VOLUME SEVENTY

ADVANCES IN CATALYSIS

Catalysis for Enabling Carbon Dioxide Utilization

VOLUME SEVENTY

Advances in
CATALYSIS

Catalysis for Enabling Carbon Dioxide Utilization

Edited by

MONTSERRAT DIÉGUEZ

Departament de Química Física i Inorgànica, Universitat Rovira i Virgili, Tarragona, Spain

ARJAN W. KLEIJ

Institute of Chemical Research of Catalonia (ICIQ), The Barcelona Institute of Science & Technology (BIST), Av. Països Catalans 16, Tarragona; Catalan Institute of Research and Advanced Studies (ICREA), Pg. Lluis Companys 23, Barcelona, Spain

ELSEVIER

ACADEMIC PRESS

An imprint of Elsevier

Academic Press is an imprint of Elsevier
50 Hampshire Street, 5th Floor, Cambridge, MA 02139, United States
525 B Street, Suite 1650, San Diego, CA 92101, United States
The Boulevard, Langford Lane, Kidlington, Oxford OX5 1GB, United Kingdom
125 London Wall, London, EC2Y 5AS, United Kingdom

First edition 2022

ISBN: 978-0-323-98935-0
ISSN: 0360-0564

For information on all Academic Press publications
visit our website at https://www.elsevier.com/books-and-journals

Publisher: Zoe Kruze
Acquisitions Editor: Sam Mahfoudh
Developmental Editor: Cindy Angelita Gardose
Production Project Manager: Sudharshini Renganathan
Cover Designer: Victoria Pearson

Typeset by STRAIVE, India

Contents

Contributors

Benjamin R. Buckley
Department of Chemistry, Loughborough University, Loughborough, LE, United Kingdom

José A. Castro-Osma
Universidad de Castilla-La Mancha, Departamento de Química Inorgánica, Orgánica y Bioquímica-Centro de Innovación en Química Avanzada (ORFEO–CINQA), Facultad de Farmacia, Albacete, Spain

Giulia Chiarioni
Chemical Engineering group, Engineering and Technology institute Groningen (ENTEG), Faculty of Science and Engineering, University of Groningen, Groningen, The Netherlands

Pegie Cool
Research Group LADCA, Department of Chemistry, University of Antwerp, Antwerp, Belgium

Shoubhik Das
ORSY Division, Department of Chemistry, University of Antwerp, Antwerp, Belgium

Felipe de la Cruz-Martínez
Universidad de Castilla-La Mancha, Departamento de Química Inorgánica, Orgánica y Bioquímica-Centro de Innovación en Química Avanzada (ORFEO–CINQA), Facultad de Ciencias y Tecnologías Químicas, Ciudad Real, Spain

Yi He
Key Laboratory of Green Chemistry & Technology of Ministry of Education, College of Chemistry, Sichuan University, Chengdu, People's Republic of China

Arjan W. Kleij
Institute of Chemical Research of Catalonia (ICIQ), The Barcelona Institute of Science & Technology (BIST), Av. Països Catalans 16, Tarragona; Catalan Institute of Research and Advanced Studies (ICREA), Pg. Lluis Companys 23, Barcelona, Spain

Agustín Lara-Sánchez
Universidad de Castilla-La Mancha, Departamento de Química Inorgánica, Orgánica y Bioquímica-Centro de Innovación en Química Avanzada (ORFEO–CINQA), Facultad de Ciencias y Tecnologías Químicas, Ciudad Real, Spain

Francesca Milocco
Chemical Engineering group, Engineering and Technology institute Groningen (ENTEG), Faculty of Science and Engineering, University of Groningen, Groningen, The Netherlands

Paolo P. Pescarmona
Chemical Engineering group, Engineering and Technology institute Groningen (ENTEG), Faculty of Science and Engineering, University of Groningen, Groningen, The Netherlands

Wei Wang
Key Laboratory of Green Chemistry & Technology of Ministry of Education, College of Chemistry, Sichuan University, Chengdu, People's Republic of China

Si-Shun Yan
Key Laboratory of Green Chemistry & Technology of Ministry of Education, College of Chemistry, Sichuan University, Chengdu, People's Republic of China

Da-Gang Yu
Key Laboratory of Green Chemistry & Technology of Ministry of Education, College of Chemistry, Sichuan University, Chengdu, People's Republic of China

Hai-Peng Zhang
Key Laboratory of Green Chemistry & Technology of Ministry of Education, College of Chemistry, Sichuan University, Chengdu, People's Republic of China

Hong Zhao
Key Laboratory of Green Chemistry & Technology of Ministry of Education, College of Chemistry, Sichuan University, Chengdu, People's Republic of China

Jian Zhu
Research Group LADCA; ORSY Division, Department of Chemistry, University of Antwerp, Antwerp, Belgium

Preface

Carbon dioxide is a readily accessible, cheap, and renewable carbon feedstock for organic synthesis. Catalysis has postulated itself as a key technology to convert CO_2 into valuable products and, as such, CO_2 catalysis has become a very active area of research. Many research groups working in this area use approaches that span the areas of homogeneous and heterogeneous catalysis with applications ranging from the sustainable synthesis of C-containing building blocks to fuels. Though remarkable results have been reported, there is still a need and desire to optimize the process efficiency of known transformations and design novel ones. At this point, we believe that contributions that discuss the key strategies and latest progress within the context of renewable carbon valorization are both attractive and stimulating to further advance this domain in sustainable development.

Following these considerations, this volume of Advances in Catalysis entitled *Catalysis for Enabling Carbon Dioxide Utilization* contains six top-quality chapters written by experts in the field with a common focus on sustainable chemical production. In Chapter 1, A. W. Kleij provides the most recent advances in the use of heterocyclic organic carbonates in synthetic chemistry and polymer development. Future directions are also identified that will prospectively spur new advancements while enforcing a prominent position of these CO_2-based heterocyclic structures in various areas of modern science and sustainable progress. In Chapter 2, P. Cool, S. Das, and coworker explain the advances made in electrochemical reduction of CO_2 into methanol over the past 10 years. The fundamental insights are provided for each kind of catalyst, which gives guidelines for the design and synthesis of novel and more efficient catalysts. In Chapter 3, B. R. Buckley focuses on electrocarboxylation methods to generate a C—CO_2 bond. The chapter is organized by the type of bond carboxylated starting with the reduction of CO_2 to oxalate and moving forward to C—C, C=C, C≡C, and C—X bond formations. The author examines in detail the breadth of carboxylation reactions nowadays available and the type of electrosynthetic cell and catalyst required to carry out these transformations. In Chapter 4, D.-G. Yu, S.-S. Yan, and coworkers focus on photocatalytic carboxylation of C—H bonds, C—(*pseudo*)halide bonds, and unsaturated substrates, among others, with CO_2 to prepare diverse carboxylic acids. In Chapter 5, P. P. Pescarmona and coworkers provide a didactic and critical overview of the main classes of heterogeneous catalysts for the conversion of CO_2 into cyclic carbonates and polycarbonates. The authors present and discuss the type of active sites, the catalytic mechanisms, and the state-of-the-art strategies to design active, selective, and stable heterogeneous catalysts. Finally, in Chapter 6, A. Lara-Sánchez, J. A. Castro-Osma, and coworker cover recent advances in the use of bio-based feedstock and carbon dioxide as a sustainable C1 building block for the synthesis of bio-sourced organic carbonates and sustainable hybrid materials.

We thank Mr. Sam Mahfoudh, who helped us initiate this project, and Ms. Cindy Angelita P. Gardose, whose help during the preparation, editing, and production process was invaluable. We thank the Spanish Ministry of Science and Innovation (PID2019-104904GB-I00 and PID2020-112684GB-100), the Catalan Government (2017/SGR/1720 and 2017/SGR/232), ICREA, and the CERCA Program/Generalitat de Catalunya. Finally, we specially thank all authors who contributed to this project.

Montserrat Diéguez
Arjan W. Kleij

CHAPTER ONE

Functional CO_2 based heterocycles as precursors in organic synthesis

Arjan W. Kleij*
Institute of Chemical Research of Catalonia (ICIQ), The Barcelona Institute of Science & Technology (BIST), Av. Països Catalans 16, Tarragona, Spain
Catalan Institute of Research and Advanced Studies (ICREA), Pg. Lluis Companys 23, Barcelona, Spain
*Corresponding author: e-mail address: akleij@iciq.es

Contents

Abstract

This chapter describes the most recent advances made in the use of functional cyclic organic carbonates in synthetic chemistry and polymer development. In particular, the type of carbonate, its functionalization and its ring size will be discussed and highlighted in detail. Cyclic carbonates are amenable to various types of ring-opening chemistry which allows to use them as precursors toward fine-chemical, advanced intermediates and monomers for poly-carbonate and poly-urethane synthesis. The synthetic concepts and their mechanistic intricacies are presented to provide a detailed context from their conception to their utilization as a means to mark the current achievements but also their limitations. Future directions are identified that will hopefully spur new advancements while enforcing a prominent position of these CO_2-based heterocyclic structures in various areas of modern science and sustainable progress.

1. General introduction

The utilization of carbon dioxide (CO_2) as an example of a renewable carbon reagent marks a stepping stone in sustainable development that preferably leads to a well-controlled carbon management through the use of

Advances in Catalysis, Volume 70
ISSN 0360-0564
https://doi.org/10.1016/bs.acat.2022.06.001

circular chemistry principles *(1,2)*. In this regard, catalysis is often considered as a key enabling tool that helps to advance the development of downstream products through ingenious approaches toward CO_2 activation and transformation *(3–5)*. Apart from reductive transformations, non-reductive conversion of CO_2 (i.e., those that do not alter the oxidation state of the carbon center of this carbon reagent upon conversion) has become prominent among the catalysis community offering various ways to increase the value of a typical waste molecule from combustion processes. Historically, the formation of heterocyclic products from CO_2 has been popular in the area of CO_2 catalysis with the formation of cyclic carbonates (and related carbamates) having a leading role *(6–9)*. (Cyclic) organic carbonates are attractive targets due to their wide application potential as synthons in preparative chemistry *(10,11)*, monomers for new polymer development *(12–14)* and as non-protic solvents and electrolytes *(15,16)*.

Five-membered cyclic carbonates are without a doubt the most sought after products in the non-reductive conversion of CO_2, and the latter can undergo a formal [2 + 3] cycloaddition reaction with epoxides to afford the target heterocyclic product. The extent to which this type of coupling process has been developed is clear from the various review articles that have appeared over the years mostly describing efforts in terms of catalyst/process intensification *(17–22)*. Their use as intermediates through ring-opening strategies has been less documented but the recent literature does show interesting perspectives and incentives how these CO_2-based rings can be utilized to forge organic compounds with an increased structural complexity.

There are several options to convert five-membered cyclic carbonates via ring-opening chemistry, while higher homologues (six- and larger-membered ring compounds) most frequently are regarded and used as suitable monomers for ring-opening polymerization (Fig. 1) *(23–26)*.

In the presence of suitable nucleophiles, five-membered cyclic carbonates can undergo ring-opening triggered by attack on one of the electrophilic positions within the five-membered ring or through *exo*-cyclic attack of a nucleophilic species following ring-opening. While not all of these manifolds are equally developed and discussed in the open literature, together with the metal-promoted transformation of specific examples of functionalized cyclic carbonates (i.e., vinyl and alkynyl substituted ones) and the use of multi-cyclic carbonates in sustainable polymer synthesis offer

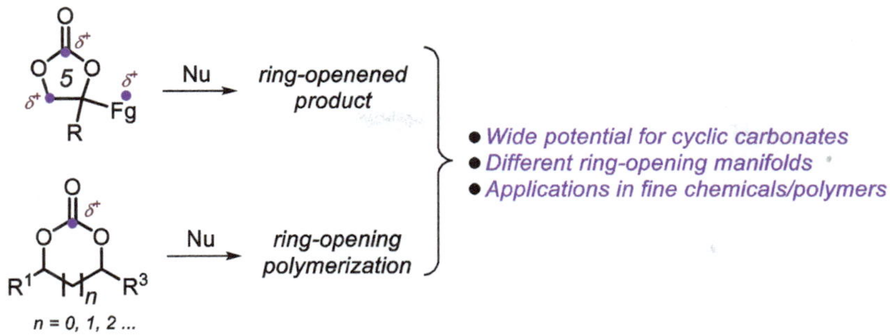

Fig. 1 Ring-opening diversity of cyclic carbonates vs polymerization potential of larger-ring carbonates. Nu stands for an external nucleophile.

a wide pallet of opportunities to valorize these CO_2-based heterocycles in a few steps into advanced molecules with a range of potentially interesting applications. This chapter is therefore divided into sections that will delineate these synthetic conversions and highlight illustrative examples of successful (catalytic) approximations toward the creation of small functionalized molecules and polymers.

2. Ring-opening functionalization of cyclic carbonates

As indicated in the introduction, there are several ways to ring-open functionalized five-membered cyclic carbonates. One of the most widely applied processes used is the aminolysis of these heterocycles using aliphatic amines producing linear hydroxy-functionalized carbamates as products *(27)*. This type of process is characterized by high chemo-selectivity, mild reaction conditions and practical simplicity while retaining all atoms in the final products. The transformation can often be carried out in the absence of a catalyst as primary (and sometimes secondary) amines are both nucleophilic and reactive enough under ambient conditions or slightly elevated temperatures. The ring-opening process is typically accompanied by formation of two regio-isomers (Fig. 2A) that arise from distinct C—O scission reactions involving the carbonate functional group. A comprehensive study in this context was performed by Caillol, Andrioletti and coworkers *(28)*, who reported a rational investigation considering variations in catalyst, solvent, reaction temperature and cyclic carbonate. In these studies, a

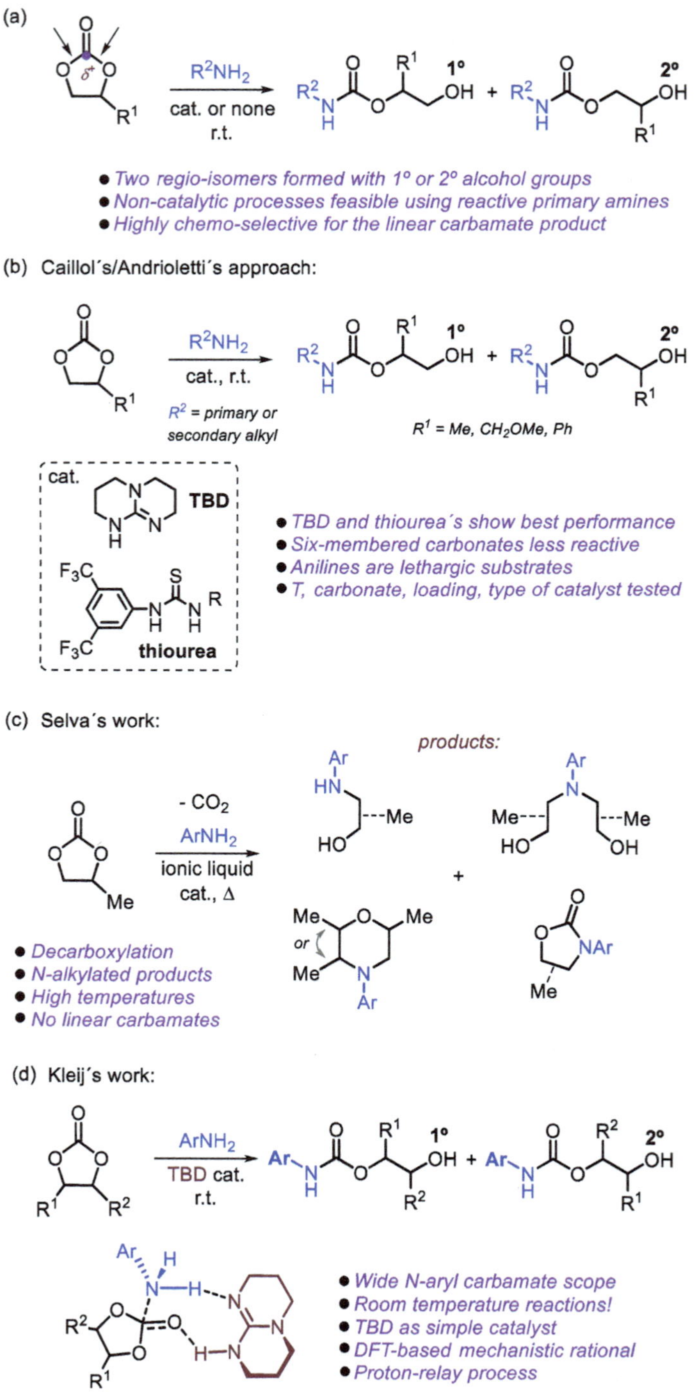

Fig. 2 (A) General features of cyclic carbonate aminolysis. (B) Studies reported by Andrioletti and Caillol. (C) Reactions of anilines and cyclic carbonates studied by Selva et al. (D) Catalyzed ambient formation of *N*-aryl carbamates and DFT rationalization.

primary amine featuring a secondary alkyl substituent was chosen (cyclohexyl amine), as the steric impediment reduces the aminolysis capacity and a catalyst is thus warranted for efficient turnover (Fig. 2B). The major findings of this work showed that hydrogen-bond activation through the use of organocatalysis is a powerful approach with thioureas and the bicyclic guanidinium compound TBD (= 1,5,7-triazabicyclodec-5-ene) demonstrating the best catalytic performance under solvent-free conditions and with a typical amine reactivity order primary alkyl amine > secondary alkyl amine > secondary amine. Another worthy observation is the apparent lower reactivity of six- vs five-membered cyclic carbonates, which may be explained by the lack of planarity (and thus steric blocking) in the six-membered congeners upon aminolysis.

The lack of any reactivity in the presence of aryl amines and absence of a catalyst inspired several groups to investigate this specific combination of reaction partners. Selva reported the treatment of alkyl-substituted cyclic carbonates such as propylene carbonate (PC) with anilines in the presence of an ionic liquid as catalyst *(29)*, and found that very high reaction temperatures (150–190 °C) were required for carbonate conversion to take place. Moreover, the chemo-selectivity of this manifold was very low and several products were noted that are formal decarboxylated *N*-alkylated compounds (Fig. 2C). However, under these demanding conditions no linear hydroxy carbamates were formed.

Later on, and inspired by the previous achievements of both Andrioletti and Selva, Kleij and coworkers reported that TBD is a suitable catalyst that allows for simple and selective aminolysis of a wide range of mono- and di-substituted alkyl cyclic carbonates in the presence of anilines affording in good yields the carbamate targets (Fig. 2D) *(30)*. More importantly, density functional theory (DFT) calculations unequivocally showed that the TBD catalysts helps to stabilize (via H-bonding) key intermediates in which both the carbonate and amine are involved, and establishing a formal proton-relay process (Fig. 2D). The computed energetic span of this process ($\Delta E = 18.1$ kcal/mol) is in good agreement with the experimentally determined feasibility of a room temperature transformation. These N-aryl carbamates are interesting synthons with potential use as precursors to *N*-aryl isocyanates through thermolysis, or as intermediates for pharmaceutically relevant structures *(31)* and non-symmetrical ureas *(30,32)*.

In a follow-up study, the Kleij group showed that TBD can also be applied as a catalyst for the regio-selective aminolysis of di-, tri- and even tetra-substituted cyclic organic carbonates *(33)*. Apart from some scattered

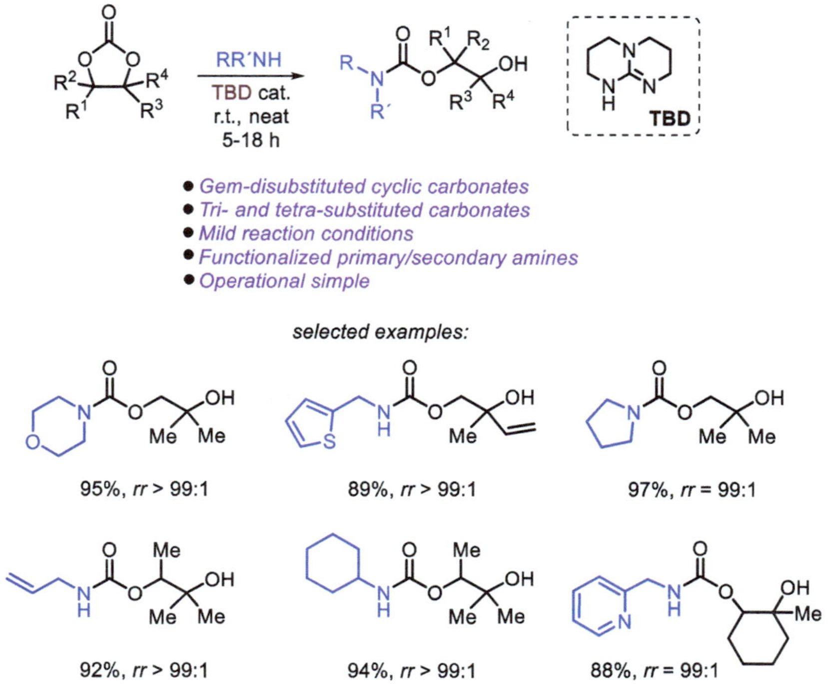

Fig. 3 TBD-catalyzed regio-selective aminolysis of substituted cyclic carbonates. The denotion *rr* stands for regioisomer ratio.

examples in the literature *(34,35)*, this latter procedure (Fig. 3) represents the first general protocol that enables the virtual complete regio-selective ring-opening of cyclic carbonates by amine nucleophiles and demonstrates the great versatility of organic carbonates in synthetic chemistry.

The Åkermark group reported an interesting variant on the aminolysis of cyclic carbonates using amino acids as the nucleophiles under aqueous conditions *(36)*. In this approach the desired carbamate formation is in competition with formal hydrolysis, and fine-tuning of the reaction conditions was required allowing to isolate functionalized linear carbamates in typically good yields of 75–95%.

While aminolysis of cyclic carbonates is well-documented, the use of other nucleophiles able to ring-open these heterocycles is much less studied. A rather simple though potentially useful synthetic transformation is the controlled hydrolysis of *cis*-configured cyclic carbonates affording their 1,2-*syn* diols, and highlights the use of CO_2 as a temporary protecting group. In 2014, Laserna et al. reported the in situ formation of complex multi-cyclic carbonates through Al-catalysis, and their subsequent transformation into *cis*-diols (Fig. 4A) *(37)*, that are useful synthons in naturally product

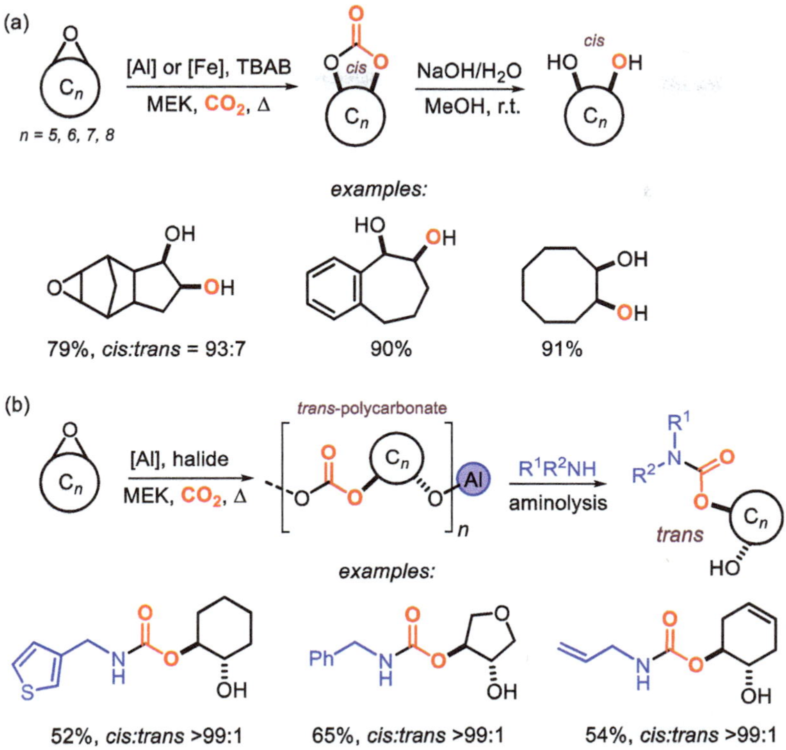

Fig. 4 (A) *Cis*-diol formation from cyclic carbonates by controlled hydrolysis. (B) In situ aminolysis of polycarbonate intermediates giving *trans*-configured linear carbamates.

synthesis. Key to the stereo-control is a double inversion pathway taking place provoked by the experimental conditions allowing to preserve the original configuration of the epoxide precursors. This is distinct from epoxide hydrolysis that would favor the formation of *trans*-diols *(38)*. In situ trapping of bicyclic organic carbonates has also been reported to control *cis/trans* stereochemistry in the formation of linear carbamates *(39)*, which is enabled by pre-formation of either cyclic carbonate or polycarbonate intermediates that react productively with amine reagents toward the desired carbamates (Fig. 4B).

Other less frequently used nucleophiles for ring-opening of five-membered ring carbonates are cyanide and thiols. Recently, Kleij et al. communicated that cyanide reacts in a rather unusual way with vinyl-substituted cyclic carbonates at elevated temperatures *(40)*. In this manifold, the CN anion preferentially attacks the methylene carbon (and thus not the carbonate C-center) leading to a cascade of steps and subsequent extrusion of CO_2 and MeCN (Fig. 5A), and labeling studies supported the proposed

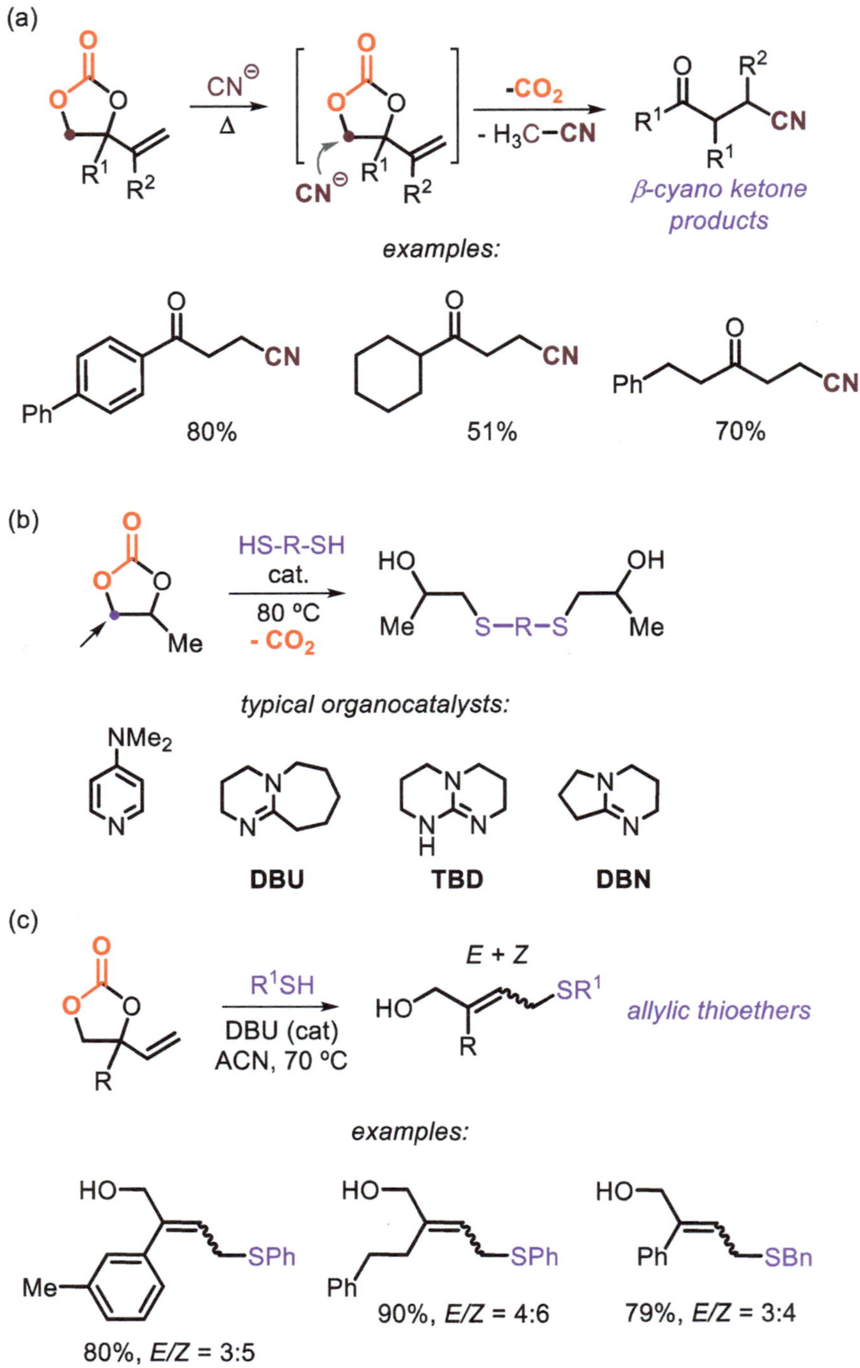

Fig. 5 (A) CN-promoted formation of β-cyano ketones from vinyl cyclic carbonates. (B) Use of thiols in cyclic carbonate ring-opening reactions. [Al] and [Fe] stand for aminotriphenolate metal complexes. (C) DBU-catalyzed formal allylic thio-etherification using vinyl cyclic carbonates and RSH based reagents.

mechanism. Unexpectedly, the formation of β-cyano ketones was favored in yields up to 80%. This a rare example of "methylene" attack by an external nucleophile, and similar notions were previously done by Selva using aryl amine nucleophiles *(29)*.

Finally, Detrembleur and coworkers used thiol reagents to widen the scope of polyurethane materials *(41)*. In a model study, propylene carbonate was treated with a dithiol reagent thereby creating a bis β-hydroxy thioether (Fig. 5B). This process resembles (in terms of the site-preference of the nucleophilic attack) the CN- and aryl amine promoted ring-opening *(29,40)*. Both the formal *S*-alkylation and aminolysis reactions could be united to produce self-blown isocyanate-free polyurethane foams by utilizing chemo- and regio-selective additions of diamines and dithiols to five-membered cyclic carbonates.

When vinyl cyclic carbonates are combined with alkyl or aryl thiols, attack of the S-based nucleophile takes place on the *exo*-cyclic double bond resulting into a decarboxylative formal allylic substitution (Fig. 5C) under DBU catalysis *(42)*. This process represents a unique attack of a hard nucleophilic onto a different, *exo*-cyclic position of the involved cyclic carbonate with the base catalyst most likely activating the RSH reagent thereby increasing the nucleophilic character of the S-atom. This allylic substitution process could be regarded as cheaper and more sustainable alternative to previously reported formation of allylic thioethers under Pd catalysis *(43)*.

This section shows clearly that nucleophilic attack onto (functional) cyclic carbonates represents a rich chemistry with softer nucleophiles preferring to engage with the carbonate carbon center and hard(er) nucleophiles inclined to react at a methylene carbon of the carbonate ring or an *exo*-cyclic C=C bond. In the next section, focus will shift to the activation of *exo*-cyclic functionalities by transition metal catalysts to bring about stereo/enantioselective formation of allylic derivatives.

3. Conversion of vinyl cyclic carbonates by transition metal catalysis

In the preceding section the ring-opening functionalization of cyclic carbonates is discussed using various nucleophiles with a certain site-preference. In this section, the conversion of vinyl-substituted cyclic carbonates through the use of transition metal catalysis will be presented that allows to create a diverse portfolio of allylated compounds under high stereocontrol *(10,11,44)*. Key in these protocols is the presence of a metal

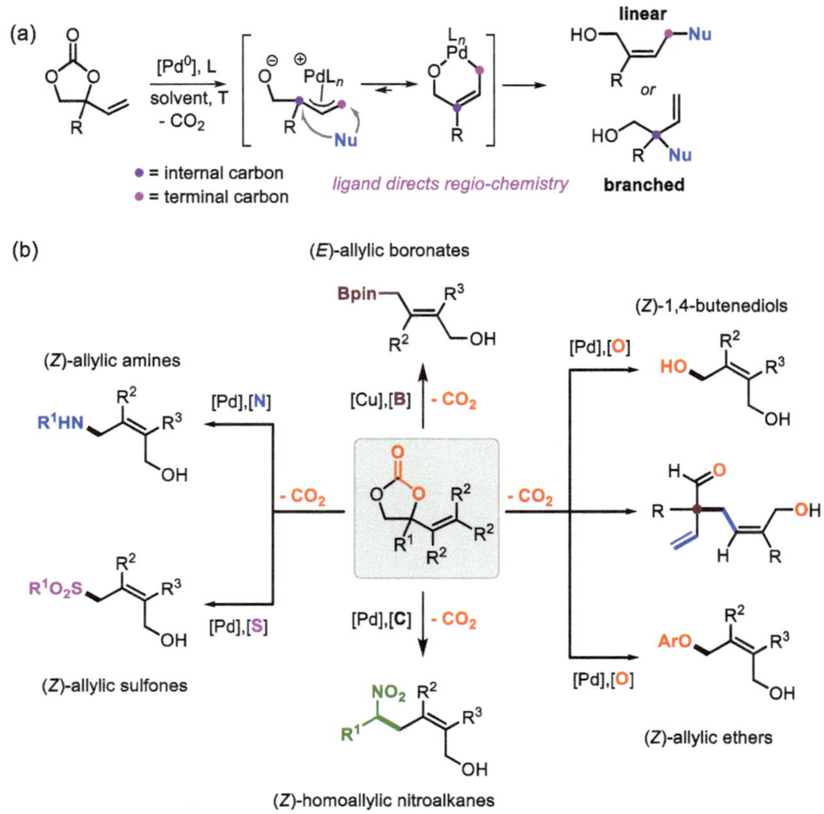

Fig. 6 (A) Decarboxylative allylation through the use of vinyl cyclic carbonates under Pd catalysis. (B) Scope of reported transformation to stereoselective access to "linear" allylic compounds.

complex that has an appropriate ligand associated (i.e., ligand engineering is crucial) enabling both efficient, selective and versatile conversion of vinyl carbonates and selected (pro)nucleophiles.

Vinyl cyclic carbonates undergo facile decarboxylation in the presence of transition metal complexes of Pd and Co (among others). The resultant intermediates are defined as reactive metal-allyl species (Fig. 6A) that can undergo allylic substitution in the presence of various nucleophiles. Depending on the regio-chemistry and the ligand, either stereoselective (*E*/*Z*, Fig. 6B) or enantioselective transformations (Fig. 7) can be designed. A wide range of N-, O-, S-, C- and B-based pronucleophiles can be employed in concert with a wide variety of substituted and thus modular vinyl carbonate precursors. This area of research has been blossoming for

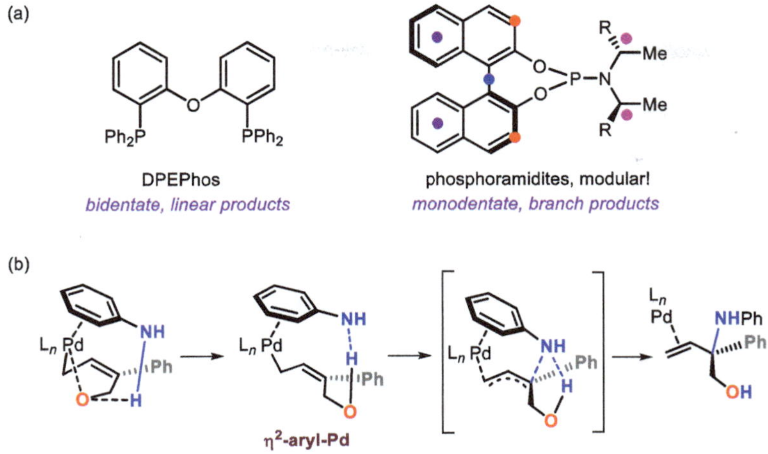

Fig. 7 (A) Privileged bidentate vs mono-dentate ligands in Pd-catalyzed allylic substitution reactions. (B) A unique inner-sphere computed mechanism that accounts for "branch" type regio-selectivity in allylic amination.

many years *(10,11,44,45)* with seminal contributions from various groups including those from Zhang et al. *(46–49)*, Khan and coworkers *(50–52)*, the Kleij laboratory *(53–57)*, Zhao *(58–60)*, among others *(61–65)*.

While bidentate phosphines (such as DPEPhos, Fig. 7A) tend to favor the formation of linear allylic compounds with high degrees of stereo-control ($Z/E > 99$:1) *(54)*, the presence of highly modular phosphoramidites (Fig. 7A; note that the colored dots represent sites where both steric and electronic effect can be exerted) *(66)* directs the nucleophilic attack toward the internal carbon of the intermediate metal allyl. The rationale behind these features was both experimentally and computationally studied for the allylic amination using aryl amines under Pd-catalysis *(67)*, with the steric bulk of the ligand favoring the presence of competitive tricoordinate Pd species and a unique inner-sphere pathway during the nucleophilic attack (Fig. 7B). In this manifold, after decarboxylation of the vinyl cyclic carbonate the system advances to a Pd(allyl) intermediate featuring a η^2-aryl-Pd interaction that drives the regio-selectivity toward the inner-carbon of the allyl fragment.

More recently other types of inner-sphere nucleophilic attack in Pd-mediated allylic substitutions have been uncovered *(68)*, showing the versatility of this chemistry and new potential to create sterically encumbered carbon stereocenters.

Apart from the reaction of vinyl cyclic carbonates with externally added nucleophiles, Kleij and coworkers reported a unusual cross-coupling between two molecules of the starting carbonate in the presence of a Pd catalyst *(55)*. In this process, one molecule of the carbonate precursor is in situ converted into a nucleophilic dienolate species that subsequently and rather efficiently is cross-coupled with an electrophilic Pd(allyl) to afford complex scaffold that feature highly functional quaternary carbon stereocenters (Fig. 8). Depending on the substrate and reaction conditions, also "simple" aldehyde products may be obtained through a β-hydride elimination process.

The procedure thus involves a formal umpolung of one molecule of vinyl carbonate following a "cross-electrophile" coupling step with a second one activated by the Pd complex. Other examples of the use of similar dienolates derived from these vinyl carbonates were also reported by other groups using allylic acetates as reaction partners *(69)*, and in principle this provides access to a wider molecular diversity.

To conclude this section, in an attempt to replace precious Pd by an economical more attractive metal, several reports have shown that Co-based allylic alkylation reactions can be designed making use of a photocatalyst to reduce the Co center at various stages of the process (Fig. 9). These approaches, better known as dual metal/photo catalysis (PC) *(70)*, enable mild single-electron photo-reduction of the Co(II) precursor thereby producing a Co(I) intermediate that facilely undergoes oxidative addition in the presence of suitable allylic precursors. The resultant Co(III)allyl species is then again reduced by the photocatalyst affording a formal nucleophilic Co(II)allyl when then reacts with carbon electrophiles such as aldehydes. This manifold was recently exploited to stereo-selectively generate both *syn*-configured 1,3-diols (in the presence of DPEPhos as ligand) *(56)* and homo-allylic alcohols (in the presence of bipyridines as ligand) *(57)* that comprise of quaternary carbon centers. In the latter work, an attractive conversion of the vinyl carbonates into 1,3-dienes was reported that in situ can be 1,2-dialkylated marking new potential for these CO_2-based precursors.

Full details of the intricacies of the mechanism that involves the Co-mediated formation of *syn*-1,3-diols has just been released *(71)*, showing the importance of the type and strength of the base additive and energy-dissipating side-reactions.

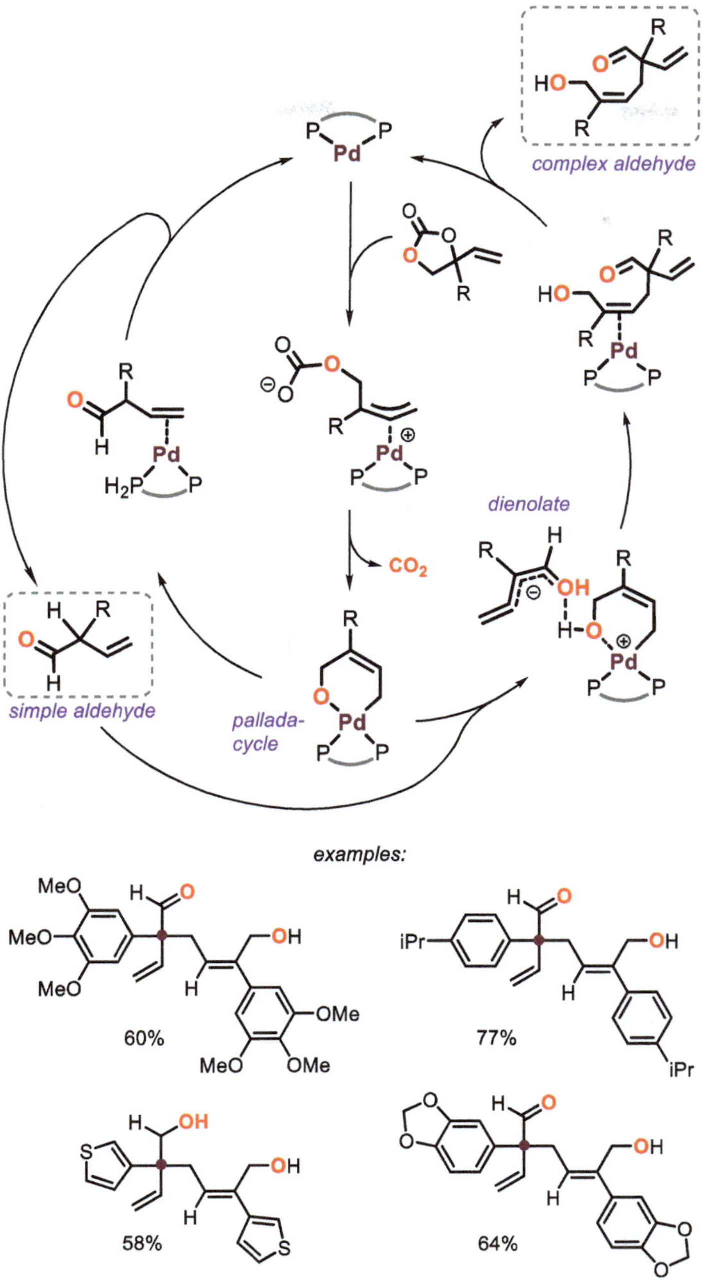

Fig. 8 Formal cross-electrophile coupling between vinyl cyclic carbonates producing compounds with highly decorated quaternary carbon centers.

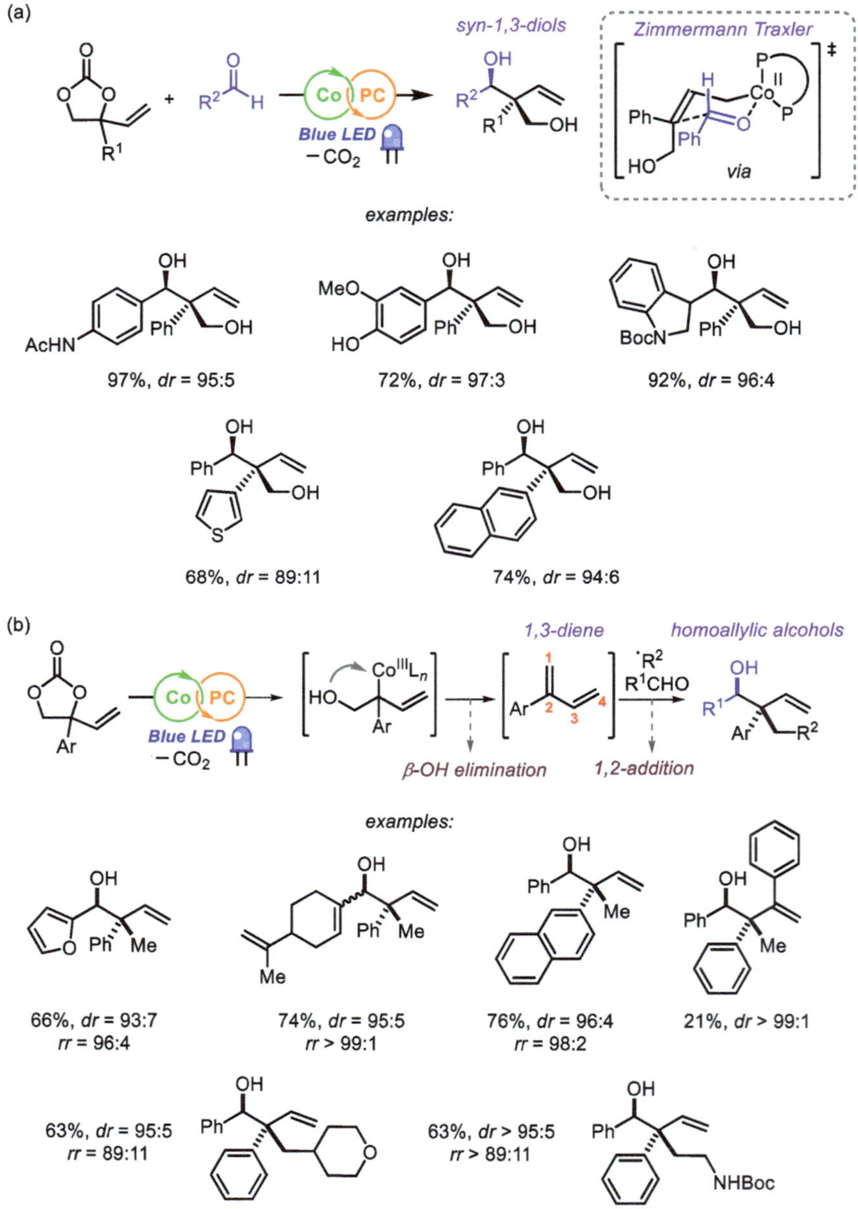

Fig. 9 (A) Stereoselective dual Co/photo-catalytic approach toward *syn*-1,3-diol scaffolds comprising a quaternary carbon center. (B) Formation of homo-allylic alcohols by dual Co/PC catalysis, *rr* stands for regioisomeric ratio.

4. Transformation of alkynyl cyclic carbonates

While vinyl cyclic carbonates have been often used in metal-mediated synthesis, the utilization of other functionalized cyclic carbonates is less prominent. Alkyne-substituted carbonates are attractive since they represent propargylic surrogates that arise after in situ decarboxylation and offer thus a means to diversify both C—C and C—X (X = heteroatom) bonds *(72)*. Previous examples of the use of alkynyl cyclic carbonates involved the Cu-mediated synthesis of chiral β-amino alcohols and dihydrofurans *(73,74)*. Soon after these results appeared, Gómez et al. reported the use of S-centered nucleophiles (i.e., sulfinate salts) in the asymmetric preparation of propargylic sulfones that incorporate tetrasubstituted tertiary carbon centers *(75)*. This Cu-promoted transformation shows the versatile use of these alkynyl carbonates while preparing functional chiral scaffolds with ambivalent functionality (Fig. 10A). In this transformation a clear positive non-linear effect (NLE) between the *ee* of the Cu catalyst and the *ee* of the propargylic sulfone product was observed. While Nishibayashi found that such NLEs may be explained by the intermediacy of dinuclear complexes *(76,77)*, further mechanistic research will be necessary to elucidate the details of this propargylic sulfonylation.

Other examples of the use of alkynyl carbonates in synthetic chemistry have been reported, for instance Yuan et al. developed a decarboxylative [3 + 2] annulation reaction between alkynyl carbonates and azlactones using Cu-catalysis and in the presence of a base *(78)*. Through this approach, a large set of γ-butyrolactones featuring vicinal quaternary carbon centers could be prepared in good to high yields with high levels of diastereocontrol (>95:5 *dr*). Gong and Song were able to extend the use of these alkyne precursors to interesting enantioselective [3 + 3] and [3 + 4] annulation reactions between five-membered alkynyl cyclic carbonates or six-membered alkynyl cyclic carbamates, and isatin-derived enals under cooperative Cu/NHC (*N*-heterocyclic carbene) catalysis providing spirooxindole derivatives (Fig. 10B) *(79)*.

Formal propargylic substitution reactions require the substrate to have a terminal alkyne group. In 2020, Guo et al. used cyclic carbonates equipped with terminally capped alkyne groups and studied the Cu-mediated conversion of these substrates in the presence of silylboranes *(80)*. These alternative alkynyl carbonates have a different reactivity profile and allow to prepare tetrasubstituted, silylated 2,3-allenols (Fig. 11A). In a different embodiment,

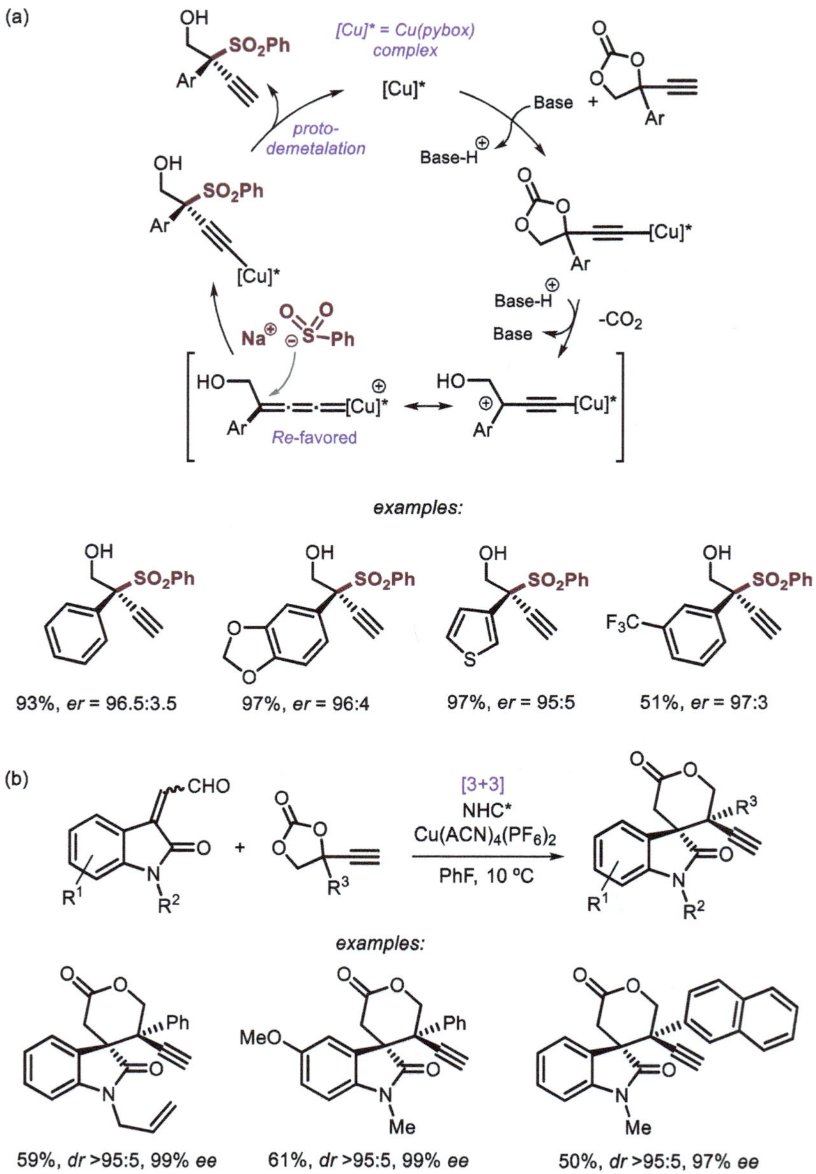

Fig. 10 (A) Propargylic sulfonylation empowered by Cu-catalysis using alkynyl cyclic carbonates as substrates. (B) [3+3] cycloaddition reactions with alkynyl cyclic carbonates.

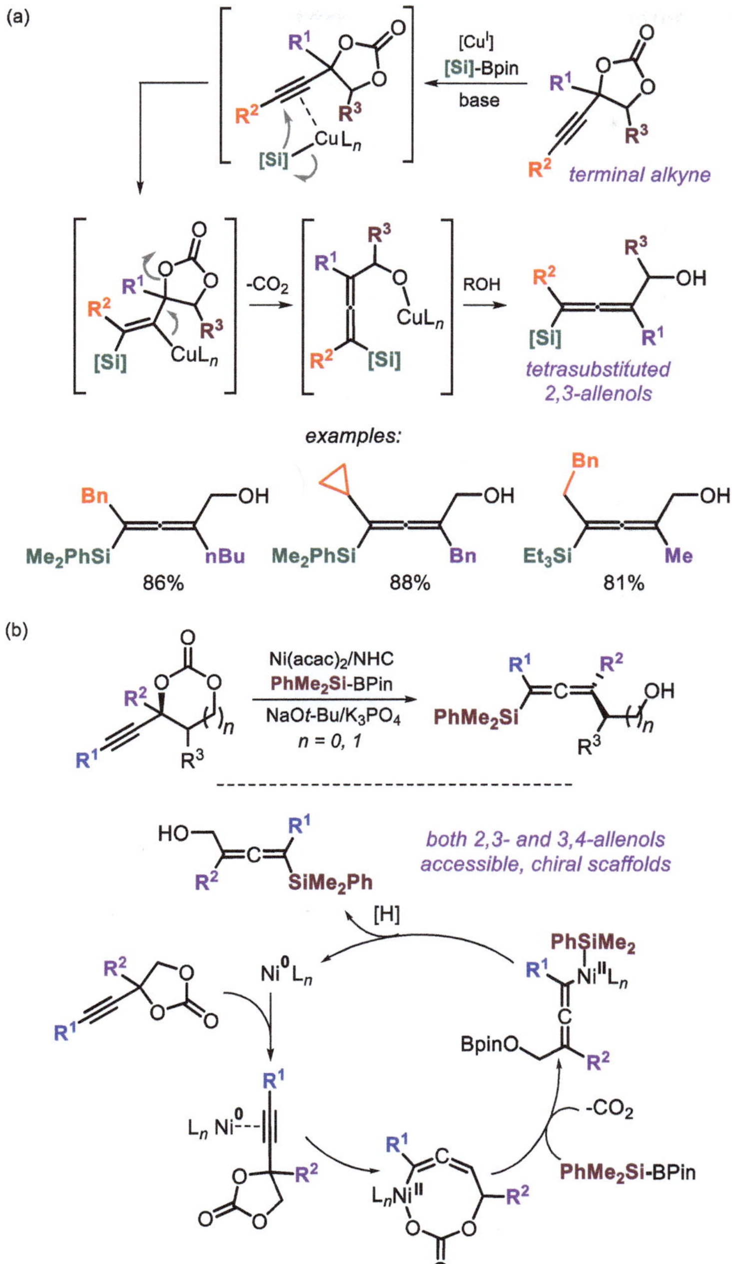

Fig. 11 (A) Cu-mediated formation of silylated 2,3-allenols using alkynyl carbonates. (B) Ni-promoted enantiospecific transformations of alkynyl carbonates affording chiral 2,3- and 3,4-allenols.

the same authors also demonstrated that Ni-catalysis is fruitful to design an enantiospecific version of this protocol giving access to both chiral 2,3- and 3,4-allenols through five- and six-membered alkynyl carbonate substrates, respectively (Fig. 11B) *(81)*. In this latter manifold (Fig. 11B), the mechanism is believed to start at Ni(0), with the first step featuring the activation of the alkyne group through a metal-pi interaction. After oxidative addition, a *C,O*-chelated larger ring nickelacycle is formed which, after decarboxylation, isomerizes to a Ni(allenyl) species that has a silyl ligand as ligand. Reductive elimination of the product from the Ni complex concludes the process. The entire scheme takes advantage of the use of Ni catalysis that has been shown to be both versatile and efficient in enantiospecific transformations *(82,83)*, which is not easily achieved under Cu or Pd catalysis.

Later on, Guo et al. found that similar alkynyl carbonates can be converted divergently under suitable Cu catalysis by a judicious choice over the reaction partner (Fig. 12) *(84)*. More specifically, a dichotomic borylation of alkyne carbonates was observed with the use of B_2pin_2 [bis(pinacolato) diborane] giving stereoselective access to (*E*)-configured 1,2-diborylated 1,3-dienes, while the utilization of $B_2(neop)_2$ [bis(neopentyl-glycolato) diboron] producing α-hydroxyallenes. The mechanistic intricacies were

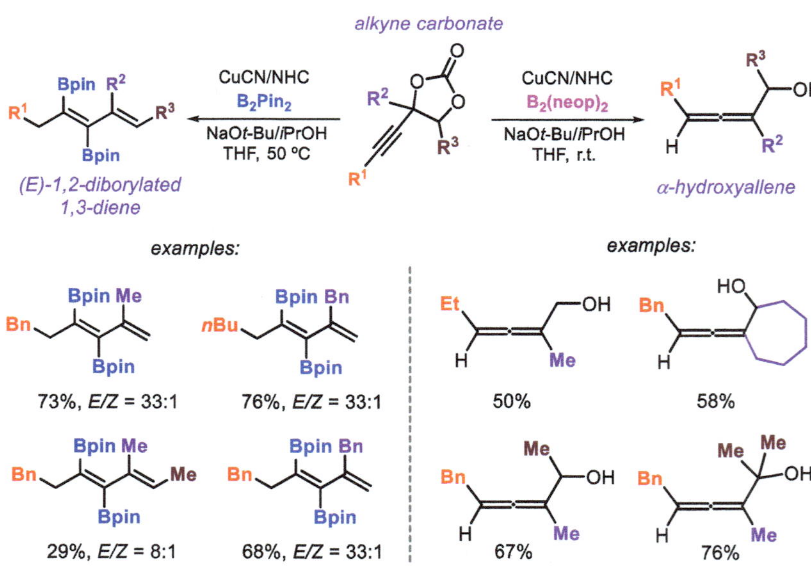

Fig. 12 Divergent transformation of alkynyl cyclic carbonates under Cu-catalysis through a dichotomic behavior of diboron reagents.

explained on the basis of various control experiments and literature precedents. The basis for the dichotomous behavior (with an apparent small structural difference between both boron reagents) and it was proposed that there is a clear dependence on the relative rate of borylation and the ease with which the Cu catalyst can activate a C-Bneop bond with respect to C-Bpin ones. Altogether, this divergent protocol allows to quickly diversify the molecular complexity from easily attained alkynyl carbonates under mild operating conditions. Along these lines, Volla and coworkers developed a Rh(I)-promoted decarboxylative arylation of alkynyl cyclic carbonates that allowed for a similar divergent synthesis of either substituted α-allenols or 1,3-butadienes depending on whether either internal or terminal alkyne-functionalized carbonate precursors were employed *(85)*.

5. Miscellaneous uses of cyclic carbonates

The preceding sections have demonstrated that functional cyclic carbonates can be effectively activated toward advanced organic transformations including stereoselective/enantioselective conversions. In this section, some recent other less frequent uses of functional carbonates are discussed. The Guo group recently reported the use of unusual allene-appended cyclic carbonates as precursors for the Pd-mediated decarboxylative [3 + 3] annulation with nitrile oxides furnishing 5,6-dihydro-1,4,2-dioxazine derivatives that feature allenyl-substituted quaternary carbon centers (Fig. 13) *(86)*. The developed protocol has ample scope in heterocyclic products that can be produced, and Xantphos and $Pd(dmdba)_2$ (dmdba = 3,5,3′,5′-dimethoxydibenzylidene acetone) were shown to be the most effective Pd/ligand combination under the experimental conditions. It is interesting to note that the allene functional group is retained and is rather robust, which enables the post-synthetic modulation of these building blocks. Furthermore, a preliminary attempt to develop an asymmetric variant of the procedure was also reported (53% *ee* in the benchmark conversion) changing the ligand to a chiral ferrocene-based phosphine ligand.

Another recent development is the use of alcohol functionalized cyclic carbonates that show unique potential to access other carbonate scaffolds that are difficult to prepare through conventional epoxide-CO_2 coupling reactions. Kleij et al. reported previously that epoxy alcohols/amines possess dual reactivity and if properly activated by suitable catalysts, can be converted through a new manifold that involves CO_2 activation by the alcohol/amine group *(7,87)*. Through this alternative pathway, the in situ preparation of a

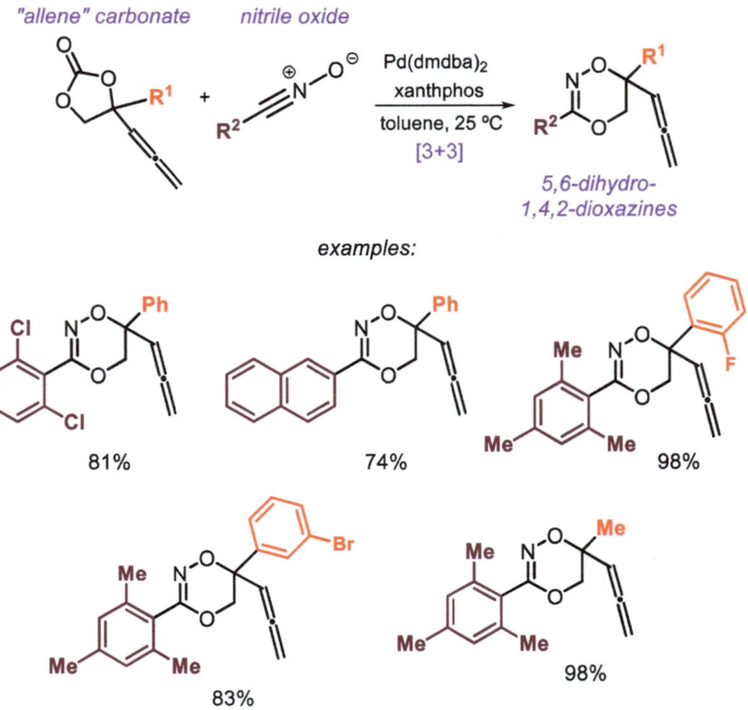

Fig. 13 Allene-functionalized cyclic carbonates in the formation of six-membered heterocycles through [3 + 3] annulation under Pd catalysis.

nucleophilic intermediate can be accomplished and therefore the addition of a nucleophilic halide (which is typically done) can be avoided. Soon hereafter other transformations were developed building on this new reactivity paradigm, paving the way for novel opportunities in the area of cyclic organic carbonate chemistry. In a mechanistic study, the same authors showed that an Al(III) aminotriphenolate complex can act as a bifunctional system, with the phenolate O-atoms acting as a Brønsted base while the Al center functions as a Lewis acid. Both fragments of the catalyst structure work in concert in the activation of glycidol and its conversion into glycerol carbonate with proton relay, the involvement of H_2O during key stages of the mechanism and intramolecular S_Ni chemistry playing a crucial role *(88)*.

With this new reactivity principle established, Kleij and coworkers then envisioned that the based-regulated Payne rearrangement of epoxy alcohols *(89)* may also be projected onto hydroxy-methyl cyclic carbonates, which can be easily accessed from epoxy alcohols and CO_2. Once hydroxy-methyl cyclic carbonates are formed they can be isomerized under basic conditions to a second type of cyclic carbonate. As both heterocyclic structures differ in the type of alcohol that is present (primary vs tertiary), selective protection of the primary alcohol based-carbonate can be achieved (being the most substituted one) and the scope of complex CO_2-based scaffolds be amplified *(90)*. Using this approach, highly functional and highly substituted tri- and tetra-substituted heterocycles were designed showing the value of new conceptual approaches in this area of research (Fig. 14A).

At a later stage, the use of β-hydroxy epoxides was proposed as a prelude to larger-ring type cyclic carbonates *(91)*. Larger-ring carbonates, and particularly the family of six-membered ones, are useful toward new CO_2-based polymer development as they can be easily ring-open polymerized *(14)*. This potential makes six-membered cyclic carbonate synthesis worthwhile but only very few general and catalytic methods have been reported to date *(92–94)*. When β-hydroxy epoxides are treated with CO_2 under binary catalytic conditions, five-membered carbonates are produced (Fig. 14B) that can be further advanced to six-membered ones by alcohol-induced isomerization. Since the six-membered carbonate is thermodynamically unstable, it needs protection at the primary alcohol site thereby providing a means to control the five-to-six-membered carbonate equilibrium. This methodology is rather unique in the sense that it provides larger-ring heterocycles from smaller-ring analogues with non-typical substitution degrees in the target product.

One drawback noted for the six-membered carbonates of Fig. 14B was the difficulties encountered exploring their ring-opening polymerization (ROP) capacity. Under standard conditions (i.e., catalytic TBD/BnOH), de-acylation occurs, which causes back-equilibration of the six- to a five-membered carbonate, which are known to be much more stable. Therefore, an alternative catalytic strategy was planned that took into account the rigidity of the β-hydroxy epoxide precursors. Utilization of more rigid cyclic β-epoxy alcohols proved to be a viable strategy for the direct synthesis of bicyclic six-membered carbonates with a free alcohol substitution *(24)*. Crucial in this transformation is the configuration of the epoxy precursor, with anti-isomers providing clean access to the desired products whereas syn-configured substrates only gave five-membered carbonate products. A wide-spanning scope of complex bicyclic carbonates could be prepared this way (Fig. 15) and, furthermore, their ROP potential was studied. Upon using a representative carbonate, room-temperature ROP was possible providing a new poly-carbonate with a number average molecular weight (M_n) of 8.6 kg/mol and a glass transition temperature almost

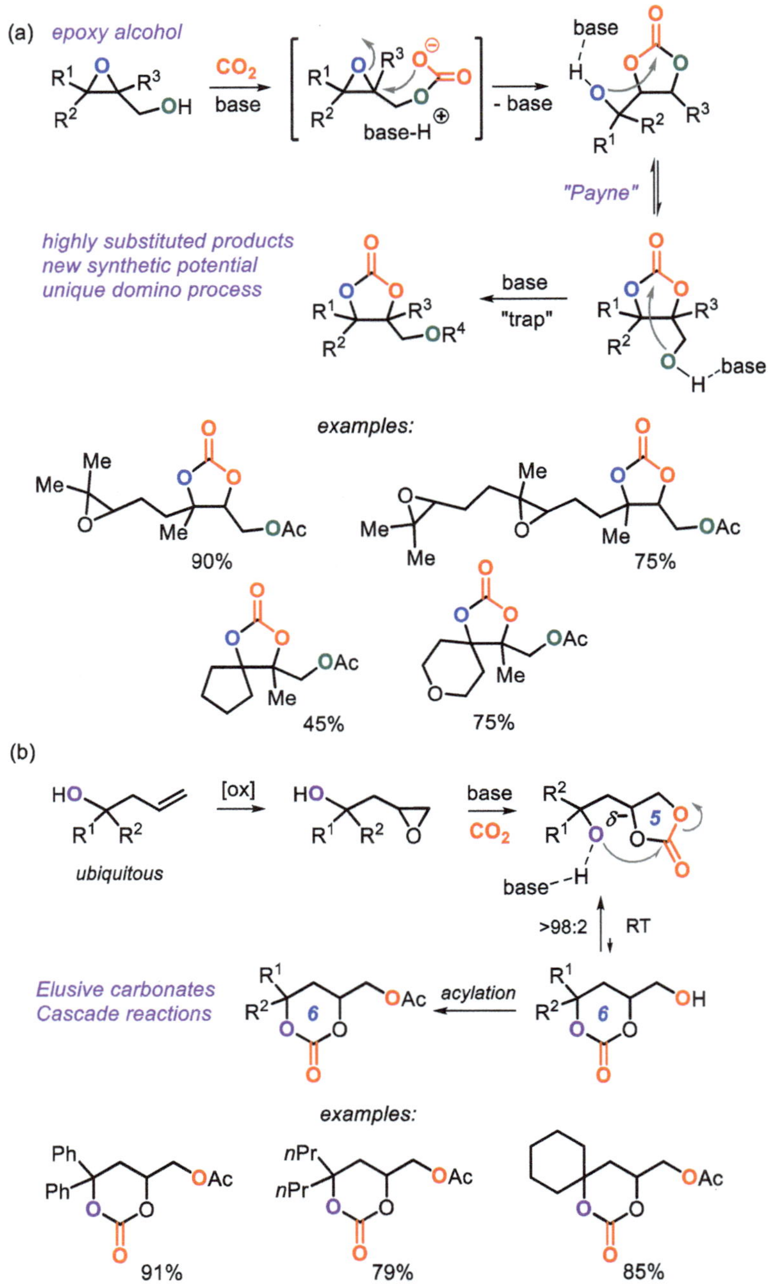

Fig. 14 (A) Payne-type rearrangement to produce highly substituted cyclic carbonates. (B) Sequential conversion of β-epoxy alcohols into six-membered cyclic carbonates.

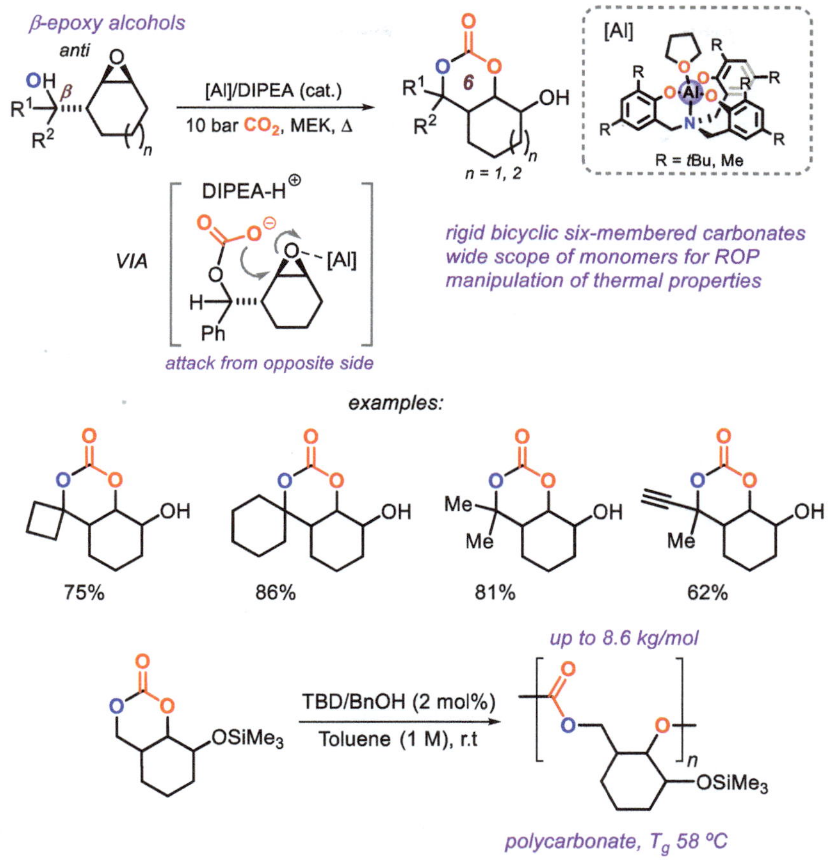

Fig. 15 Rigid bicyclic carbonate monomer synthesis through catalytic β-epoxy alcohol/ CO_2 coupling reactions.

80 °C higher than a known non-substituted polycarbonate of this type having a similar molecular weight. These data suggest that macroscopic manipulation of polycarbonates can be achieved by subtle variations in the backbone substitution and possible through changes in the O-functionalization (Fig. 15).

As a final example of the use of cyclic carbonates as synthetic precursors, their application in the formation of non-isocyanate polyurethanes (NIPUs) is highlighted here *(95,96)*. This development started as to answer to a growing need for more sustainable polyurethane preparation, which currently in industry is still dominated by the use of (toxic) aryl isocyanate reagents *(97)*. Cyclic carbonates are less reactive but their aminolysis by suitable amine reagents (Section 2) does represent a potentially useful alternative to create urethane (= linear carbamate) bonds. Typically, polymers that contain multiple double bonds can be converted into their cyclic carbonate derivatives by a sequence that involves epoxidation and catalytic [3+2] cycloaddition in the presence of CO_2.

The cyclic carbonate functionalized polymers can then be treated with polyamine reagents to create urethane type cross-links that help to cure or rigidify the polymer matrix *(98)*. Alternative approaches build on the use of more simple, low-molecular weight di- or multi-cyclic carbonates and di- or multi-amines to forge NIPUs with a wide range of structural, thermal and mechanical properties *(99)*. Of special current interest is the use of renewable feedstock to construct these NIPUs with recent examples including terpenes *(100)*, fatty acids *(101)* and sugars *(102)*.

6. Conclusion and outlook

This chapter demonstrates that cyclic carbonates have conquered a prominent position as CO_2-based heterocycles in the area of synthetic chemistry and polymer development. Their ease of preparation from simple and relatively cheap starting materials makes them attractive as intermediates toward more complex or higher-molecular weight architectures. Over the last decade, intense research efforts have provided access to a wide portfolio of cyclic carbonates with variations in their ring-size, substitution degree and functionality. Their rich ring-opening chemistry, decarboxylative functionalization, versatility in complex heterocyclic synthesis and suitability as monomers and intermediates in new polyurethane development has significantly increased their value within a diverse combination of academic and commercial communities. What holds their future? First of all, higher-order analogues of five-membered cyclic carbonates are still much desired and new developments ought to be initiated to create such targets from new, cheap and readily available resources as this will undoubtedly advance their usage and application. Further to this, so far a limited amount of functionalities have been explored as "substituents" of cyclic carbonates (vinyl, alkyne and alcohol groups for the vast majority of the cases) and their activation and subsequent utilization in the creation of complex products. Surely a wider set of functional groups could further amplify their use in other sub-areas of synthetic and polymer chemistry, for instance in cases when functional groups are introduced that can undergo reversible chemistry (imine formation). Such reversibility may of great value in the creation of responsive materials and fully degradable polymers. A key aspect to allow for new developments centered around the use of functional cyclic carbonates is the availability of synthetic methods that permit the introduction of functionality and controls the molecular properties of these fascinating molecules. Catalysis has been and will be of primary importance to enable a bright future for these modular building blocks.

References

1. Keijzer, T.; Bakker, V.; Slootweg, C. *Nat. Chem.* **2019**, *11*, 190–195.
2. Jehanno, C.; Alty, J.; Roosen, M.; de Meester, S.; Dove, A.; Chen, E. Y.-X.; Leibfarth, F.; Sardon, H. *Nature* **2022**, *603*, 803–814.
3. Liu, Q.; Wu, L.; Jackstell, R.; Beller, M. *Nat. Commun.* **2015**, *6*, 5933.
4. Dabral, S.; Schaub, T. *Adv. Synth. Catal.* **2019**, *361*, 223–246.
5. Klankermayer, J.; Wesselbaum, S.; Beydoun, K.; Leitner, W. *Angew. Chem., Int. Ed.* **2016**, *55*, 7296–7343.
6. Yu, B.; He, L.-N. *ChemSusChem* **2015**, *8*, 52–62.
7. Rintjema, J.; Epping, R.; Fiorani, G.; Martín, E.; Escudero-Adán, E. C.; Kleij, A. W. *Angew. Chem., Int. Ed.* **2016**, *55*, 3972–3976.
8. Wang, S.; Chanjuan, X. *Chem. Soc. Rev.* **2019**, *48*, 382–404.
9. Wang, J.-L.; Miao, C.-X.; Dou, X.-Y.; Gao, J.; He, L.-N. *Curr. Org. Synth.* **2011**, *15*, 621–646.
10. Guo, W.; Gómez, J. E.; Cristòfol, À.; Xie, J.; Kleij, A. W. *Angew. Chem., Int. Ed.* **2018**, *57*, 13735–13747.
11. Allen, B. D. W.; Lakeland, C. P.; Harrity, J. P. A. *Chem. – Eur. J.* **2017**, *23*, 13830–13857.
12. Maquilón, C.; Della Monica, F.; Limburg, B.; Kleij, A. W. *Adv. Synth. Catal.* **2021**, *363*, 4033–4040.
13. McGuire, T. M.; Pérale, C.; Castaing, R.; Kociok-Köhn, G. I.; Buchard, A. *J. Am. Chem. Soc.* **2019**, *141*, 13301–13305.
14. Yu, W.; Maynard, E.; Chiaradia, V.; Arno, M. C.; Dove, A. P. *Chem. Rev.* **2021**, *121*, 10865–10907.
15. Ouhib, F.; Meabe, L.; Mahmoud, A.; Grignard, B.; Thomassin, J.-M.; Boschini, F.; Zhu, H.; Forsyth, M.; Mecerreyes, D.; Detrembleur, C. *ACS Appl. Polym. Mater.* **2020**, *2*, 922–931.
16. Sua, C.-C.; Hea, M.; Amine, R.; Chen, Z.; Sahore, R.; Dietz Rago, N.; Amine, K. *Energy Storage Mater.* **2019**, *17*, 284–292.
17. Shaikh, R. R.; Pornpraprom, V.; V. D'Elia. *ACS Catal.* **2018**, *8*, 419–450.
18. Kamphuis, A. J.; Picchioni, F.; Pescarmona, P. P. *Green Chem.* **2019**, *21*, 406–448.
19. Comerford, J. W.; Ingram, I. D. V.; North, M.; Wu, X. *Green Chem.* **2015**, *17*, 1966–1987.
20. Della Monica, F.; Kleij, A. W. *Catal. Sci. Technol.* **2020**, *10*, 3483–3501.
21. Schäffner, B.; Schäffner, F.; Verevkin, S. P.; Börner, A. *Chem. Rev.* **2010**, *110*, 4554–4581.
22. Martín, C.; Fiorani, G.; Kleij, A. W. *ACS Catal.* **2015**, *5*, 1353–1370.
23. Guerin, W.; Diallo, A. K.; Kirilov, E.; Helou, M.; Slawinski, M.; Brusson, J.-M.; Carpentier, J.-F.; Guillaume, S. M. *Macromolecules* **2014**, *47*, 4230–4235.
24. Qiao, C.; Shi, W.; Brandolese, A.; Benet-Buchholz, J.; Escudero-Adán, E. C.; Kleij, A. W. *Angew. Chem., Int. Ed.* **2022**, *61*, e202205053.
25. Song, Y.; Yang, X.; Shen, Y.; Dong, M.; Lin, Y.-N.; Hall, M. B.; Wooley, K. L. *J. Am. Chem. Soc.* **2020**, *142*, 16974–16981.
26. Gregory, G. L.; Kociok-Köhn, G.; Buchard, A. *Polym. Chem.* **2017**, *8*, 2093–2104.
27. Quienne, B.; Poli, R.; Pinaud, J.; Caillol, S. *Green Chem.* **2021**, *23*, 1678–1690.

28. Blain, M.; Jean-Gérard, L.; Auvergne, R.; Benazet, D.; Caillol, S.; Andrioletti, B. *Green Chem.* **2014**, *16*, 4286–4291.
29. Selva, M.; Fabris, M.; Lucchini, V.; Perosa, A.; Noè, M. *Org. Biomol. Chem.* **2010**, *8*, 5187–5198.
30. Guo, W.; González-Fabra, J.; Bandeira, N. A. G.; Bo, C.; Kleij, A. W. *Angew. Chem., Int. Ed.* **2015**, *54*, 11686–11690.
31. Ghosh, A. K.; Brindisi, M. *J. Med. Chem.* **2015**, *58*, 2895–2940.
32. Laserna, V.; Guo, W.; Kleij, A. W. *Adv. Synth. Catal.* **2015**, *357*, 2849–2854.
33. Sopeña, S.; Laserna, V.; Guo, W.; Martin, E.; Escudero-Adán, E. C.; Kleij, A. W. *Adv. Synth. Catal.* **2016**, *358*, 2172–2178.
34. Tomita, H.; Sanda, F.; Endo, T. *J. Polym. Sci.: Part A: Polym. Chem.* **2001**, *39*, 3678–3685.
35. Lombardo, V. M.; Dhulst, E. A.; Leitsch, E. K.; Wilmot, N.; Heath, W. H.; Gies, A. P.; Miller, M. D.; Torkelson, J. M.; Scheidt, K. A. *Eur. J. Org. Chem.* **2015**, 2791–2795.
36. Olsén, P.; Oschmann, M.; Johnston, E. V.; Åkermark, B. *Green Chem.* **2018**, *20*, 469–475.
37. Laserna, V.; Fiorani, G.; Whiteoak, C. J.; Martin, E.; Escudero-Adán, E.; Kleij, A. W. *Angew. Chem., Int. Ed.* **2014**, *53*, 10416–10419.
38. Wang, Z.; Cui, Y.-T.; Xu, Z.-B.; Qu, J. *J. Org. Chem.* **2008**, *73*, 2270–2274.
39. Guo, W.; Laserna, V.; Martin, E.; Escudero-Adán, E. C.; Kleij, A. W. *Chem. – Eur. J.* **2016**, *22*, 1722–1727.
40. Ni, J.; Cristòfol, À.; Kleij, A. W. *Org. Chem. Front.* **2021**, *8*, 4520–4526.
41. Monie, F.; Grignard, B.; Thomassin, J.-M.; Mereau, R.; Tassaing, T.; Jerome, C.; Detrembleur, C. *Angew. Chem., Int. Ed.* **2020**, *59*, 17033–17041.
42. Ni, J.; Lanzi, M.; Kleij, A. W. Organocatalytic Formal Allylic Thioetherification of Vinyl Cyclic Carbonates, *Org. Chem. Front.* **2022**, in preparation.
43. Gómez, J. E.; Guo, W.; Kleij, A. W. *Org. Lett.* **2016**, *18*, 6042–6045.
44. Zuo, L.; Liu, T.; Chang, X.; Guo, W. *Molecules* **2019**, *24*, 3930.
45. You, Y.; Li, Q.; Zhang, Y.-P.; Zhao, J.-Q.; Wang, Z.-H.; Yuan, W.-C. *ChemCatChem* **2022**, *14*, e202101887.
46. Khan, A.; Khan, S.; Khan, I.; Zhao, C.; Mao, Y.; Chen, Y.; Zhang, Y. J. *J. Am. Chem. Soc.* **2017**, *139*, 10733–10741.
47. Khan, A.; Zheng, R.; Kan, Y.; Ye, J.; Xing, J.; Zhang, Y. J. *Angew. Chem., Int. Ed.* **2014**, *53*, 6439–6442.
48. Khan, A.; Yang, L.; Xu, J.; Jin, L. Y.; Zhang, Y. J. *Angew. Chem., Int. Ed.* **2014**, *53*, 11257–11260.
49. Liu, K.; Khan, I.; Cheng, J.; Hsueh, Y. J.; Zhang, Y. J. *ACS Catal.* **2018**, *8*, 11600–11604.
50. Khan, S.; Wang, Y.; Zhang, M.-N.; Perveen, S.; Zhang, J.; Khan, A. *Org. Chem. Front.* **2022**, *9*, 456–461.
51. Xu, Y.; Salman, M.; Khan, S.; Zhang, J.; Khan, A. *J. Org. Chem.* **2020**, *85* (17), 11501–11510.
52. Khan, A.; Zhang, J.; Khan, S. *Green Chem.* **2020**, *22*, 4116–4120.
53. Cai, A.; Guo, W.; Martínez-Rodríguez, L.; Kleij, A. W. *J. Am. Chem. Soc.* **2016**, *138*, 14194–14197.
54. Guo, W.; Martínez-Rodríguez, L.; Kuniyil, R.; Martin, E.; Escudero-Adán, E. C.; Maseras, F.; Kleij, A. W. *J. Am. Chem. Soc.* **2016**, *138*, 11970–11978.
55. Guo, W.; Kuniyil, R.; Gómez, J. E.; Maseras, F.; Kleij, A. W. *J. Am. Chem. Soc.* **2018**, *140*, 3981–3987.
56. Cristòfol, À.; Limburg, B.; Kleij, A. W. *Angew. Chem., Int. Ed.* **2021**, *60*, 15266–15270.

57. Xue, S.; Cristòfol, À.; Limburg, B.; Zeng, Q.; Kleij, A. W. *ACS Catal.* **2022**, *12*, 3651–3659.
58. Yang, G.; Ke, Y.-M.; Zhao, Y. *Angew. Chem., Int. Ed.* **2021**, *60*, 12775–12780.
59. Huang, Y.; Ma, C.; Liu, S.; Yang, L.-C.; Lan, Y.; Zhao, Y. *Chem* **2021**, *7*, 812–826.
60. Yang, L.-C.; Tan, Z. Y.; Rong, Z.-Q.; Liu, R.; Wang, Y.-N.; Zhao, Y. *Angew. Chem., Int. Ed.* **2018**, *57*, 7860–7864.
61. Xu, Y.; Chen, L.; Yang, Y.-W.; Zhang, Z.; Yang, W. *Org. Lett.* **2019**, *21*, 6674–6678.
62. Ming, S.; Qurban, S.; Du, Y.; Su, W. *Chem. – Eur. J.* **2021**, *27*, 12742–12746.
63. Ke, M.; Liu, Z.; Huang, G.; Wang, J.; Tao, Y.; Chen, F. *Org. Lett.* **2020**, *22*, 4135–4140.
64. Feng, W.; Xu, L.; Li, D.-Y.; Liu, P.-N. *Org. Lett.* **2020**, *22*, 5094–5098.
65. Zhao, H.-W.; Du, J.; Guo, J.-M.; Feng, N.-N.; Wang, L.-R.; Ding, W.-Q.; Song, X.-Q. *Chem. Commun.* **2018**, *54*, 9178–9181.
66. Teichert, J. F.; Feringa, B. L. *Angew. Chem., Int. Ed.* **2010**, *49*, 2486–2528.
67. Hu, L.; Cai, A.; Wu, Z.; Kleij, A. W.; Huang, G. *Angew. Chem., Int. Ed.* **2019**, *58*, 14694–14702.
68. Wu, H.; Hu, L.; Shi, Y.; Shen, Z.; Huang, G. *ACS Catal.* **2022**, *12*, 2722–2728.
69. Wang, H.; Qiu, S.; Wang, S.; Zhai, H. *ACS Catal.* **2018**, *8*, 11960–11965.
70. Skubi, K. L.; Blum, T. R.; Yoon, T. P. *Chem. Rev.* **2016**, *116*, 10035–10074.
71. Limburg, B.; Cristòfol, À.; Kleij, A. W. *J. Am. Chem. Soc.* **2022**, *144*. https://doi.org/10.1021/jacs.2c03692.
72. For a recent account: Y. Nishibayashi. *Chem. Lett.* **2021**, *50*, 1282–1288.
73. Tian, L.; Gong, L.; Zhang, X. *Adv. Synth. Catal.* **2018**, *360*, 2055–2059.
74. Zhang, Y.-C.; Zhang, B.-W.; Geng, R.-L.; Song, J. *Org. Lett.* **2018**, *20*, 7907–7911.
75. Gómez, J. E.; Cristòfol, À.; Kleij, A. W. *Angew. Chem., Int. Ed.* **2019**, *58*, 3903–3907.
76. Tsuchida, K.; Senda, Y.; Nakajima, K.; Nishibayashi, Y. *Angew. Chem., Int. Ed.* **2016**, *55*, 9728–9732.
77. Nakajima, K.; Shibata, M.; Nishibayashi, Y. *J. Am. Chem. Soc.* **2015**, *137*, 2472–2475.
78. Lu, W.-Y.; Wang, Y.; You, Y.; Wang, Z.-H.; Zhao, J.-Q.; Zhou, M.-Q.; Yuan, W.-C. *J. Org. Chem.* **2021**, *86*, 1779–1788.
79. Zhang, Z.-J.; Zhang, L.; Geng, R.-L.; Song, J.; Chen, X.-H.; Gong, L.-Z. *Angew. Chem., Int. Ed.* **2019**, *58*, 12190–12194.
80. Guo, K.; Kleij, A. W. *Org. Lett.* **2020**, *22*, 3942–3945.
81. Guo, K.; Zeng, Q.; Villar-Yanez, A.; Bo, C.; Kleij, A. W. *Org. Lett.* **2022**, *24*, 637–641.
82. Tollefson, E. J.; Hanna, L. E.; Jarvo, E. R. *Acc. Chem. Res.* **2015**, *48*, 2344–2353.
83. Lucas, E. L.; Jarvo, E. R. *Nat. Rev. Chem.* **2017**, *1*, 0065.
84. Guo, K.; Kleij, A. W. *Angew. Chem., Int. Ed.* **2021**, *60*, 4901–4906.
85. Sontakke, G. S.; Shukla, R. K.; Volla, C. M. R. *Adv. Synth. Catal.* **2022**, *364*, 565–573.
86. Pan, T.; Gao, X.; Yang, S.; Wang, L.; Hu, Y.; Liu, M.; Wang, W.; Wu, Y.; Zheng, B.; Guo, H. *Org. Lett.* **2021**, *23*, 5750–5754.
87. Rintjema, S. J.; Kleij, A. W. *Synthesis* **2016**, *48*, 3863–3878.
88. Huang, R.; Rintjema, J.; González-Fabra, J.; Martín, E.; Escudero-Adán, E. C.; Bo, C.; Urakawa, A.; Kleij, A. W. *Nat. Catal.* **2019**, *2*, 62–70.
89. Payne, G. B. *J. Org. Chem.* **1962**, *27*, 3819–3822.
90. Sopeña, S.; Cozzolino, M.; Maquilón, C.; Escudero-Adán, E. C.; Martínez Belmonte, M.; Kleij, A. W. *Angew. Chem., Int. Ed.* **2018**, *57*, 11203–11207.
91. Qiao, C.; Villar-Yanez, A.; Sprachmann, J.; Limburg, B.; Bo, C.; Kleij, A. W. *Angew. Chem., Int. Ed.* **2020**, *59*, 18446–18451.
92. Honda, M.; Tamura, M.; Nakao, K.; Suzuki, K.; Nakagawa, Y.; Tomishige, K. *ACS Catal.* **2014**, *4*, 1893–1896.

93. Rintjema, J.; Guo, W.; Martin, E.; Escudero-Adán, E. C.; Kleij, A. W. *Chem. – Eur. J.* **2015**, *21*, 10754–10762.
94. Diao, Z.-F.; Zhou, Z.-H.; Guo, C.-X.; Yu, B.; He, L.-N. *RSC Adv.* **2016**, *6*, 32400–32404.
95. Grignard, B.; Thomassin, J.-M.; Gennen, S.; Poussard, L.; Bonnaud, L.; Raquez, J.-M.; Dubois, P.; Tran, M.-P.; Park, C. B.; Jerome, C.; Detrembleur, C. *Green Chem.* **2016**, *18*, 2206–2215.
96. Schimpf, V.; Asmacher, A.; Fuchs, A.; Bruchmann, B.; Mülhaupt, R. *Macromolecules* **2019**, *52*, 3288–3297.
97. Engels, H.-W.; Pirkl, H.-G.; Albers, R.; Albach, R. W.; Krause, J. *Angew. Chem., Int. Ed.* **2013**, *52*, 9422–9441.
98. Dechent, S.; Kleij, A. W.; Luinstra, G. A. *Green Chem.* **2020**, *22*, 969–978.
99. Bizet, B.; Grau, E.; Asua, J. M.; Cramail, H. *Macromol. Chem. Phys.* **2022**, *223*, e2100437.
100. Bähr, M.; Bitto, A.; Mülhaupt, R. *Green Chem.* **2012**, *14*, 1447–1454.
101. Estelle Rix, E.; Grau, G.; Chollet, H. C. *Eur. Polym. J.* **2016**, *84*, 863–872.
102. Dannecker, P.-K.; Meier, M. A. R. *Sci. Rep.* **2019**, *9*, 9858.

About the author

Arjan W. Kleij is an ICREA professor and ICIQ Group Leader at the Institute of Chemical Research of Catalonia (Tarragona, Spain). After his Ph.D. with Gerard van Koten at the University of Utrecht (the Netherlands), he completed two postdoctoral positions at the Autonomous University of Madrid (with Javier de Mendoza) and the University of Amsterdam (with Joost Reek), and two industrial appointments in the Netherlands at Avantium Technologies (Amsterdam) and Hexion Specialty Chemicals (Rotterdam). Since October 2006, he is based at ICIQ and focuses on the use of new catalytic processes for CO_2 and biomass valorization, the creation of biobased polymers and metal-mediated stereoselective transformations.

CHAPTER TWO

Recent strategies for the electrochemical reduction of CO_2 into methanol

Jian Zhu[a,b], Shoubhik Das[b,*], and Pegie Cool[a,*]

[a]Research Group LADCA, Department of Chemistry, University of Antwerp, Antwerp, Belgium
[b]ORSY Division, Department of Chemistry, University of Antwerp, Antwerp, Belgium
[*]Corresponding authors: e-mail address: shoubhik.das@uantwerpen.be; pegie.cool@uantwerpen.be

Contents

Abstract

Excessive CO_2 emission caused by the overuse of fossil fuels and other industrial activities have threatened toward climate change. In order to avoid further CO_2 accumulation, converting atmospheric CO_2 into value-added products is the viable solution. In this respect, electrochemical conversion of CO_2 is the most attractive way due to the fact that it can be powered by environmentally friendly energies, conducted under mild reaction conditions, and is an easy to scale-up process. However, introducing selectivity could be one of the major issues in this procedure and to improve the selectivity, an effective, durable, and sustainable catalyst is highly desired. It should be also mentioned that among all the reduced products from CO_2, methanol contains a high energy density; such that it can be directly applied to fuel cells and in other industries. Considering all these, in this chapter, we summarized the specific advances in the electrochemical reduction of CO_2 into methanol developed in the last 10 years. The fundamental insights are suggested for each kind of catalyst, which provides the guideline for the design and synthesis of novel catalysts. Finally, the perspectives and conclusions are presented for future explorations in this direction.

Advances in Catalysis, Volume 70
ISSN 0360-0564
https://doi.org/10.1016/bs.acat.2022.04.004

1. Introduction

Carbon dioxide (CO_2) emissions have been drastically increased since the end of World War II, which are due to the rapid development of petrochemical, transport, metallurgical, cement industries, and human activities *(1,2)*. As a result, the imbalance between CO_2 consumption and production has led to the greenhouse effect and climate change *(3)*. For this reason, mitigating the atmospheric CO_2 level and consequently, developing alternative strategies to avoid these CO_2 emissions, utilization of CO_2 should be the urgent task for the sustainable development of our society.

There are two different ways to tackle this problem: carbon capture and storage (CCS) and carbon capture and utilization (CCU). The CCS route requires a material with extremely high CO_2 adsorption capacity and is limited by gas separation, transport cost, and compression technology *(4–6)*. In contrast, reducing CO_2 concentration by CCU is a more appealing method, by which CO_2 can be utilized directly for other industries or can be converted into other valuable chemicals such as urea *(7)*, acrylic acid derivatives *(8)*, starch *(9)*, formic acid *(10)*, and others *(11,12)*. Recently, several technologies have been developed for the utilization of CO_2 induced by photo, thermal, plasma, electricity, and enzymes *(13)*. Among these technologies, utilization and conversion of CO_2 in an electrochemical way achieves the carbon neutrality and inter-conversions between intermittent renewable energies and CO_2 reduction products, as illustrated in Fig. 1. In addition, this process occurs under mild conditions, preferably at room temperature, and at atmospheric pressure and can be scaled up easily *(14,15)*. Since the first demonstration by Hori *(16)*, tremendous efforts have been devoted for the electrochemical CO_2 reduction reaction (ECO_2RR) by keeping focusing on designing novel catalysts *(17)*, reactors *(18)*, and gas diffusion electrodes *(19)*.

Product selectivity is of significant importance in an ECO_2RR system, which varies vastly with the alteration of employed catalysts and charge transfer numbers. For example, Kuhl et al. have detected 16 products ranging from C_1 products like formic acid (HCOOH), carbon monoxide (CO), methanol (CH_3OH), formaldehyde (CH_2O), and methane (CH_4) to C_2 and C_3 products such as oxalic acid ($H_2C_2O_4$), ethylene (C_2H_4), ethanol (C_2H_6O), propanol (C_3H_8O), and so on *(20)*. Among these reduction products, methanol is a relatively environmentally-friendly product with a high energy density of 15.6 MJ L^{-1} that is half value of the gasoline fossil fuel

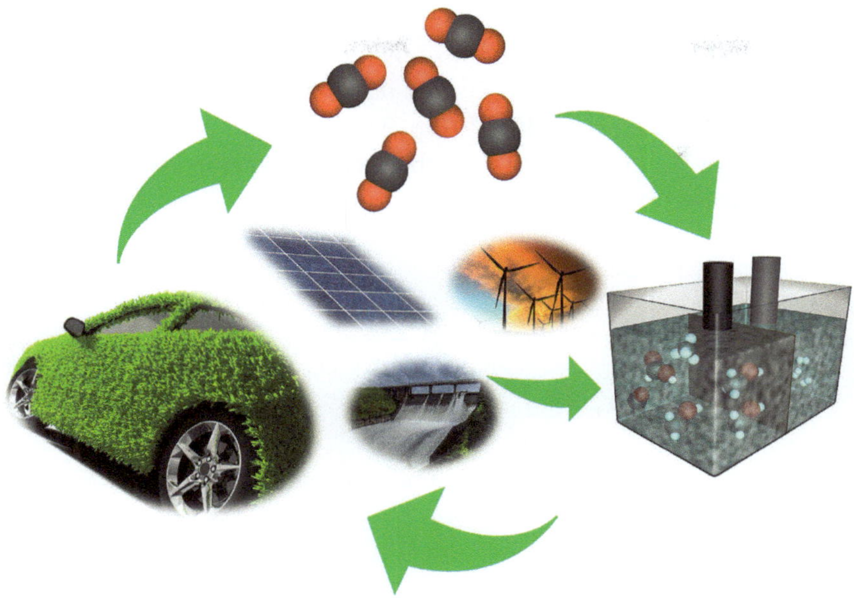

Fig. 1 Schematic illustration of inter-conversions between CO_2 and environmentally friendly energies.

(21–23). In addition, methanol is the most powerful product that can be used directly as a potential alternative in internal combustion engines and fuel cells without further modification *(24)*. Moreover, methanol is not only among the most vital intermediate for the resins and paint industry but can also be stored at atmospheric pressure *(25)*. For the reasons mentioned above, selective electroreduction of CO_2 to methanol has been an interesting research topic. However, the selective formation of methanol is highly challenging due to the extremely complex charge transfer procedure during the electrochemical CO_2 reduction progress, thus the yield and selectivity are still insufficient for commercial applications.

2. Mechanism of electroreduction of CO_2 to methanol

2.1 Illustration of the reaction mechanism

The mechanism of electrochemical reduction of CO_2 to methanol involves six electrons and protons transfer reactions, proceeding via the corresponding redox reactions below. According to the differences in the formed intermediates, two different reaction pathways have been proposed (Scheme 1) *(21)*. The first mechanism involves the generation of CO intermediate from

HOCO*, as is depicted in the equations of (1) and (2), while the other pathway involves the generation of formate intermediate (HCOO*), described in the equations of (4) and (5). Further reaction by accepting one electron generates the intermediate HCO* (Eqs. 3 and 6) which is finally reduced to CH_3OH (Eqs. 7 and 8). In summary, the overall reduction procedure (Eq. 9) requires six electrons with the formation of water as a by-product.

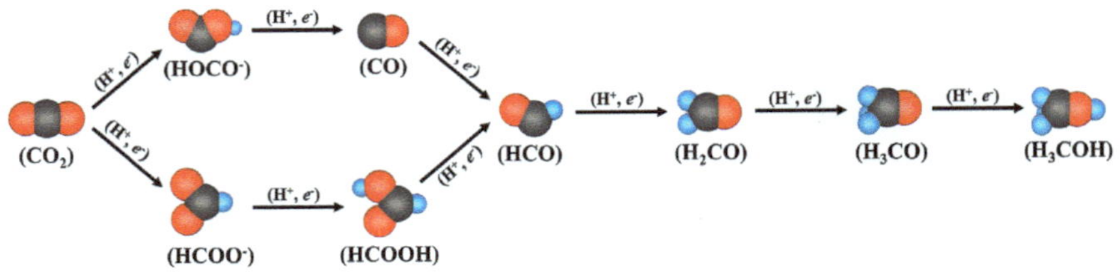

Scheme 1 The proposed two reaction pathways toward methanol formation.

Reaction pathway 1:

$$CO_2 + e^- + H^+ = HOCO^* \quad (1)$$

$$HOCO^* + e^- + H^+ = H_2O + CO \quad (2)$$

$$CO + e^- + H^+ = HCO^* \quad (3)$$

Reaction pathway 2:

$$CO_2 + e^- + H^+ = HCOO^* \quad (4)$$

$$HCOO^* + e^- + H^+ = HCOOH \quad (5)$$

$$HCOOH + e^- + H^+ = HCO^* + H_2O \quad (6)$$

Further reactions from pathway 1 and pathway 2:

$$HCO^* + e^- + H^+ = CH_2O \quad (7)$$

$$HCHO + 2e^- + 2H^+ = CH_3OH \quad (8)$$

Overall reaction:

$$CO_2 + 6e^- + 6H^+ = CH_3OH + H_2O \quad (9)$$

However, the theoretical reaction potential for the formation of methanol is only 20 mV positive than the standard hydrogen electrode (SHE). Accordingly, there is a high possibility for the hydrogen evolution reaction (HER) while the electrochemical CO_2 reduction procedure occurs *(26)*. This triggers to form a mixture of components instead of a single species during electrochemical reduction of CO_2. For example, Jaramillo et al. demonstrated 16 different reduction products, which was consistent with the results conducted experimentally and theoretically by Hori and Norskov et al *(20,27,28)*. Moreover, the multistep charge transfer procedure indicates kinetically slower reduction process. Thus, to attain high selectivity and efficiency, H_2 evolution in the system must be restricted, and thermo-kinetically feasible catalysts with high selectivity, and production rate are strongly desired.

2.2 Performances' indicators

2.2.1 Onset potential and overpotential

In order to evaluate and compare the electrochemical performances of the CO_2 reduction in a general protocol, several technical parameters are introduced, which not only relates to the catalyst itself but also influenced by external factors like electrolyte and impurities *(29)*. Regarding the specific products of an electrochemical system, the *onset potential* is defined as the starting point of the reaction, while the *overpotential* refers to the absolute difference value between the onset potential and standard reduction potential.

2.2.2 Partial current density and faradic efficiency

Partial current density is an indicator of product selectivity, which can be obtained by multiplying the normalized total current density regarding the geometric electrode area, the mass of the catalysts, or the electrochemical surface area (ECSA) with *Faradic efficiency* of the product at a given potential. *Faradic efficiency* can also be used to describe the selectivity of a given catalyst, which can be obtained according to Eqs. (10) and (11).

For gaseous product:

$$FE_{prod}\,(\%) = \frac{Q_{prod}}{Q_{total}} \times 100\% = \frac{\alpha n F}{Q_{total}} \times 100\% \tag{10}$$

For liquid product:

$$FE_{prod}\,(\%) = \frac{Q_{prod}}{Q_{total}} \times 100\% = \frac{\alpha n c F}{Q_{total}} \times 100\% \tag{11}$$

Where Q_{prod} relates to the charges for the formation of a specific product and Q_{total} is the total passed charges during the reduction process. α is the required electrons for generation of a specific product (e.g., $\alpha = 6$ for methanol). c, n, and F are the concentration, mole number of a desired product, and Faraday's constant (96,485C mol^{-1}), respectively.

2.2.3 Electrochemical surface area (ECSA)

The catalytic behaviors indicated by electrochemical surface area (ECSA) is a more direct and meaningful way, which is applicable to evaluate the electrochemical properties among a wide variety of reported different electrocatalysts *(30)*. The ECSA can be obtained by measuring the double-layer capacitance at the interface of catalyst and electrolyte, as suggested in Eq. (12).

$$ESCA = R_f \times S = \frac{C_{dl}}{C_{ref.}} \times S \qquad (12)$$

where R_f is the roughness factor, which is highly dependent on the texture of the catalyst. S is the specific surface area of a smooth metal electrode. C_{dl} represents the double-layer capacitance, C_{ref} is a benchmark value reported previously.

3. Evaluation of catalysts

3.1 Cu-based catalysts

3.1.1 Pure Cu electrode

Nowadays, Cu composited with ZnO/Al_2O_3 and their modifications have been regarded as ideal catalyst to convert CO_2 into methanol on a large scale via Fischer-Tropsch pathway *(31–33)*. Although high selectivity can be obtained, such reaction routes are limited by the poor stability of catalysts and require high energy input and hydrogen. In the early 1980s', Hori et al. reported that Cu is the only metal that could electrochemically convert CO_2 to hydrocarbons, such as CO, CH_4, C_2H_4, CH_3COOH, and C_2H_5OH *(27,34,35)*. Later, Noda et al. identified further reduced products including C_2H_5CHO, C_2H_6, C_3H_8, C_3H_6 and C_3H_7OH that were formed on the Cu electrode by regulating the applied potential *(36)*. Experiments and DFT calculation demonstrated that methane and ethylene were usually the main reduced products on the Cu electrode. While no CH_3OH was detected, the rate-determining step that produced the key intermediates such as CO and CHO played the vital role during the ECO_2RR process *(37,38)*. Hence, understanding the bonding mechanism of the preferred O—C bond or competitive H—C bond on the Cu surface provides fundamental insight into the generation of methanol.

Product selectivity is highly related to the electrode surface conditions, which displays different adsorption capacity for the generated intermediates.

Schouten et al. explored the possible intermediates of CO_2 reduction on Cu foil by reducing C_1 and C_2 organics containing oxygen coupled with in-situ mass spectrometry. They suggested methanol as the main product could be produced through the direct protonation of the $^*CH_2OH$ intermediate *(39)*. Peterson and co-workers suggested a different reaction route, which produced methanol on Cu(211) as follows: First, the formyl (CHO) intermediate was reduced to formaldehyde (CH_2O) and followed by the formation of a methoxy species (CH_3O) consequently. Then, methanol was produced by the hydrogenation reaction between the CH_3O intermediate and protons *(28)*. Very recently, Nie et al. described a new reaction mechanism by elucidating the role of kinetics of elementary steps in CO_2 electroreduction with the introduction of H_2O molecule to the model. Their research proposed a COH (—C—O—H) intermediate with a lower activation barrier than the CHO (H—C=O) intermediate toward the rate-determining step, which could be converted to CHOH and CH_2OH subsequently. Finally, methanol was formed by reducing CH_2OH intermediate, as is shown in Fig. 2 *(40)*.

Although the feasibility of electrochemical CO_2 reduction illustrated by theoretical calculation studies, there were only few reports on the methanol formation on a pure Cu electrode. For instance, as much as 16 kinds of products were detected when a Cu foil electrode was employed by Kuhl et al *(41)*. Nevertheless, only very small amount of methanol was observed

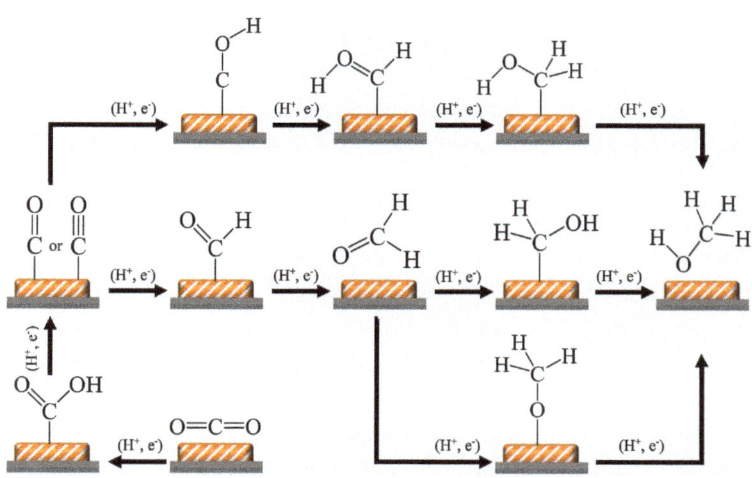

Fig. 2 The proposed reaction pathways on pure Cu electrode.

despite the detection sensitivity was maximized by enlarging the ratio of the electrode geometric surface to the counter electrode with a small electrolyte volume. The low methanol production rate was related to the variations of CO binding energy on various metals surface and surface conditions *(42–45)*.

Later, Xiao et al. explored pH dependent routes to obtain the major products and identified the key intermediate for ECO_2RR on the Cu (111) electrode when solvent was taken into consideration *(46)*. As can be seen in Fig. 3A, they demonstrated the critical role of surface-bound H_2O, which provided the hydroxyl group in the key intermediates (HCOH and H_2COH) with proton directly and then the hydroxyl group in adsorbed H_2O was reduced to adsorbed OH^-, which made ΔG as pH dependent. Whereas, the formed H_2COH intermediate thermodynamically favored the formation of methane and ethylene over the methanol formation with a barrier slightly lower by 0.05 eV when pH 1, as displayed in Fig. 3B. In addition, Le Duff et al. certified that the OH species that were covered on the electrode surface played a role that interacted with the adsorbed CO, while the electrode structure and surface roughness of the Cu cathode had little influence on the formation of methanol *(47)*. Of interest, Lum et al. proposed evidences of different product distributions on oxide-derived Cu, which was in contrast to the oriented Cu (100) and Cu (111) surface. They elucidated that the variation of binding strength regarding the intermediates at different grain boundaries was responsible for the product selectivity, although the same intermediate was obtained *(48)*. Later, Gao et al. revealed the effects of the oxygen atom under the surface of the plasma-treated Cu (100) toward ECO_2RR performances, and indicated that current density and faradic efficiency changed with the variation of the oxygen concentration as displayed in Fig. 3C and D *(49)*. Ma and co-workers determined the effect of Cu nanowire morphology with different lengths on ECO_2RR performances. They found that HER activity was strictly suppressed with the increase of nanowire length, and the product selectivity highly relied on the electrolyte pH *(50)*. Besides, other factors such as particle size *(51)*, grain boundaries *(52)*, and defects *(53)* also had significant influence on the selectivity and distribution of the electrochemical reduction products.

In the following sections, several defected Cu and its derivative-based catalysts for instance single atom Cu catalysts and Cu deposited onto a support (Cu@support) will be discussed regarding improving the methanol selectivity.

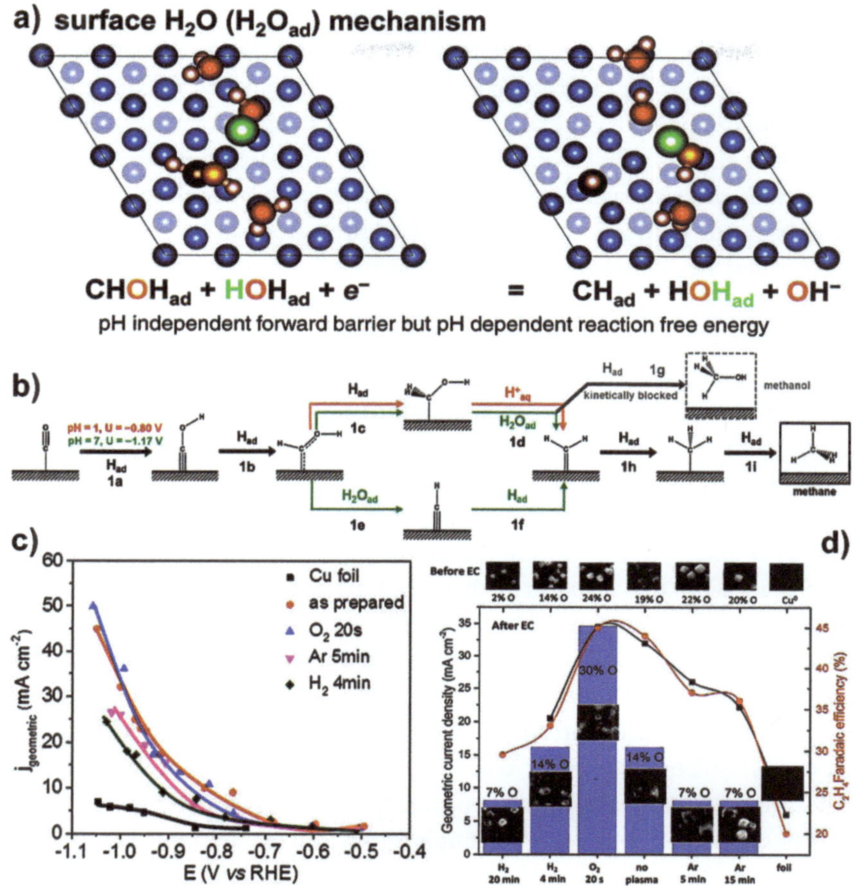

Fig. 3 (A) The mechanism for electrochemical dehydration reduction illustrated with the $CHOH_{ad}$ case: surface H_2O (H_2O_{ad}). (B) Predicted complete pH dependent pathways for C1 products. (C) Potential-dependent geometric current density after 1 h of electrochemical reaction in a CO_2-saturated 0.1 M $KHCO_3$ solution. (D) Geometrical current density (left axis) and Faradaic efficiency for C_2H_4 (right axis) after 1 h of electrochemical reaction at −1.0 V vs RHE in CO_2-saturated 0.1 M $KHCO_3$ for different plasma treatments as well as for a reference electropolished Cu foil. The oxygen content of the nanocubes after the reaction (from EDX) is indicated with column bars (blue). The oxygen content of the cubes before the reaction, as well as typical scanning electron microscope (SEM) images (lateral image size: 1.7 μm) before (top row) and after the reaction (images inside the blue columns), are shown as insets. The cubes contain ~24 at.% Cl before the reaction, which is depleted during the reaction. *Panel (A and B) Reprinted with permission from Xiao, H.; Cheng, T.; Goddard, W.A. 3rd.* J. Am. Chem. Soc. ***2017,*** 139 *(1), 130–136. Copyright 2017 American Chemical Society. Panel (C and D) Reprinted with permission from Gao, D.; Zegkinoglou, I.; Divins, N.J.; Scholten, F.; Sinev, I.; Grosse, P.; Roldan Cuenya, B.* ACS Nano ***2017,*** 11 *(5), 4825–4831. Copyright 2017 American Chemical Society.*

3.1.2 Single Cu atom catalysts

The strategy of downsizing catalysts has been widely applied to electrochemical catalysis for improving conversion efficiency and maximizing utilization of catalysts. In general, the smaller the size of the catalysts, the higher geometric activity will be obtained due to the increase of electrochemically active sites when compared with bulk materials and planar electrodes. As the particle size is reduced, the strength of adsorption of H*, OH*, and CO* species will be increased, which is comparable to some noble metals for electrocatalysis *(54,55)*. The improved electrochemical performances can be attributed to low-coordinated surface atoms and high curvature as compared to the catalysts with a larger size *(51)*.

In terms of ECO_2RR for methanol formation, there are a few reports on single atom Cu catalyst. The theoretical study on single atom catalysts for ECO_2RR was first conducted by Back et al. *(56)* As is shown in Fig. 4A, they suggested the feasibilities of methanol formation through both *OCHO and *COOH intermediate on single atom catalysts, which was consistent with the reaction path on a Cu surface reported by Peterson et al. as mentioned above *(28)*. When single-atom Cu was decorated onto ZnS substrate, the ECO_2RR followed the same procedure with a *CHO intermediate, where the positively charged Cu remained as the active site and can capture adsorbate. It is worthy to note that the bond length of CO was enlarged, which was helpful for further protonation *(57)*. However, Lee and co-workers reported a different reaction path involving a *HCOO intermediate on a single atom Cu catalyst based on First-principle *(58)*. As illustrated in Fig. 4B, CO_2 was protonated to form HCOO* followed by a series of reduction steps to H_2COO*, H_2COOH*, CH_2O*, CH_3O*, and CH_3OH subsequently, which was thermodynamically more favorable than the CHO* pathway as presented in Fig. 4C. Recently, Yang et al. prepared self-supported through-hole carbon nanofibers decorated with isolated single Cu atom by an electrospinning method and subsequent calcination (Fig. 4D) *(59)*. When evaluated as the catalyst for electrochemical CO_2 reduction, the CuSAs/TCNFs (through-hole carbon nanofibers) demonstrated a high methanol faradic efficiency of 44% and long-time catalytic stability of 50 h at $-0.9\ V_{RHE}$ as depicted in Fig. 4E and F. Fig. 5A displays the preparation of single Cu atom catalyst that was immobilized on a MXene layer (large family of 2D metal carbides and carbonitrides) by selective etching one vulnerable element and leaving the other rigid Cu element onto the MXene *(60)*. As presented in Fig. 5B, the SA-Cu-MXene delivered a maximum methanol FE of

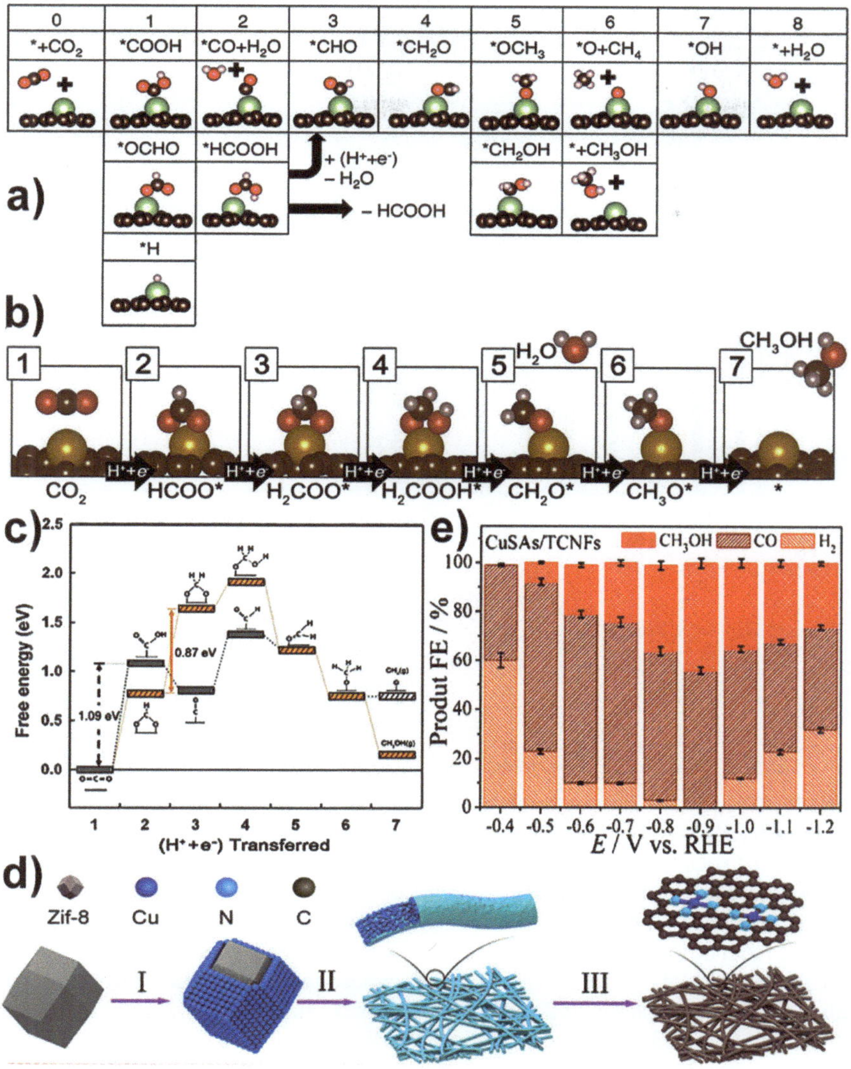

Fig. 4 (A) Key reaction intermediate species for the CO_2 reduction reaction. The numbers on top of each column are the numbers of proton-electron pairs ($H^+ + e^-$) transferred to CO_2. (B) Adsorbed configurations of intermediates in the most favorable CO_2 reduction mechanism on the Cu/G catalyst. Orange, red, brown, and white spheres represent Cu, O, C, and H atoms, respectively. (C) Comparison of most favorable mechanisms corresponding to the COOH and HCOO pathways on the Cu/G catalyst. The COOH and HCOO paths are represented by gray and orange hatched bars, respectively. The two paths coincide from CH_2O* onwards. (D) Synthesis procedure of CuSAs/THCF: I, adsorption of Cu ions; II, electrospinning of polymer fibers; III, carbonization and etching,

(Continued)

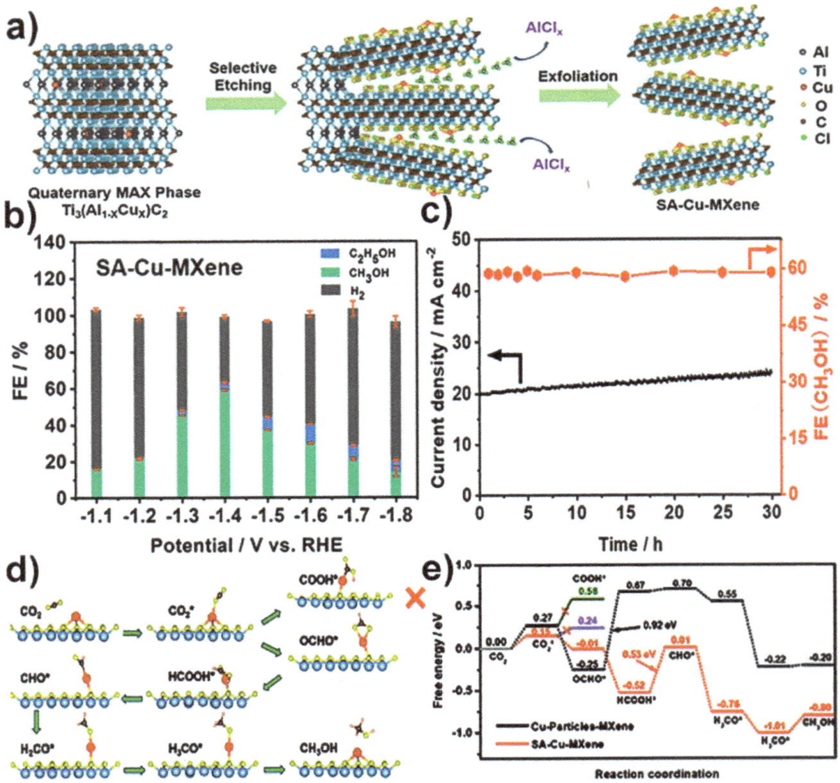

Fig. 5 (A) Schematic illustration of the fabrication of SA-Cu-MXene via selective etching quaternary MAX-$Ti_3(Al_{1-x}Cu_x)C_2$. Gray, blue, red, yellow, brown, and green balls represent Al, Ti, Cu, O, C, and Cl atoms, respectively. (B) FEs of SA-Cu-MXene. (C) Current-time responses and corresponding FEs (CH_3OH) of SA-Cu-MXene for CO_2 reduction at an overpotential of −1.4 V for 30 h. (D) DFT calculations proposed a reaction pathway for the functionalization of CO_2 to methanol on isolated Cu of SA-Cu-MXene. (E) Free energy diagram of CO_2 to CH_3OH on Cu—O_3 structure. *Reprinted with permission from Zhao, Q.; Zhang, C.; Hu, R.; Du, Z.; Gu, J.; Cui, Y.; Chen, X.; Xu, W.; Cheng, Z.; Li, S.; Li, B.; Liu, Y.; Chen, W.; Liu, C.; Shang, J.; Song, L.; Yang, S.* ACS Nano ***2021,*** 15 *(3), 4927–4936. Copyright 2021 American Chemical Society.*

Fig. 4—Cont'd (E) Faradaic efficiencies of all products at CuSAs/TCNFs. *Panel (A) Reprinted with permission from Back, S.; Lim, J.; Kim, N.Y.; Kim, Y.H.; Jung, Y.* Chem. Sci. ***2017,*** 8 *(2), 1090–1096. Copyright 2017 Royal Society of Chemical. Panel (B and C) Reprinted with permission from Lee, C.-M.; Senthamaraikannan, T.G.; Shin, D.Y.; Kwon, J.A.; Lim, D.-H.* Comput. Theor. Chem. ***2021,*** 1201, *113277. Copyright 2021 Elsevier. Panel (D and E) Reprinted with permission from Yang, H.; Wu, Y.; Li, G.; Lin, Q.; Hu, Q.; Zhang, Q.; Liu, J.; He, C.* J. Am. Chem. Soc. ***2019,*** 141 *(32), 12717–12723. Copyright 2019 American Chemical Society.*

59.1% at $-1.4\ V_{RHE}$, which is much higher than the Cu particle counterpart and pure MXene catalyst. Moreover, it showed high electrolysis durability regarding both the current density and methanol FE in Fig. 5C. DFT calculation revealed that CO_2 reduction on SA-Cu-MXene experienced a downhill type route with an *OCHO intermediate in terms of thermodynamics, which was more favorable than on a Cu-particle-MXene catalyst or through a COOH* intermediate, as suggested in Fig. 5D and E.

3.1.3 Cu_xO catalysts

Cu-based catalysts in the form of oxides have shown that they can react with CO_2 much easier than metallic Cu catalysts because of the relatively higher electron density *(61,62)*. Moreover, previous studies have shown that Cu(I) active sites were especially active to promote catalytic CO_2 activity and selectivity toward oxygenates, which allowed valance band electrons to participate in and improved CO adsorption. Further, Cu(I) sites were believed to stabilize reaction intermediates to methanol such as formate (*COOH) and methoxy adsorbates (H_3CO^*) *(63)*. For example, Le et al. suggested a higher methanol production rate of 43 μmol $cm^{-2}\ h^{-1}$ on the electrodeposited Cu_2O thin film electrode than on the air oxidized and anodized CuO electrode *(64)*. However, such catalysts were easily reduced while performing CO_2 reduction at negative potentials, which led to the poor products selectivity and catalytic stability *(65)*.

In order to further improve the catalytic activity and target product selectivity, several strategies have been adapted to stabilize Cu_xO by optimizing the surface morphology with specific crystal structure and tuning the electronic structure. For instance, Liu and co-workers prepared a series of Cu_2O nanocubes that were highly oriented to different crystal facets of Cu_2O, as displayed in Fig. 6A *(66)*. Electrochemical performances, as well as the DFT calculations, demonstrated in Fig. 6B–D that product distribution was highly correlated to the crystal facets, where C_1 alcohol is preferred to be formed at the (110) facet. In addition, the difference in bond length, surface reconstruction, and oxygen-vacancy concentration also contributed to the variation products distribution. Chang et al. achieved a high methanol production rate of 53.6% in a Cu/Cu_2O photo-electrocatalytic cell by tuning the interfaces length *(67)*. Yang and co-workers obtained a porous Metal Organic Framework-derived $Cu@Cu_2O$ catalyst and demonstrated the existence of Cu/Cu_2O interface by tuning the calcination temperature under aerobic atmosphere as displayed in Fig. 6E and F *(68)*. The optimized sample delivered a high methanol FE of 45% at $-0.7\ V_{RHE}$ in Fig. 6G and H that was higher than other samples, which could be ascribed to the

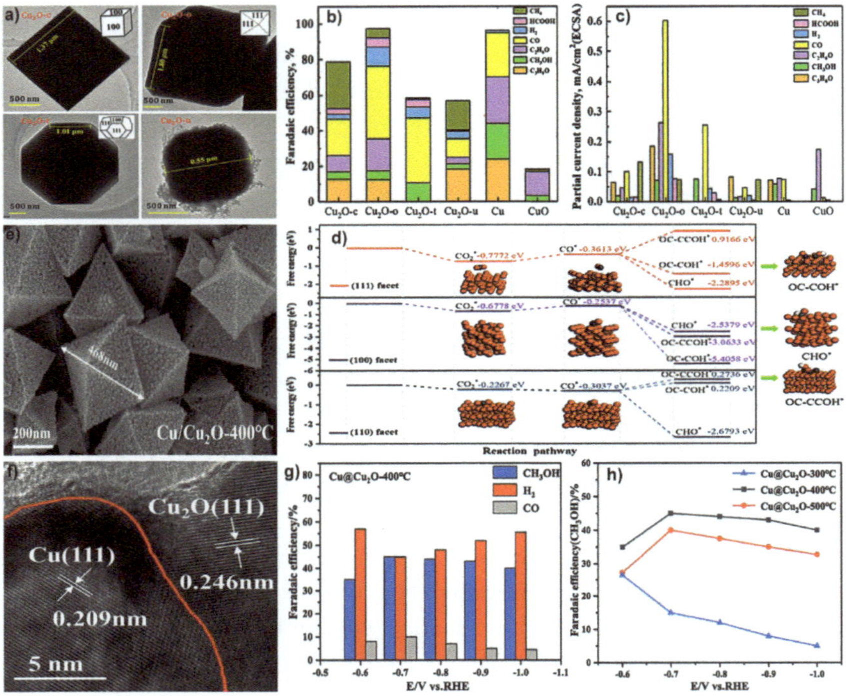

Fig. 6 (A) Transmission electron microscope (TEM) images of the different Cu_2O catalysts: Cu_2O-c, Cu_2O-o, Cu_2O-t, and Cu_2O-u. (B) FEs of the products of the different Cu_2O catalysts, bulk Cu and CuO. (C) The ECSA-normalized partial current densities for the varied products measured at −0.3 V vs RHE. (D) Free energy of the intermediates that appear in CO_2RR on (111), (100), and (110) crystal facets of Cu_2O catalysts. (E) SEM images of Cu@Cu_2O electrocatalysts derived from Cu-BTC (1,3,5-benzenetricarboxylic acid) pyrolysis at 400 °C. (F) High-resolution transmission electron microscope (HRTEM) image of Cu@Cu_2O-400 °C. (G) Faradic efficiencies of all products on Cu@Cu_2O-400 °C. (H) CH_3OH over Cu@Cu_2O-T electrocatalysts at various applied potentials in CO_2-saturated 0.5 M $KHCO_3$. *Panel (A–D) Reprinted with permission from Liu, B.; Yao, X.; Zhang, Z.; Li, C.; Zhang, J.; Wang, P.; Zhao, J.; Guo, Y.; Sun, J.; Zhao, C.* ACS Appl. Mater. Interfaces ***2021,*** 13 *(33), 39165–39177. Copyright 2021 American Chemical Society. Panel (E–H) Reprinted with permission from Yang, X.; Cheng, J.; Yang, X.; Xu, Y.; Sun, W.; Zhou, J.* Chem. Eng. J. ***2022,*** 431, *134171. Copyright 2022 Elsevier.*

interactions between the moderate bound CO^* and adsorbed H^* at the Cu/Cu^+ interface site. Although the improved methanol selectivity on the Cu_2O/Cu electrode, the catalytic activity was suffering from degradation because of the inevitable reduction of the metal oxide.

In order to stabilize Cu_xO based catalysts, improving the electrochemical performances by constructing various interfaces with Cu_2O stood out to be one of the most effective ways. For example, Prakash et al. constructed a Cu_2O/Polypyrrole (Cu_2O/Ppy) interface by coating the Cu_2O electrode with Ppy polymer, which delivered an extremely high methanol FE of 93% at -0.85 V_{RHE} and excellent catalytic stability of 15 h with negligible FE decay. The Ppy coating layer in the composite not only prevented the Cu_2O from reducing but facilitated mass transport and CO_2 adsorption due to the formed intermediates concentration gradient *(69)*. In another study, the maximum methanol FE of 66.4% was accompanied with a high production yield of 1.27×10^{-4} mol m^{-2} s^{-1} which was obtained with a CuO electrode containing partial Cu_2O, while the improved catalytic durability should contribute to the introduction of TiO_2 atom layer *(70)*.

Besides, ECO_2RR performances at the interfaces of Cu/CuO, Cu_2O/ZnO, CuO/ZnO, and Cu/Zn-foil, Cu/Cu_2O/ZnO were also explored and displayed selective relationships between the products distribution and applied potentials *(71–77)*. These researchers elucidated that the simultaneous reduction of Cu_2O and HER can be suppressed after the introduction of the second element. Furthermore, the established interfaces led to richer active sites for CO_2 adsorption and conversion, which facilitated the formation and stabilization of diverse intermediates that were essential for methanol generation.

3.1.4 Cu@substrate

3.1.4.1 Metal-free catalysts

Metal-free materials are also appealing as potential catalysts for ECO_2RR for their excellent stability, high efficiency, renewability, and low cost. Although the active sites and mechanisms are still under debate, various explorations have certified that improving the performances of metal-free catalysts by heteroatom installation is an effective way. For example, Wu et al. illustrated that ECO_2RR toward methanol through hydrogenation of *CHO with H* moieties that absorbed on the N sites is the easiest way that required much lower activation energy *(78)*. Yuan and co-workers synthesized N-doped carbon materials with mosses as the precursor as displayed in Fig. 7A *(79)*. They demonstrated that the main products were methane and ethanol upon the nitrogen doping but low methanol FE, which showed high ECO_2RR activity and production rate in Fig. 7B and C due to the high concentration of pyridinic-/graphitic-N accompanied and induced defects that could lead to the occurrences of ECO_2RR. Pyrazinic N also had vital effects on product selectivity of ECO_2RR, which could protect the active pyridinic-N from been consumption and decomposition *(80)*.

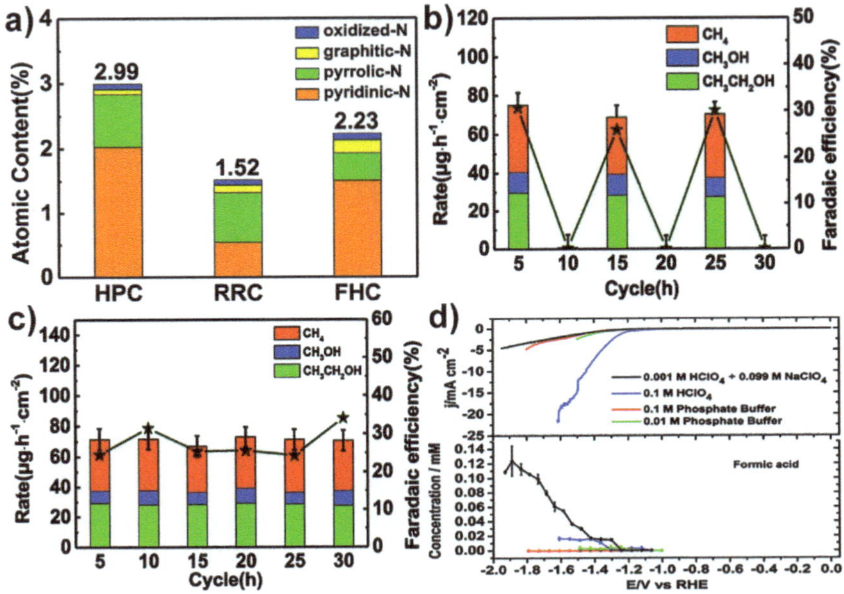

Fig. 7 (A) Comparison of the atomic concentrations of different structural configurations of doped N in the HPC, RRC, and FHC samples, as obtained via the XPS test. (B) Production yields and Faradaic efficiencies of the HPC sample at −1.2 V (vs Ag/AgCl) in 6 cycles (each for 5 h) of alternating operation in the CO_2-saturated and Ar-saturated electrolytes. (C) Production yields and Faradaic efficiencies at −1.2 V in each of these 6 cycles. (D) Formic acid formation during reduction of formaldehyde in perchloric acid (pH 1 and pH 3), 0.1 M phosphate buffer pH 6.8, and 0.01 M phosphate buffer pH 6.8. Scan rate: 1 mV s^{-1}. *Panel (A–C) Reprinted with permission from Yuan, H.; Qian, X.; Luo, B.; Wang, L.; Deng, L.; Chen, Y.* Sci. Total Environ. ***2020,*** 739, *140340. Copyright 2020 Elsevier. Panel (D) Reprinted with permission from Birdja, Y.Y.; Koper, M.T.* J. Am. Chem. Soc. ***2017,*** 139 *(5), 2030–2034. Copyright 2017 American Chemical Society.*

Boron-doped diamond (BDD) has also been considered as one of the most potential metal-free alternatives for ECO_2RR because of its wide potential window, high stability, mechanical durability, and low back current *(81)*. Jiwanti and co-workers reported a BDD catalyst with a high methanol FE of 24.3% in an ammonia solution at the potential of $-1.3\ V_{Ag/AgCl}$ *(82)*. Birdja et al. referred to this base-catalyzed ECO_2RR to Cannizzaro-type reactions, which was strongly influenced by the local pH at the electrode-electrolyte interface and buffer capacity as displayed in Fig. 7D *(83)*. They observed that HCOOH and HCHO were the main liquid products accompanied with only a trace amount of methanol in an acid electrolyte, which demonstrated a superior product selectivity to CH_3COOH on the BDD electrode. Mou et al. synthesized a boron phosphide (BP) catalyst through a vacuum-sealing process followed by heat treatment. The as-prepared BP electrode displayed a superior ECO_2RR activity with a high methanol FE of 92%, excellent cyclibility, and strong resistance to pH due to the synergistic effects of the B and P atoms that can promote the binding and activation of CO_2 *(84)*.

3.1.4.2 Cu@support catalyst

Dispersing catalysts onto support materials such as carboneous materials, TiO_2, and so on by physically interacting or coordinating with heteroatom is a promising strategy for ECO_2RR. Thanks to the synergistic effects of high electronic conductivity, physical stability, and tunability of substrates, catalysts decorated on substrate usually display exceptional performances *(85–87)*. Marepally et al. prepared an oxygen functionalized carbon nanotube (CNT) which was decorated with various amounts of Cu nanoparticles. They found that oxygenated functional groups on CNT could anchor Cu nanoparticles and could act as an electron trapping site that could stimulate the reduction of the adsorbed species *(88)*. Albo et al. reported that CuO/ZnO loaded gas diffusion electrode (GDE) produced methanol as the main product with a high conversion efficiency of 27.5%. The gas diffusion electrode applied to the system could not only promote mass transport due to its porosity but also enabled much easier CO_2 availability *(89)*.

3.1.4.3 Cu@TiO_2 catalyst

Recently, TiO_2 has been found as an ideal catalyst for the reduction of CO_2 using photocatalysis owing to its relatively low conduction band *(90)*. Several studies have identified that $Ti^{3+}O^-$ formed on the surface of TiO_2 was the catalytically active site that was favorable for charge transfer and stabilization of $CO_2^{\cdot -}$ intermediate *(91,92)*. However, the mechanism of using TiO_2 as the catalyst for CO_2 electrochemical reduction is still unclear. Ramesha et al. reported the conductivity of one-electron reduction of CO_2 to form $CO_2^{\cdot -}$ in a nonaqueous solution. They proved that Ti^{3+} played an important role in bonding with CO_2, which could reduce the reduction potential *(93)*. Teh et al. further studied tandem electroreduction of CO_2 to methanol starting with HCOOH that can be hydrogenated directly at the Ti^{3+} sites and oxygen vacancy sites (TOV) rather than through an $^*OCH_2$ intermediate as the previous reports. They demonstrated a positive linear relationship between the methanol production rate and the concentration of oxygen vacancies, and the highest methanol FE of

12.6% was obtained at $-1.0\ V_{RHE}$ *(94)*. Yuan et al. studied the ECO_2RR performances by combining Cu/TiO_2 nanoparticles on a N-doped graphene. A maximum methanol FE of 19.5% was obtained, which was relatively higher than Cu/NG, TiO_2/NG, Cu/TNT, and Cu cluster on ZnO *(95–97)*. Nevertheless, the current densities obtained are still too low to scale-up production.

3.2 Precious metal-based catalysts and their alloys

3.2.1 Precious metals and oxides

Precious metals such as Au, Pt, Ag, Pd, Ru, Ir, and so on, have generally been regarded as the third group that produces CO and H_2 as the main products while showing less reactivity for the further reduced products during the ECO_2RR process because of the relatively weaker CO^* binding strength *(98)*. For example, Jaramillo et al. have investigated the feasibility of ECO_2RR on Ag and Au surface, while only less than 0.1% production rate of methanol was obtained *(99,100)*. Although some catalysts with high methanol FE were obtained by modifying the substrates with doped nitrogen, adenine and histamine, the current densities were extremely low *(101–103)*. Zhang and co-workers synthesized a hierarchy Pd/SnO_2 by loading SnO_2 onto 2D Pd nanosheets. They demonstrated superior CO_2 electrochemical reduction properties because of the abundant optimized Pd-O-Sn interfaces produced by tuning the ratio of Pd to Sn, which improved the adsorption capacities of CO_2 and modulated the CO binding ability by partially exposed Pd *(104)*.

DFT calculations conducted on RuO_2 (110) that have the same rutile phase structure as IrO_2 (110) displayed a superior methanol selectivity to that of either methane or CO. The HER activity can be suppressed while using CO as the spectator and taking low CO^* coverage at the bridge site or coordinatively unsaturated (cus) sites into consideration *(105–108)*. Recently, Zhang et al. prepared a multi-active site catalyst Ag-Co_3O_4-CeO_2/LGC with rich oxygen vacancy defects in CeO_2 *(109)*. They tuned the electronic structure by introducing transition metals to low graphitized carbon and achieved the smallest activation barrier as shown in Fig. 8A. Accordingly, the Ag-Co_3O_4-CeO_2/LGC electrode showed a small onset potential in Fig. 8B, long catalytic stability in Fig. 8C, and a high alcohol selectivity of 77.6% at $-0.85\ V_{RHE}$ although the complex product distribution as presented in Fig. 8D.

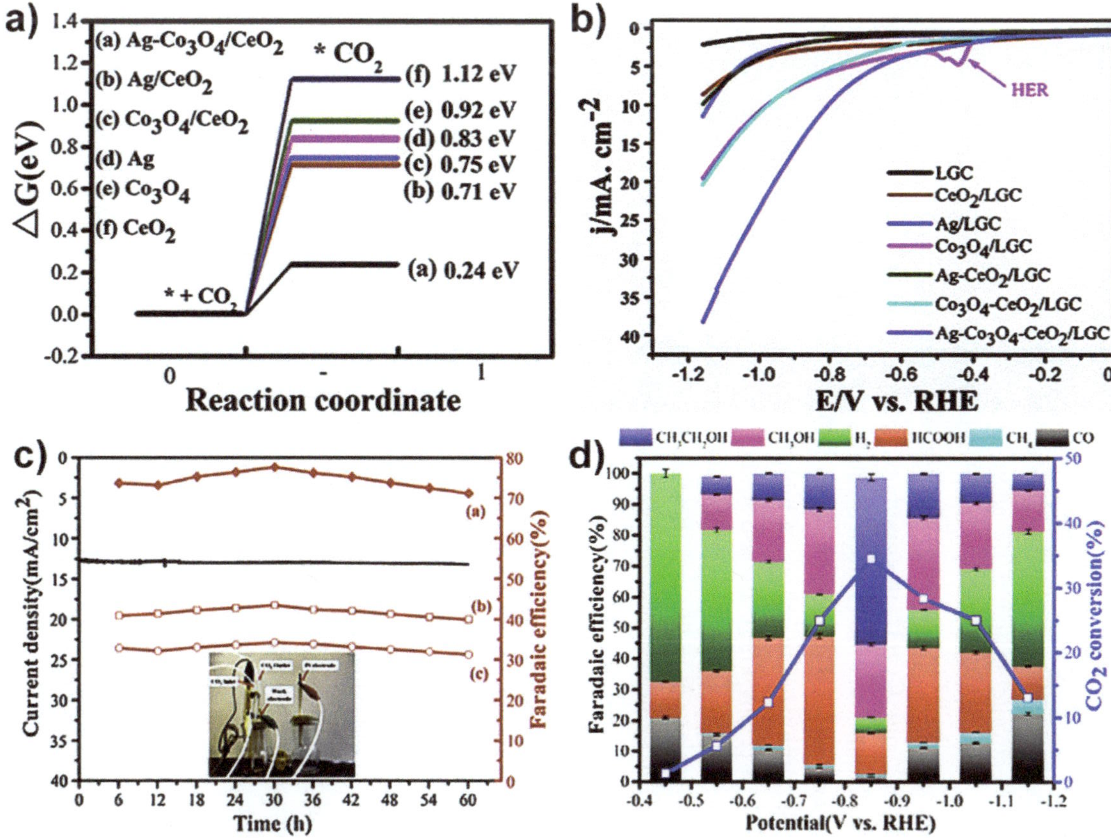

Fig. 8 See figure legend on next page.

3.2.2 Precious metal-based alloys

To improve the methanol conversion rate, as discussed above, various catalysts have been extensively discussed. Alloys consisting of at least two metals are also appealing catalytic materials for electrochemically converting CO_2 to methanol, where one of the metallic components acted as an active site while the other one usually worked as a stabilizer because of the differences in binding ability with various intermediates. Thus, modulating the electronic structure by alloying is also a promising way *(110,111)*. DFT calculations have demonstrated that the key of methanol selectivity was CH_2OH^* intermediate that could be produced by the reduction of HCO^* and COH^* intermediate, which preferred to be protonated at the C atom over the O atom on alloys *(112)*. They concluded that the binding energies and concentrations of CO^*, H^*, and OH^* on alloy surfaces had to be optimized to achieve a high FE and selectivity during the ECO_2RR process *(113)*. There are also studies that have suggested that the compositions, morphologies, oxidation states, and defects have a significant influence on product selectivity and distribution *(114,115)*.

Low et al. explored the electrochemical performances of Zn dendrites that were deposited on various substrates by a pulse-deposition method. The highest methanol FE of 10.5% was obtained on a Zn dendrite electrode with Ag foam substrate due to the 3D porous structure and low coordination numbers of Zn in Ag/Zn alloy *(116)*. Payra and co-workers studied the ECO_2RR performances of Pt/Zn nano-alloys by varying synthesis temperature *(117)*. A series of alloys were obtained by changing Pt/Zn ratios, and the optimized Pt_xZn/C electrode indicated a high methanol FE of 81.4% and superior durability of 16 h at the potential of $-0.9\ V_{RHE}$, as is shown in Fig. 9A–C. Lu and co-workers developed a series of Pt_xCo_y bimetallic aerogels by a template-free self-assembly process *(118)*. They demonstrated that electrochemical performances highly depended on the Pt_xCo_y

Fig. 8 (A) The adsorption free energy of $^*CO_2^{\delta-}$ on CeO_2, Co_3O_4, Ag, Co_3O_4/CeO_2, Ag/CeO_2 and Ag-Co_3O_4/CeO_2 at 0 V. (B) Linear sweep voltammetry (LSV) scans of different samples with CO_2 in 0.1 M $KHCO_3$. (C) The CO_2ER stability test (60 h) of Ag-Co_3O_4-CeO_2/LGC catalyst at −0.85 V vs RHE in CO_2-saturated 0.1 M $KHCO_3$. (Left axis): total geometric current density, (right axis): FE, (a) total lower alcohols, (b) CH_3OH, (c) CH_3CH_2OH. (D) CO_2 conversion (right axis) and FE (left axis) of CO_2ER at different potential with steady CO_2 flow (3 mL min^{-1}) on Ag-Co_3O_4-CeO_2/LGC. *Reprinted (adapted) with permission from Zhang, Q.; Du, J.; He, A.; Liu, Z.; Tao, C. J.* CO_2 Util. ***2019,*** 34, *635–645. Copyright 2019 Elsevier.*

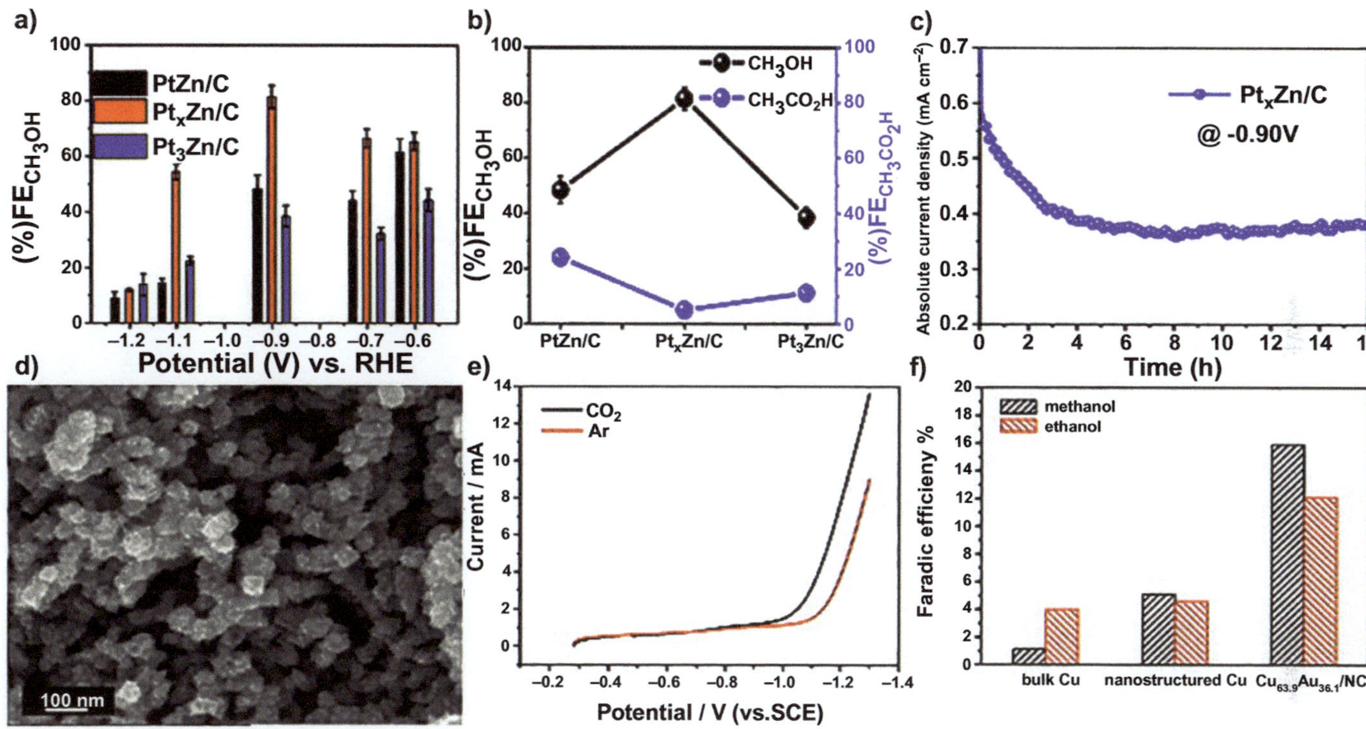

Fig. 9 See figure legend on next page.

compositions, and a high methanol FE of 80% with ultra-stability on the $Pd_{87}Co_{13}$ electrode was achieved. Jia et al. prepared nanostructured Cu/Au alloys as ECO_2RR catalysts by an electrochemical deposition method with a nanoporous Cu film as template as displayed in Fig. 9D *(119)*. They suggested high reductivity in the CO_2 saturated phosphate buffered solution (PBS) in Fig. 9E and an improved FE was obtained on $Cu_{63.9}Au_{36.1}$/NCF in Fig. 9F, where the alloys could facilitate alcohol production through catalyzing the conversion of CO intermediate.

3.3 Transition metal-based composites

Transition metals and their derivatives are among the best catalysts for electrocatalysis that have been studied for decades owing to their tunable *d* orbitals that have different spin-polarizations. The slight differences in electronic structure among elements may lead to vast changes toward ECO_2RR selectivity and reaction path, which may result from the different priorities of binding strength to key intermediates *(120,121)*. Therefore, it will be of significance to develop a catalytic system with a tunable and moderate adsorption ability to key intermediates. Several strategies have been adopted to modulate the electronic structure of electrodes. For example, Sun et al. obtained a high methanol FE of 71.2% by doping Bi into MoS_2 electrode, where the doped Bi atoms could drive CO generation, while CO could bind with Mo atoms and could be further reduced to methanol with H_2 that was produced on Mo sites *(122,123)*. Sun et al. studied electrochemical performances of FeP(211) array on Ti mesh by exposing abundant reactive sites with enhanced adsorption and reduction abilities of CO_2 *(124)*. The obtained electrode suggested an extremely high total alcohol FE of 94.3% with 80.2% methanol at the potential of $-0.2\ V_{RHE}$

Fig. 9 (A) Faradaic efficiencies of CH_3OH in CO_2RR over the intermetallic nano-alloys as a function of potentials. (B) Comparison Faradaic efficiencies for CH_3OH and CH_3CO_2H at −0.90 V over the three intermetallic nano-alloys. (C) A 16 h long-term CPE at −0.90 V over Pt_xZn/C show the remarkable stability of current density. (D) SEM images of Cu/Au sample deposited on NCF. (E) LSV curves of $Cu_{63.9}Au_{36.1}$/NCF in 0.2 M PBS solution saturated with CO_2 (black line) and Ar (red line). (F) Faradic efficiencies of methanol and ethanol by use of different electrodes in 0.5 M $KHCO_3$ solution. *Panel (A–C) Reprinted with permission from Payra, S.; Shenoy, S.; Chakraborty, C.; Tarafder, K.; Roy, S.* ACS Appl. Mater. Interfaces ***2020,*** 12 *(17), 19402–19414. Copyright 2020 American Chemical Society. Panel (D–F) Reprinted with permission from Jia, F.; Yu, X.; Zhang, L.* J. Power Sources ***2014,*** 252, *85–89. Copyright 2014 Elsevier.*

and displayed superior durability and cyclability at all the measured potentials. Zhao et al. achieved a FeS_2/NiS nanocomposite by a hydrothermal method, and the TEM image in Fig. 10A proved the FeS_2/NiS interface *(125)*. When assessed as the catalyst for ECO_2RR, the obtained FeS_2/NiS catalyst displayed superior ECO_2RR activity to the single composite as displayed in Fig. 10B. They obtained a high methanol FE of 64% with an unprecedented overpotential of 280 mV (Fig. 10C) and stable catalysis (Fig. 10D) owing to the synergistic effects of nanostructured FeS_2/NiS and corresponding interfaces. Moreover, modulating ECO_2RR selectivity toward methanol by decorating active materials onto modified substrates *(126,127)*, and morphologies control *(128,129)* have been demonstrated and summarized in Table 1.

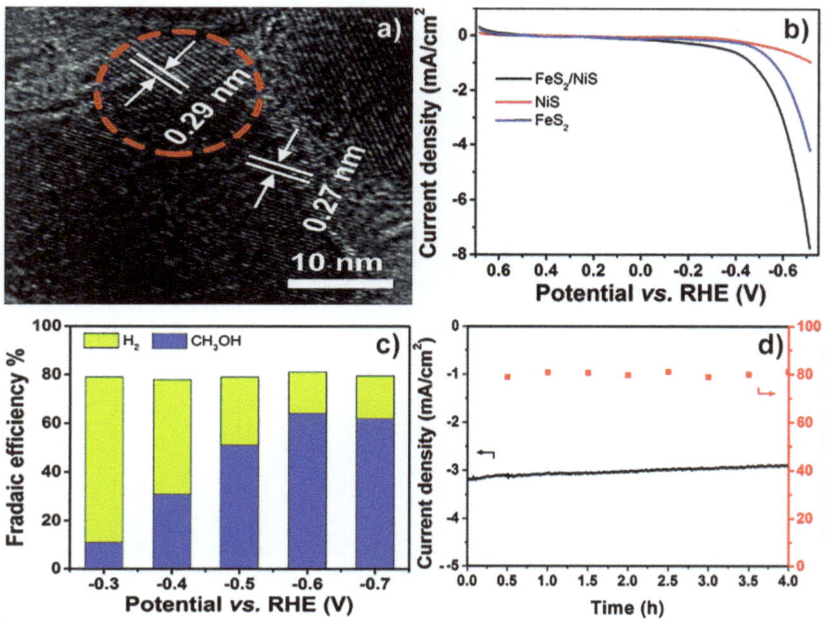

Fig. 10 (A) The HRTEM image of FeS_2/NiS nanocomposite. (B) LSVs of FeS_2/NiS nanocomposite (black trace), NiS (red trace), and FeS_2 (blue trace) in 0.5 M $KHCO_3$ aqueous solution under CO_2 atmosphere. (C) Faradaic efficiencies vs potential of CH_3OH and H_2 for FeS_2/NiS nanocomposite. (D) Stability of FeS_2/NiS nanocomposite for CO_2 reduction operated at a potential of −0.6 V vs RHE for 4 h. Geometric current density vs time (left axis) and total FE for production (CH_3OH and H_2) vs time (right axis). *Reprinted with permission from Zhao, S.; Guo, S.; Zhu, C.; Gao, J.; Li, H.; Huang, H.; Liu, Y.; Kang, Z.* RSC Adv. ***2017,*** *7 (3), 1376–1381. Copyright 2017 Elsevier.*

Table 1 Summary of heterogeneous catalysts and related current densities and Faradic efficiencies for CH_3OH production via ECO_2RR.

Catalyst	Electrolyte	Current density	Faradic efficiency (%)	Stability	Reference
SA-Cu-Mxene	0.1 M $KHCO_3$	21.3 mA cm^{-2}@−1.4 V_{RHE}	59.1%@−1.4 V_{RHE}	30 h@−1.6 V_{RHE}	*(60)*
Cu_2O NPs	0.5 M $KHCO_3$	6.5 mA cm^{-2}@-0.3 V_{RHE}	4.9%@-0.3 V_{RHE}	400 min@−0.3 V_{RHE}	*(66)*
Cu_2O/CuO	0.5 M $KHCO_3$ + 10 mM pyridine	46 mA cm^{-2}@−1.4 $V_{Ag/AgCl}$	6.46%@−1.3 $V_{Ag/AgCl}$	2 h@−0.3 $V_{Ag/AgCl}$	*(71)*
CuO nanowires	0.1 M $KHCO_3$	56.8 mA cm^{-2}@−1.8 $V_{Ag/AgCl}$	66.4%@−1.4 $V_{Ag/AgCl}$	2 h@−1.8 $V_{Ag/AgCl}$	*(70)*
CuSAs/TCNFs	0.1 M $KHCO_3$	–	44%@0.9 V_{RHE}	50 h@0.9 V_{RHE}	*(59)*
Cu_2O nano-catalyst	× M $KHCO_3$	–	47%@2.0 V_{SHE}	50 min @−2.0 V_{SHE}	*(63)*
Cu/Cu_2O-ZnO-7 h	0.1 M $KHCO_3$	−0.15 mA cm^{-2}@−1.0 $V_{Ag/AgCl}$	67.22%@−1.7 $V_{Ag/AgCl}$	0.5 h@−1.2 $V_{Ag/AgCl}$	*(72)*
Cu@Cu_2O-400 °C	0.5 M $KHCO_3$	–	45@−0.7 V_{RHE}	10 h	*(68)*
$Cu_2O_{(OL-MH)}$/Ppy LT paper	0.5 M $KHCO_3$	0.223 mA cm^{-2}@−0.85 V_{RHE}	93 ± 1.2%@0.85 V_{RHE}	15 h@−0.85 V_{RHE}	*(69)*
Cu_2O-ZnO(1:1)	0.5 M $KHCO_3$	–	17.7%@−1.3 $V_{Ag/AgCl}$	5 h@−1.3 $V_{Ag/AgCl}$	*(74)*
Cu(core)/CuO (Shell)	1 M $KHCO_3$	17.3 mA cm^{-2}@−1.35 $V_{Ag/AgCl}$	2.5%@−1.35 $V_{Ag/AgCl}$	–	*(75)*

$Zn_{0.05}$@Cu-30	1.14 M $NaHCO_3$	–	45%@−1.27 V_{SCE}	6 h@−1.27 V_{SCE}	*(73)*
Cu_2O@CC	0.5 M $NaHCO_3$	−5.57 mA cm^{-2}@−1.7 V_{SCE}	–	4 h @−1.7 V_{SCE}	*(86)*
Cu_2O/ZnO	0.5 M $KHCO_3$	−1.16 $V_{Ag/AgCl}$@10 mA cm^{-2}	27.5%@ 10 mA cm^{-2}	20 h	*(89)*
HPC	0.5 M $KHCO_3$	50 mA cm^{-2}@−1.6 $V_{Ag/AgCl}$	10.5%@−1.2 $V_{Ag/AgCl}$	30 h@−1.2 $V_{Ag/AgCl}$	*(79)*
BP	0.1 M $KHCO_3$	–	92%@−0.5 V_{RHE}	18 h@−0.6 $V_{Ag/AgCl}$	*(84)*
OD Cu/C-1000	0.1 M $KHCO_3$	–	43.2%@−0.3 V_{RHE}	15 h@−0.7 V_{RHE}	*(87)*
$Ti_{an,\ 3\ min}$	0.1 M K_2SO_4 + 0.5 M HCOOH	−2 mA cm^{-2}@−1 V_{RHE}	12.6%@−1 V_{RHE}	6 h@−1 V_{RHE}	*(94)*
Cu/TiO_2/NG	0.2 M KI	0.061 mA cm^{-2}@−0.2 V_{RHE}	19.5%@−0.2 V_{RHE}	20 h@−0.2 V_{RHE}	*(95)*
10% Cu-TNT	0.5 M $KHCO_3$	0.0055 A cm^{-2}@−1.5 V_{SCE}	4%@−2.5 V_{SCE}	100 min @−1.7 V_{SCE}	*(96)*
3 ML Cu/ZnO	0.1 M $KHCO_3$	12 mA cm^{-2}@−1.4 $V_{Ag/AgCl}$	2.8%@−1.4 $V_{Ag/AgCl}$	–	*(97)*
$Pd_{83}Cu_{17}$	25 mol% [Bmim] BF_4 in water	31.8 mA cm^{-2}	80%@−2.1 $V_{Ag/AgCl}$	24 h@−2.1 $V_{Ag/AgCl}$	*(118)*
Cu/Zn-foil	0.1 M $KHCO_3$	−30 mA cm^{-2}@−1.4 V_{RHE}	48%@−1.0 V_{RHE}	2000 s@−1.9 V_{RHE}	*(76)*

Continued

Table 1 Summary of heterogeneous catalysts and related current densities and Faradic efficiencies for CH_3OH production via ECO_2RR.—cont'd

Catalyst	Electrolyte	Current density	Faradic efficiency (%)	Stability	Reference
PD-Zn/Ag foam	0.1 M $KHCO_3$	-2.1 mA cm^{-2}@-1.38 V_{RHE}	8.1%@-1.38 V_{RHE}	8 h@-1.38 V_{RHE}	*(116)*
C-Py-Sn-Zn	0.1 M $KHCO_3$	–	59.9% @-0.5 V_{RHE}	26 h @-0.5 V_{RHE}	*(115)*
Ni_3Al	0.1 M K_2SO_4	-2.1 mA cm^{-2}@-1.38 $V_{Ag/AgCl}$	1%@-1.38 $V_{Ag/AgCl}$	–	*(110)*
Mo-Bi BMC/CP	0.5 M [Bmim]BF_4 in MeCN	12.1 mA cm^{-2}@-0.7 V_{SHE}	71.2%@-0.7 V_{SHE}	5 h@-0.7 V_{SHE}	*(122)*
Mo-Bi BMC@CNT-15%	60 wt% [Emim]BF_4	5.6 mA cm^{-2}@-0.3 V_{SCE}	81%@-0.3 V_{SCE}	12 h@-0.3 V_{SCE}	*(123)*
Pt_xZn/C	0.1 M $KHCO_3$	–	81.4%@-0.9 V_{RHE}	16 h@-0.9 V_{RHE}	*(117)*
Pt@Adenine-rGO	0.1 M KNO_3	0.5 mA cm^{-2}@-0.3 $V_{Ag/AgCl}$	85 ± 2.1%@-0.3 $V_{Ag/AgCl}$	3 h@-0.3 $V_{Ag/AgCl}$	*(102)*
Pd/SnO_2 Nss	0.1 M $KHCO_3$	-1.45 mA cm^{-2}@-0.24 V_{RHE}	54.8 ± 2%@-0.24 V_{RHE}	24 h@-0.24 V_{RHE}	*(104)*
Ag-Co_3O_4-CeO_2/LGC	0.1 M $KHCO_3$	12.8 mA cm^{-2}@-0.85 V_{RHE}	23.4%@-0.85 V_{RHE}	60 h@-0.85 V_{RHE}	*(109)*
FeS_2/NiS	0.5 M $KHCO_3$	7.8 mA cm^{-2}@-0.68 V_{RHE}	64%@-0.6 V_{RHE}	4 h@-0.6 V_{RHE}	*(125)*

FeP NA/TM	0.5 M $KHCO_3$	–	80.2%@−0.2 V_{RHE}	36 h@−0.2 V_{RHE}	*(124)*
Fe_2P2S_6	0.5 M $KHCO_3$	–	65.2%@−0.2 V_{RHE}	30 h@−0.5 V_{RHE}	*(128)*
CoO/CN/Ni	0.5 M $KHCO_3$	10.6 mA cm^{-2}@−0.7 V_{RHE}	70.7%@−0.7 V_{RHE}	10 h@−0.7 V_{RHE}	*(127)*
$Co(CO_3)_{0.5}(OH)\cdot 0.11H_2O$	0.1 M $KHCO_3$	0.59 mA cm^{-2}@−0.98 V_{SCE}	97%@−0.98 V_{SCE}	10 h@−0.98 V_{SCE}	*(129)*
PO-5 nm Co/ SL-NG	0.1 mol dm^{-3} KCO_3	4 mA cm^{-2}@−0.9 V_{SCE}	71.4%@−0.9 V_{SCE}	10 h@−1.2 V_{SCE}	*(126)*

4. Future perspectives and conclusions

CO_2 alleviation and conversion in order to reach carbon neutrality coupled with the assistance of intermittent clean energy have been highly significant following the Kyoto Protocol and the Paris Agreement. The assessed catalysts so far reported indicated that formic acid, H_2, and CO are the main reduction products and exhibited less reactivity and selectivity toward the further reduced products, especially to methanol. The main obstacles for electrochemical CO_2 reduction originate from the slow kinetics of multistep electron-proton transport and variations of binding preference with key intermediates on the electrode surface. From this perspective, the rational design and synthesis of catalysts are mainly focused on modulating conversion rate and selectivity by exposing more active sites on electrode surfaces. Therefore, the major topic for controlling product selectivity in a methanol electrosynthesis system is to precisely tune surface structure that has moderate adsorption ability to key intermediates.

In this chapter, we have stressed on the Cu-based catalysts, started with discussing possible reaction paths and followed by manipulating strategies on Cu electrodes. Then, we have summarized three other typical groups of catalysts based on the designing strategies on electrode surface for facilitating methanol generation in CO_2 electroreduction with high selectivity and low overpotential. Such designing strategies aim to modify the surface properties by controlling catalyst morphology and size, controlling crystal facets, introducing interfaces, alloying with a second metal, and others, which may alter the electronic structure of the as-discussed catalysts and decrease the reaction energy barrier. Finally, we summarized research results obtained over the last 10 years.

Although some progress has been made over the past decades, the reaction mechanism varies with the intrinsic nature of electrode surface and post-treatments. Most catalysts process with a COOH* intermediate while RuO_2 (110) usually results in an OCHO* intermediate toward the same product with different potential limiting step (PLS). In addition, the product selectivity and conversion rate for methanol production are still too low to meet the requirements for large-scale applications. It still remains the biggest challenge to precisely regulate surface structure toward establishing a dependence relationship between adsorption energy and diverse intermediates. Therefore, the specific reaction mechanisms should be carefully clarified by coupling with advanced characterization and theoretical simulations, which should provide constructive insights for the designing of catalysts and their modifications. Advanced technologies such as operando spectroscopic measurement help to quantify and detect the active sites and key intermediates dynamically at atom/molecular scale. The development of theoretical simulations provides an in-depth understanding of possible reaction feasibilities from the perspectives of thermodynamics and reaction energetics. With the rapid development of these advanced technologies, certainly, electrochemical CO_2 reduction will be one of the very important technologies to reach a sustainable global economy.

Acknowledgments

Jian Zhu acknowledges the funding from the China Scholarship Council (CSC No. 202008440279) for a PhD stipend.

References

1. Zheng, T.; Jiang, K.; Wang, H. *Adv. Meter.* **2018**, *30* (48), e1802066.
2. Jiang, Y.; Long, R.; Xiong, Y. *Chem. Sci.* **2019**, *10* (31), 7310–7326.
3. Roy, A.; Jadhav, H. S.; Park, S. J.; Seo, J. G. *J. Alloys Compd.* **2021**, *887*, 161449.
4. Navarro-Jaén, S.; Virginie, M.; Bonin, J.; Robert, M.; Wojcieszak, R.; Khodakov, A. Y. *Nat. Rev. Chem.* **2021**, *5* (8), 564–579.
5. Nitopi, S.; Bertheussen, E.; Scott, S. B.; Liu, X.; Engstfeld, A. K.; Horch, S.; Seger, B.; Stephens, I. E. L.; Chan, K.; Hahn, C.; Norskov, J. K.; Jaramillo, T. F.; Chorkendorff, I. *Chem. Rev.* **2019**, *119* (12), 7610–7672.
6. Rubin, E. S.; Davison, J. E.; Herzog, H. J. *Int. J. Greenhouse Gas Control* **2015**, *40*, 378–400.
7. Peterson, S. L.; Stucka, S. M.; Dinsmore, C. J. *Org. Lett.* **2010**, *12* (6), 1340–1343.
8. Wang, X.; Wang, H.; Sun, Y. *Chem* **2017**, *3* (2), 211–228.
9. Cai, T.; Sun, H.; Qiao, J.; Zhu, L.; Zhang, F.; Zhang, J.; Tang, Z.; Wei, X.; Yang, J.; Yuan, Q.; Wang, W.; Yang, X.; Chu, H.; Wang, Q.; You, C.; Ma, H.; Sun, Y.; Li, Y.; Li, C.; Jiang, H.; Wang, Q.; Ma, Y. *Science* **2021**, *373* (6562), 1523–1527.
10. Yang, H.; Kaczur, J. J.; Sajjad, S. D.; Masel, R. I. J. *J. CO_2 Util.* **2017**, *20*, 208–217.
11. Ren, J.; Li, F. F.; Lau, J.; Gonzalez-Urbina, L.; Licht, S. *Nano Lett.* **2015**, *15* (9), 6142–6148.
12. Mita, T.; Chen, J.; Sugawara, M.; Sato, Y. *Org. Lett.* **2012**, *14* (24), 6202–6205.
13. Saravanan, A.; Senthil Kumar, P.; Vo, D.-V. N.; Jeevanantham, S.; Bhuvaneswari, V.; Anantha Narayanan, V.; Yaashikaa, P. R.; Swetha, S.; Reshma, B. *Chem. Eng. Sci.* **2021**, *236*, 116515.
14. Birdja, Y. Y.; Pérez-Gallent, E.; Figueiredo, M. C.; Göttle, A. J.; Calle-Vallejo, F.; Koper, M. T. M. *Nat. Energy* **2019**, *4* (9), 732–745.
15. Han, N.; Ding, P.; He, L.; Li, Y.; Li, Y. *Adv. Energy Mater.* **10, 2020**, 1902338.
16. Yoshio, H.; Katsuhei, K.; Akira, M.; Shin, S. *Chem. Lett.* **1986**, *15* (6), 897–898.
17. Kuo, L.; Dinh, C.-T. *Curr. Opin. Electrochem.* **2021**, *30*, 100807.
18. Liu, X.; Li, B.-Q.; Ni, B.; Wang, L.; Peng, H.-J. *J. Energy Chem.* **2022**, *64*, 263–275.
19. Nguyen, T. N.; Dinh, C.-T. *Chem. Soc. Rev.* **2020**, *49* (21), 7488–7504.
20. Kuhl, K. P.; Cave, E. R.; Abram, D. N.; Jaramillo, T. F. *Energy Environ. Sci.* **2012**, *5* (5), 7050–7059.

21. Albo, J.; Alvarez-Guerra, M.; Castaño, P.; Irabien, A. *Green Chem.* **2015**, *17* (4), 2304–2324.
22. Al-Rowaili, F. N.; Jamal, A.; Ba Shammakh, M. S.; Rana, A. *ACS Sustain. Chem. Eng.* **2018**, *6* (12), 15895–15914.
23. Olah, G. A. *Angew. Chem., Int. Ed.* **2013**, *52* (1), 104–107.
24. Li, X.; Faghri, A. *J. Power Sources* **2013**, *226*, 223–240.
25. Omae, I. *Coord. Chem. Rev.* **2012**, *256* (13–14), 1384–1405.
26. Liu, Y.; Li, F.; Zhang, X.; Ji, X. *Curr. Opin. Green Sustain. Chem.* **2020**, *23*, 10–17.
27. Hori, Y.; Murata, A.; Takahashi, R. *J. Chem. Soc., Faraday Trans. 1* **1989**, *85* (8), 5022–5023.
28. Peterson, A. A.; Abild-Pedersen, F.; Studt, F.; Rossmeisl, J.; Norskov, J. K. *Energy Environ. Sci.* **2010**, *3* (9), 1311–1315.
29. Zhang, L.; Zhao, Z. J.; Gong, J. *Angew. Chem., Int. Ed.* **2017**, *56* (38), 11326–11353.
30. Clark, E. L.; Resasco, J.; Landers, A.; Lin, J.; Chung, L.-T.; Walton, A.; Hahn, C.; Jaramillo, T. F.; Bell, A. T. *ACS Catal.* **2018**, *8* (7), 6560–6570.
31. Pastor-Pérez, L.; Baibars, F.; Le Sache, E.; Arellano-García, H.; Gu, S.; Reina, T. R. *J. CO_2 Util.* **2017**, *21*, 423–428.
32. Liao, F.; Huang, Y.; Ge, J.; Zheng, W.; Tedsree, K.; Collier, P.; Hong, X.; Tsang, S. C. *Angew. Chem., Int. Ed.* **2011**, *50* (9), 2162–2165.
33. Zhang, Y.; Zhong, L.; Wang, H.; Gao, P.; Li, X.; Xiao, S.; Ding, G.; Wei, W.; Sun, Y. *J. CO_2 Util.* **2016**, *15*, 72–82.
34. Hori, Y.; Kikuchi, K.; Murata, A.; Suzuki, S. *Chem. Lett.* **1986**, *15* (6), 897–898.
35. Hori, Y.; Kikuchi, K.; Suzuki, S. J. C. L. *Chem. Lett.* **1985**, *14* (11), 1695–1698.
36. Hidetomo, N.; Shoichiro, I.; Yoshiyuki, O.; Kaname, I. *Chem. Lett.* **1989**, *18* (2), 289–292.
37. Schouten, K. J.; Qin, Z.; Perez Gallent, E.; Koper, M. T. *J. Am. Chem. Soc.* **2012**, *134* (24), 9864–9867.
38. Hori, Y.; Takahashi, I.; Koga, O.; Hoshi, N. *J. Phys. Chem. B* **2002**, *106* (1), 15–17.
39. Schouten, K. J. P.; Kwon, Y.; van der Ham, C. J. M.; Qin, Z.; Koper, M. T. M. *Chem. Sci.* **2011**, *2* (10), 1902–1909.
40. Nie, X.; Esopi, M. R.; Janik, M. J.; Asthagiri, A. *Angew. Chem., Int. Ed.* **2013**, *52* (9), 2459–2462.
41. Kuhl, K. P.; Cave, E. R.; Abram, D. N.; Jaramillo, T. F. *Energy Environ. Sci.* **2012**, *5* (5), 7050–7059.
42. Kuhl, K. P.; Hatsukade, T.; Cave, E. R.; Abram, D. N.; Kibsgaard, J.; Jaramillo, T. F. *J. Am. Chem. Soc.* **2014**, *136* (40), 14107–14113.
43. Hori, Y.; Wakebe, H.; Tsukamoto, T.; Koga, O. *Electrochim. Acta* **1994**, *39* (11 – 12), 1833–1839.
44. Hori, Y.; Murata, A.; Yoshinami, Y. *J. Chem. Soc., Faraday Trans.* **1991**, *87* (1), 125–128.
45. Hori, Y.; Murata, A.; Takahashi, R.; Suzuki, S. *J. Am. Chem. Soc.* **1987**, *109* (16), 5022–5023.
46. Xiao, H.; Cheng, T.; Goddard, W. A., 3rd. *J. Am. Chem. Soc.* **2017**, *139* (1), 130–136.
47. Le Duff, C. S.; Lawrence, M. J.; Rodriguez, P. *Angew. Chem., Int. Ed.* **2017**, *56* (42), 12919–12924.
48. Lum, Y.; Ager, J. W. *Nat. Catal.* **2018**, *2* (1), 86–93.
49. Gao, D.; Zegkinoglou, I.; Divins, N. J.; Scholten, F.; Sinev, I.; Grosse, P.; Roldan Cuenya, B. *ACS Nano* **2017**, *11* (5), 4825–4831.
50. Ma, M.; Djanashvili, K.; Smith, W. A. *Angew. Chem., Int. Ed.* **2016**, *55* (23), 6680–6684.
51. Reske, R.; Mistry, H.; Behafarid, F.; Roldan Cuenya, B.; Strasser, P. *J. Am. Chem. Soc.* **2014**, *136* (19), 6978–6986.

52. Li, C. W.; Ciston, J.; Kanan, M. W. *Nature* **2014**, *508* (7497), 504–507.
53. Liu, X.; Xiao, J.; Peng, H.; Hong, X.; Chan, K.; Norskov, J. K. *Nat. Commun.* **2017**, *8*, 15438.
54. Jin, H.; Liu, X.; Chen, S.; Vasileff, A.; Li, L.; Jiao, Y.; Song, L.; Zheng, Y.; Qiao, S.-Z. *ACS Energy Lett.* **2019**, *4* (4), 805–810.
55. Mukerjee, S.; McBreen, J. *J. Electroanal. Chem.* **1998**, *448* (2), 163–171.
56. Back, S.; Lim, J.; Kim, N. Y.; Kim, Y. H.; Jung, Y. *Chem. Sci.* **2017**, *8* (2), 1090–1096.
57. Kour, G.; Mao, X.; Du, A. *ChemNanoMat* **2020**, *6* (12), 1806–1811.
58. Lee, C.-M.; Senthamaraikannan, T. G.; Shin, D. Y.; Kwon, J. A.; Lim, D.-H. *Comput. Theor. Chem.* **2021**, *1201*, 113277.
59. Yang, H.; Wu, Y.; Li, G.; Lin, Q.; Hu, Q.; Zhang, Q.; Liu, J.; He, C. *J. Am. Chem. Soc.* **2019**, *141* (32), 12717–12723.
60. Zhao, Q.; Zhang, C.; Hu, R.; Du, Z.; Gu, J.; Cui, Y.; Chen, X.; Xu, W.; Cheng, Z.; Li, S.; Li, B.; Liu, Y.; Chen, W.; Liu, C.; Shang, J.; Song, L.; Yang, S. *ACS Nano* **2021**, *15* (3), 4927–4936.
61. Li, C. W.; Kanan, M. W. *J. Am. Chem. Soc.* **2012**, *134* (17), 7231–7234.
62. Lee, S.; Kim, D.; Lee, J. *Angew. Chem., Int. Ed.* **2015**, *54* (49), 14701–14705.
63. Hazarika, J.; Manna, M. S. *Electrochim. Acta* **2019**, *328*, 135053.
64. Le, M.; Ren, M.; Zhang, Z.; Sprunger, P. T.; Kurtz, R. L.; Flake, J. C. *J. Electrochem. Soc.* **2011**, *158* (5), E45–E49.
65. Nie, X.; Griffin, G. L.; Janik, M. J.; Asthagiri, A. *Catal. Commun.* **2014**, *52*, 88–91.
66. Liu, B.; Yao, X.; Zhang, Z.; Li, C.; Zhang, J.; Wang, P.; Zhao, J.; Guo, Y.; Sun, J.; Zhao, C. *ACS Appl. Mater. Interfaces* **2021**, *13* (33), 39165–39177.
67. Chang, X.; Wang, T.; Zhao, Z. J.; Yang, P.; Greeley, J.; Mu, R.; Zhang, G.; Gong, Z.; Luo, Z.; Chen, J.; Cui, Y.; Ozin, G. A.; Gong, J. *Angew. Chem., Int. Ed.* **2018**, *57* (47), 15415–15419.
68. Yang, X.; Cheng, J.; Yang, X.; Xu, Y.; Sun, W.; Zhou, J. *Chem. Eng. J.* **2022**, *431*, 134171.
69. Periasamy, A. P.; Ravindranath, R.; Senthil Kumar, S. M.; Wu, W. P.; Jian, T. R.; Chang, H. T. *Nanoscale* **2018**, *10* (25), 11869–11880.
70. Azenha, C.; Mateos-Pedrero, C.; Alvarez-Guerra, M.; Irabien, A.; Mendes, A. *Electrochim. Acta* **2020**, *363*, 137207.
71. Roy, A.; Jadhav, H. S.; Gil Seo, J. *Electroanalysis* **2020**, *33* (3), 705–712.
72. Munir, S.; Varzeghani, A. R.; Kaya, S. *Sustainable Energy Fuels* **2018**, *2* (11), 2532–2541.
73. Zhang, W.; Zhou, Q.; Qi, J.; Li, N. *React. Kinet. Mech. Catal.* **2021**, *134* (1), 243–257.
74. Albo, J.; Sáez, A.; Solla-Gullón, J.; Montiel, V.; Irabien, A. *Appl. Catal. B* **2015**, *176–177*, 709–717.
75. Lan, Y.; Ma, S.; Lu, J.; Kenis, P. J. *Int. J. Electrochem. Sci.* **2014**, *9* (12), 7300–7308.
76. Ashok, A.; Kumar, A.; Saad, M. A. S.; Al-Marri, M. J. *J. CO_2 Util.* **2021**, *53*, 101749.
77. Dongare, S.; Singh, N.; Bhunia, H.; Bajpai, P. K. *Electrochim. Acta* **2021**, *392*, 138988.
78. Wu, H.-Z.; Bandaru, S.; Liu, J.; Li, L.-L.; Jin, L. *Carbon* **2020**, *169*, 73–81.
79. Yuan, H.; Qian, X.; Luo, B.; Wang, L.; Deng, L.; Chen, Y. *Sci. Total Environ.* **2020**, *739*, 140340.
80. Li, W.; Fechler, N.; Bandosz, T. J. *Appl. Catal., B* **2018**, *234*, 1–9.
81. Liu, Y.; Zhang, Y.; Cheng, K.; Quan, X.; Fan, X.; Su, Y.; Chen, S.; Zhao, H.; Zhang, Y.; Yu, H.; Hoffmann, M. R. *Angew. Chem., Int. Ed.* **2017**, *56* (49), 15607–15611.
82. Jiwanti, P. K.; Natsui, K.; Nakata, K.; Einaga, Y. *RSC Adv.* **2016**, *6* (104), 102214–102217.
83. Birdja, Y. Y.; Koper, M. T. *J. Am. Chem. Soc.* **2017**, *139* (5), 2030–2034.

84. Mou, S.; Wu, T.; Xie, J.; Zhang, Y.; Ji, L.; Huang, H.; Wang, T.; Luo, Y.; Xiong, X.; Tang, B.; Sun, X. *Adv. Mater.* **2019**, *31* (36), e1903499.
85. Irfan Malik, M.; Malaibari, Z. O.; Atieh, M.; Abussaud, B. *Chem. Eng. Sci.* **2016**, *152*, 468–477.
86. Chang, T.-Y.; Liang, R.-M.; Wu, P.-W.; Chen, J.-Y.; Hsieh, Y.-C. *Mater. Lett.* **2009**, *63* (12), 1001–1003.
87. Zhao, K.; Liu, Y.; Quan, X.; Chen, S.; Yu, H. *ACS Appl. Mater. Interfaces* **2017**, *9* (6), 5302–5311.
88. Marepally, B. C.; Ampelli, C.; Genovese, C.; Tavella, F.; Veyre, L.; Quadrelli, E. A.; Perathoner, S.; Centi, G. *J. CO_2 Util.* **2017**, *21*, 534–542.
89. Albo, J.; Irabien, A. *J. Catal.* **2016**, *343*, 232–239.
90. Low, J.; Cheng, B.; Yu, J. *Appl. Surf. Sci.* **2017**, *392*, 658–686.
91. Farkhondehfal, M. A.; Hernández, S.; Rattalino, M.; Makkee, M.; Lamberti, A.; Chiodoni, A.; Bejtka, K.; Sacco, A.; Pirri, F. C.; Russo, N. *Int. J. Hydrogen Energy* **2020**, *45* (50), 26458–26471.
92. Ma, S.; Lan, Y.; Perez, G. M.; Moniri, S.; Kenis, P. J. *ChemSusChem* **2014**, 7 (3), 866–874.
93. Ramesha, G. K.; Brennecke, J. F.; Kamat, P. V. *ACS Catal.* **2014**, *4* (9), 3249–3254.
94. Teh, W. J.; Piqué, O.; Low, Q. H.; Zhu, W.; Calle-Vallejo, F.; Yeo, B. S. *ACS Catal.* **2021**, *11* (14), 8467–8475.
95. Yuan, J.; Yang, M.-P.; Hu, Q.-L.; Li, S.-M.; Wang, H.; Lu, J.-X. *J. CO_2 Util.* **2018**, *24*, 334–340.
96. Hossain, S. K.; Saleem, J.; Rahman, S.; Zaidi, S.; McKay, G.; Cheng, C. *Catalysts* **2019**, *9* (3), 298.
97. Andrews, E.; Ren, M.; Wang, F.; Zhang, Z.; Sprunger, P.; Kurtz, R.; Flake, J. *J. Electrochem. Soc.* **2013**, *160* (11), H841–H846.
98. Zhu, W.; Kattel, S.; Jiao, F.; Chen, J. G. *Adv. Energy Mater.* **2019**, *9* (9), 1802840.
99. Cave, E. R.; Montoya, J. H.; Kuhl, K. P.; Abram, D. N.; Hatsukade, T.; Shi, C.; Hahn, C.; Nørskov, J. K.; Jaramillo, T. F. *Phys. Chem. Chem. Phys.* **2017**, *19* (24), 15856–15863.
100. Hatsukade, T.; Kuhl, K. P.; Cave, E. R.; Abram, D. N.; Jaramillo, T. F. *Phys. Chem. Chem. Phys.* **2014**, *16* (27), 13814–13819.
101. Ensafi, A. A.; Alinajafi, H. A.; Rezaei, B. *J. Electroanal. Chem.* **2016**, *783*, 82–89.
102. Alinajafi, H. A.; Ensafi, A. A.; Rezaei, B. *Int. J. Hydrogen Energy* **2018**, *43* (52), 23262–23274.
103. Zarandi, R. F.; Rezaei, B.; Ghaziaskar, H. S.; Ensafi, A. A. *Int. J. Hydrogen Energy* **2019**, *44* (59), 30820–30831.
104. Zhang, W.; Qin, Q.; Dai, L.; Qin, R.; Zhao, X.; Chen, X.; Ou, D.; Chen, J.; Chuong, T. T.; Wu, B.; Zheng, N. *Angew. Chem., Int. Ed.* **2018**, *57* (30), 9475–9479.
105. Zhao, H.; Zhu, H.; Feng, Y.; Zhao, Q.; Suo, B.; Zou, W.; Han, H.; Zhai, G.; Jiang, Z.; Song, Q.; Li, Y. *ChemElectroChem* **2020**, 7 (24), 5036–5043.
106. Karamad, M.; Hansen, H. A.; Rossmeisl, J.; Nørskov, J. K. *ACS Catal.* **2015**, *5* (7), 4075–4081.
107. Tayyebi, E.; Hussain, J.; Skulason, E. *Chem. Sci.* **2020**, *11* (35), 9542–9553.
108. Bhowmik, A.; Hansen, H. A.; Vegge, T. *J. Phys. Chem. C* **2017**, *121* (34), 18333–18343.
109. Zhang, Q.; Du, J.; He, A.; Liu, Z.; Tao, C. *J. CO_2 Util.* **2019**, *34*, 635–645.
110. Paris, A. R.; Bocarsly, A. B. *ACS Catal.* **2017**, 7 (10), 6815–6820.
111. Hatsukade, T.; Kuhl, K. P.; Cave, E. R.; Abram, D. N.; Feaster, J. T.; Jongerius, A. L.; Hahn, C.; Jaramillo, T. F. *Energy Technol.* **2017**, *5* (6), 955–961.
112. Hirunsit, P.; Soodsawang, W.; Limtrakul, J. *J. Phys. Chem. C* **2015**, *119* (15), 8238–8249.

113. Back, S.; Kim, H.; Jung, Y. *ACS Catal.* **2015**, *5* (2), 965–971.
114. Zhang, F. Y.; Sheng, T.; Tian, N.; Liu, L.; Xiao, C.; Lu, B. A.; Xu, B. B.; Zhou, Z. Y.; Sun, S. G. *Chem. Commun.* **2017**, *53* (57), 8085–8088.
115. Huang, W.; Yuan, G. *Electrochem. Commun.* **2020**, *118*, 106789.
116. Low, Q. H.; Loo, N. W. X.; Calle-Vallejo, F.; Yeo, B. S. *Angew. Chem., Int. Ed.* **2019**, *58* (8), 2256–2260.
117. Payra, S.; Shenoy, S.; Chakraborty, C.; Tarafder, K.; Roy, S. *ACS Appl. Mater. Interfaces* **2020**, *12* (17), 19402–19414.
118. Lu, L.; Sun, X.; Ma, J.; Yang, D.; Wu, H.; Zhang, B.; Zhang, J.; Han, B. *Angew. Chem., Int. Ed.* **2018**, *57* (43), 14149–14153.
119. Jia, F.; Yu, X.; Zhang, L. *J. Power Sources* **2014**, *252*, 85–89.
120. Yin, J.; Yin, Z.; Jin, J.; Sun, M.; Huang, B.; Lin, H.; Ma, Z.; Muzzio, M.; Shen, M.; Yu, C.; Zhang, H.; Peng, Y.; Xi, P.; Yan, C. H.; Sun, S. *J. Am. Chem. Soc.* **2021**, *143* (37), 15335–15343.
121. Perryman, J. T.; Ortiz-Rodríguez, J. C.; Jude, J. W.; Hyler, F. P.; Davis, R. C.; Mehta, A.; Kulkarni, A. R.; Patridge, C. J.; Velázquez, J. M. *Mater. Horiz.* **2020**, 7 (1), 193–202.
122. Sun, X.; Zhu, Q.; Kang, X.; Liu, H.; Qian, Q.; Zhang, Z.; Han, B. *Angew. Chem., Int. Ed.* **2016**, *55* (23), 6771–6775.
123. Chi, C.; Duan, D.; Zhang, Z.; Wei, G.; Li, Y.; Liu, S. *Coatings* **2020**, *10* (12), 1142.
124. Ji, L.; Li, L.; Ji, X.; Zhang, Y.; Mou, S.; Wu, T.; Liu, Q.; Li, B.; Zhu, X.; Luo, Y.; Shi, X.; Asiri, A. M.; Sun, X. *Angew. Chem., Int. Ed.* **2020**, *59* (2), 758–762.
125. Zhao, S.; Guo, S.; Zhu, C.; Gao, J.; Li, H.; Huang, H.; Liu, Y.; Kang, Z. *RSC Adv.* **2017**, 7 (3), 1376–1381.
126. Huang, J.; Guo, X.; Yue, G.; Hu, Q.; Wang, L. *ACS Appl. Mater. Interfaces* **2018**, *10* (51), 44403–44414.
127. Wang, L.; Xu, Y.; Chen, T.; Wei, D.; Guo, X.; Peng, L.; Xue, N.; Zhu, Y.; Ding, M.; Ding, W. *J. Catal.* **2021**, *393*, 83–91.
128. Ji, L.; Chang, L.; Zhang, Y.; Mou, S.; Wang, T.; Luo, Y.; Wang, Z.; Sun, X. *ACS Catal.* **2019**, *9* (11), 9721–9725.
129. Huang, J.; Hu, Q.; Guo, X.; Zeng, Q.; Wang, L. *Green Chem.* **2018**, *20* (13), 2967–2972.

About the authors

Mr. Jian Zhu obtained his M.S. degree from Central South University in 2018 and he is now a PhD candidate at the University of Antwerp. His interests mainly focus on the alloy type anode materials for secondary rechargeable batteries and transition metals for electrochemical CO2 reduction.

Shoubhik Das did his PhD under the supervision of Prof. Matthias Beller at the Leibniz Institut für Katalyse e.V., Rostock. He received his PhD degree in 2012. Subsequently, he joined the research group of Prof. Matthew Gaunt in the Cambridge University as a post-doctoral research associate. In 2013, he moved to the research group of Prof. Paul J. Dyson at the EPFL, Switzerland and joined as a scientist. In August 2015, he started his independent research career at the Georg-August-Universität Göttingen, Germany. Since November 2019 he has joined the department of Chemistry at the University of Antwerp. His main research interest is the activation of small molecules and their applications inorganic synthesis and fuel generation.

Pegie Cool received her PhD degree in Science, Chemistry in 1998 at the University of Antwerp, Belgium. During her postdoctoral fellowship funded by the Research Foundation Flanders, she was associated at the Texas A&M University, USA and at the University of Queensland, Australia. In 2004 she was appointed Assistant Professor and since 2012 she is Full Professor and head of the Laboratory of Adsorption and Catalysis at the Chemistry Department in Antwerp. Her main expertise is on the development of inorganic nanoporous materials, involving pore size modification and activation for sorption and (photo)catalysis. The main focus is on environmental applications, such as wastewater purification, CO2 conversion, and automotive exhaust gas conversion. She is Board of Directors member of the European Nanoporous Materials Institute of Excellence (ENMIX aisbl) and Board Member of the Dutch Zeolite Association (Netherlands-Flanders).

CHAPTER THREE

Electrosynthetic routes toward carbon dioxide activation and utilization

Benjamin R. Buckley*
Department of Chemistry, Loughborough University, Loughborough, LE, United Kingdom
*Corresponding author: e-mail address: b.r.buckley@lboro.ac.uk

Contents

Abstract

The development of electrochemical methods for the carboxylation of a variety of substrates to form new C–CO_2 bonds has been an area of interest for a number of years and is a powerful way to construct such a bond, often negating the use of water sensitive/pyrophoric reagents (e.g. metal halogen exchange with alkyl lithium compounds). In this review we will examine the breadth of carboxylation reactions available and the type of electrosynthetic cell and catalyst required to carry out the transformation. Traditionally many of these reactions were carried out using sacrificial magnesium/aluminum or zinc anodes, but recent developments have enabled the circumvention of these non-sustainable approaches and have opened the door toward further development of continuous flow processes. These developments have run alongside the increased availability of standardized batch and flow electrochemical systems now available to the synthetic chemist. Competition between the areas of photocatalysis and electrocatalysis through comparative reactivity of single electron transfer processes has driven many of the new innovations in this exciting area of research.

Advances in Catalysis, Volume 70
ISSN 0360-0564
https://doi.org/10.1016/bs.acat.2022.07.002

1. Introduction

Electrosynthesis has over recent years seen a renaissance in its use aided by the development of low-cost standardized equipment and the development of bench top laboratory scale flow reactors *(1)*. The use of electrocatalytic systems for the activation and/or incorporation of carbon dioxide has, however, been a topic of interest for a number of decades *(2)*. Electrochemical approaches to preparative scale carbon dioxide utilization are of particular interest since efficient chemical incorporation of carbon dioxide has, until recent years, been restricted to reactive substrates, such as epoxides *(3)* to produce cyclic carbonates and amines to produce carbamates *(4)*. Alternative technologies in addition to electrosynthesis, such as photocatalysis, have started to appear along-side traditional transition metal catalysis enabling carbon dioxide insertion into less energetic substrates.

In general terms the electrosynthesis equipment employed in carbon dioxide utilization is fairly non-specialized, however, in some specific cases high pressure cells have been employed. Typically, electrochemical reduction of either carbon dioxide or the substrate needs to occur at the cathode surface, thus in order to complete the circuit a corresponding oxidation must take place at the anode. In the majority of cases thus far reported this oxidation is achieved through the use of a sacrificial metal electrode (e.g. Mg, Al, Zn). In the cases where inert electrodes (e.g. Pt or carbon) are used an organic reducing agent or water splitting is employed. The majority of carbon dioxide incorporation reactions are carried out in simple undivided cells although more recently there are some examples that employ a divided cell system to prevent destruction of either the product or starting material (Fig. 1) *(2)*.

A major challenge in the area has been the development of electrochemical systems that do not require a sacrificial agent. This is important because the production of large quantities of inorganic or organic waste inhibit the "green" credentials of the system and the formation of solid particulate in the reaction medium prevents continuous processing. With the development of non-sacrificial electrode systems the possibility to utilize electrochemical flow reactors is now possible. This is aided by the fact that several commercial systems are now commercially available to the synthetic chemist.

The construction of a thermodynamically and kinetically stable C-C bond is perhaps one of the most desirable applications for CO_2 utilization. The direct formation of carboxylic acids from the coupling of carbon

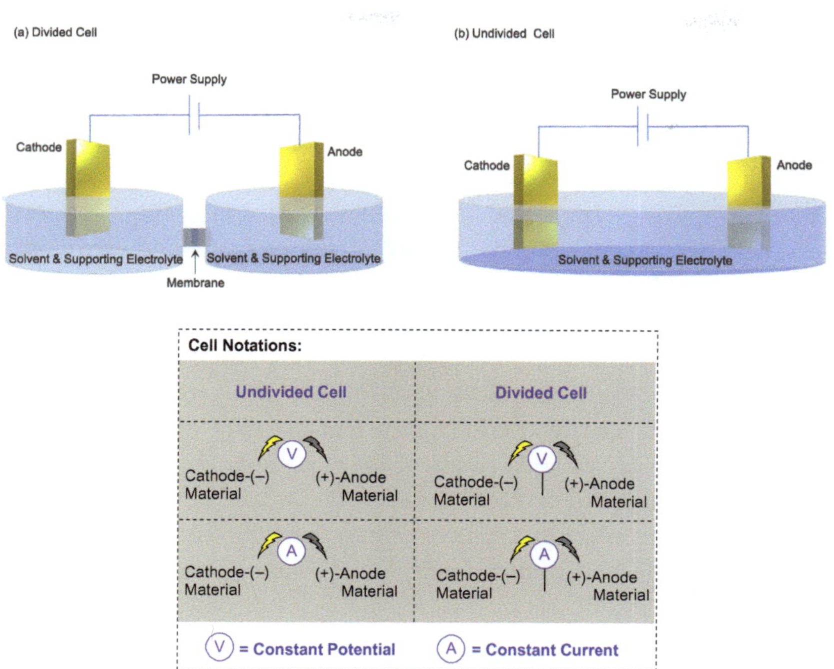

Fig. 1 Common cell set-ups used in electrochemical carbon dioxide utilization and cell notation used in this chapter.

dioxide and C—C, C=C, C≡C and C—X bonds under "green" conditions are described as one of the most challenging carbon dioxide utilization processes, with the hydrocarboxylation of ethene with CO_2 being described as a "dream process." *(5)* Traditional routes employing carbon dioxide to generate carboxylic acids concentrate on the reaction of carbon dioxide with an organometallic nucleophile, however, these reactions are particularly moisture/air sensitive and generate significant inorganic waste *(6)*. Therefore, activation of CO_2 in an alternative air/moisture tolerant process, such as in electrochemistry, is of significant interest due in part to the fact that the high thermodynamic stability of CO_2 is negated by a single-electron reduction.

Although there are many reports that now show successful and energy-efficient reduction of CO_2 to formic acid/formate the focus of this chapter will be on those processes that have been electrochemically driven to afford a C—CO_2 bond. The chapter is therefore organized by the bond carboxylated starting with the reduction of CO_2 to oxalate and moving through C—C, C=C, C≡C and C—X carboxylation reactions.

2. Reduction of CO_2 to oxalate

Oxalic acid can be prepared through the dimerization of carbon dioxide and is perhaps one of the simplest molecules to result in C—C bond formation from CO_2. Oxalic acid has been widely used in dyeing processes, in the cleaning industry, and in metallurgy. Carbon dioxides reduction potential in aprotic solution is –2.2 V vs SCE and in the absence of other reactants generally affords a mixture of oxalate, CO, and CO_3^{2-}. Savage reported several approaches to carbon dioxide reduction including routes to oxalate. Two processes are shown in Scheme 1; (a) The direct reduction of carbon dioxide to afford oxalate at constant current in DMF with Bu_4NClO_4 as electrolyte and a Hg cathode (at low CO_2 concentration and high temperature oxalate was favored over $CO + CO_3^{2-}$) *(7)*; and (b) indirect reduction utilizing aromatic esters or nitrile electrocatalysts *(8)*. Savage proposed that oxalate formation occurred through the coupling of two carbon dioxide radical anions.

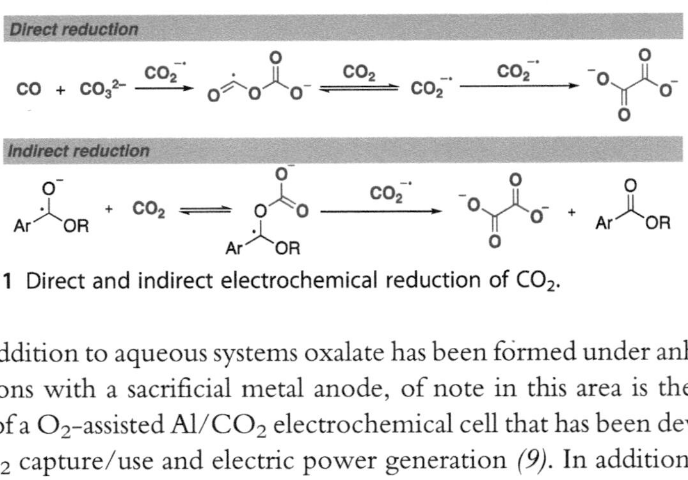

Scheme 1 Direct and indirect electrochemical reduction of CO_2.

In addition to aqueous systems oxalate has been formed under anhydrous conditions with a sacrificial metal anode, of note in this area is the recent report of a O_2-assisted Al/CO_2 electrochemical cell that has been developed for CO_2 capture/use and electric power generation *(9)*. In addition, nickel electrocatalysis and ionic liquids have been employed in oxalate synthesis *(10,11)*. However, a more sustainable approach has been developed in which a non-sacrificial electrode system has been employed to produce oxalate with current efficiencies of over 50% *(7)*.

3. Carboxylation of C(sp³)–C(sp³) bonds

Carboxylation of the carbon-carbon sp^3-sp^3 bond is a particularly challenging transformation. Ring opening of small ring carbocycles represents a viable route to this problem since ring strain relief can enable selective reactivity across the sp^3–sp^3 bond. Typically activated cyclopropanes or cyclobutanes have been treated with Lewis acids or transition metal catalysts to functionalize these small molecules *(12)*. Yu and co-workers noted that this relief of ring strain has been facilitated using photo- or electrosynthesis *(13,14)*. Utilizing a novel sacrificial electrochemical approach, they were able to selectively dicarboxylate a range of 1,2-diarylcyclopropanes in good yield (Scheme 2A) and were able to extend this methodology to 2-arylcyclopropanecarboxylates and 1,2-diarylcyclobutanes (Scheme 2B and C) *(15)*. In the absence of carbon dioxide the fully reduced ring opened product is observed in good yield indicating the formation of a radical anion, and a possible electron transfer-chemical protonation-electron transfer–chemical protonation (ECEC) mechanism. Further cyclic voltammetry studies showed the Nb cathode was able to reduce diphenylcyclopropane at –3.0 V vs SCE and CO_2 at –2.21 V vs SCE, indicating that both CO_2 and the cyclopropane can be reduced under the reaction conditions. Further exploitation of the reaction products was achieved through the preparation of a carbocyclic cleft analogous to Trögers base and a novel self-healing polymer. At present Yu's approach described above represents the only reported method for the dicarboxylation of single bonds, undoubtedly this seminal report will spur other researchers in the area to develop novel routes to sp^3–sp^3 carboxylation processes.

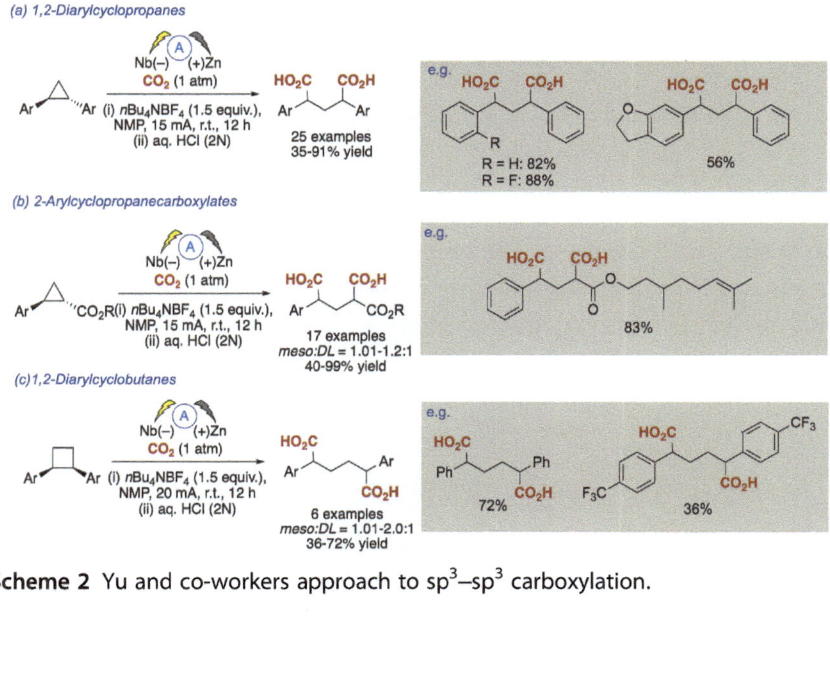

Scheme 2 Yu and co-workers approach to sp^3–sp^3 carboxylation.

4. Carboxylation of C═C bonds

4.1 Insertion of two CO_2H groups

The electrochemical carboxylation C═C bonds, particularly alkenes has been studied for several decades, however this area has seen a renaissance due to related photoredox and nickel catalyzed processes being reported, culminating in several new approaches to electrochemical C═C carboxylation. Vasil'ev and co-workers in 1991 were among the first to report electrochemical carboxylation of alkenes. Using a stainless steel cathode and an aluminum non-sacrificial anode at 5 atm CO_2 pressure afforded good selectivity to aluminum oxalate in excellent current efficiency (up to 89%) *(16)*. Vasil'ev postulated that oxalate formation proceeded through the CO_2 radical anion and in the presence of an acceptor molecule, such as ethene can form the longer chain dicarboxylic acids (Scheme 3). Thus, the CO_2 radical anion reacts with ethene and then either undergoes carboxylation or reacts with another molecule of ethene, consequently generating a mixture of oxalic (C_2), succinic acid (C_4), adipic acid (C_6) and suberic acid (C_8).

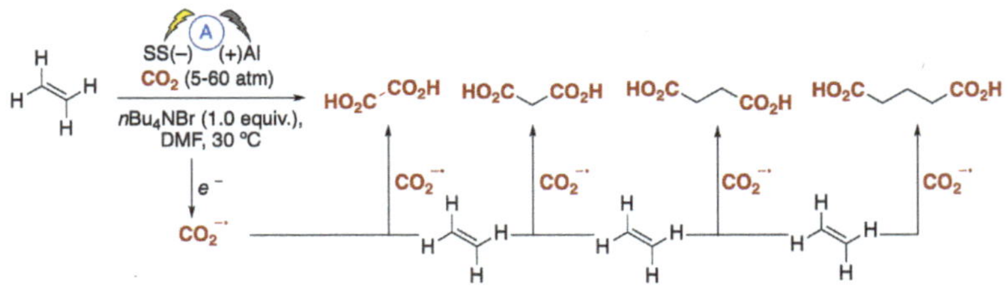

Scheme 3 Vasil'ev's carboxylation of ethene to dicarboxylated products.

Duñach, a pioneer of electrosynthetic carboxylation chemistry, reported the dicarboxylation of a range of alkenes utilizing a sacrificial magnesium anode and a nickel catalyst. The system worked for aliphatic (strained bicyclic systems), aromatic, α,β-unsaturated, hetero-substituted and silyl alkenes, with moderate to good conversions and selectivities (Scheme 4) *(17)*.

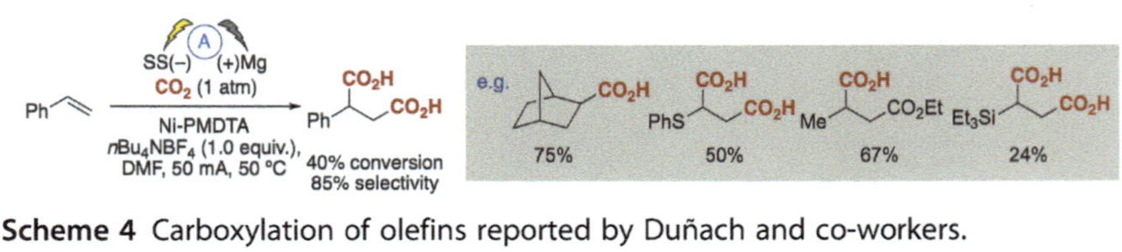

Scheme 4 Carboxylation of olefins reported by Duñach and co-workers.

Several reports utilizing a similar system have appeared, most recently Senboku and co-workers reported the electrocarboxylation of aryl-substituted alkenes without the need for an additional nickel catalyst, however, the process does rely on the use of a platinum electrode (Scheme 5) *(18)*. A limited number of aryl alkenes were carboxylated in good to excellent yield (69–99%) The proposed mechanism is similar to that reported by Vasil'ev. However, unlike CO_2 and ethene, cyclic voltammetry studies indicate that CO_2 and styrenes have similar reduction potentials: thus, two reaction pathways (path a or path b, Scheme 6) are plausible. Oxidation of the sacrificial magnesium anode enables the reduction of CO_2 or the alkene. Further analysis of the reaction parameters, such as electrolytes, temperature, electrodes, and other factors have been examined in follow-up studies *(19)*.

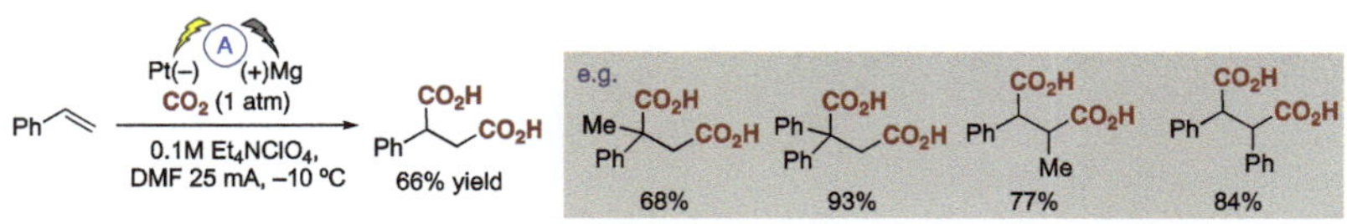

Scheme 5 Senboku's modified route to dicarboxylic acids using a Pt cathode.

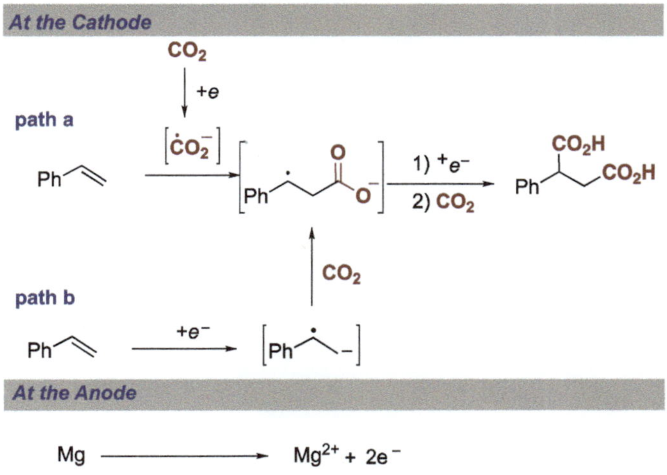

Scheme 6 Senboku's proposed mechanism for dicarboxylic acid synthesis.

Maeda and Mita have reported the dearomative dicarboxylation of a series of heterocyclic compounds using a platinum cathode and sacrificial magnesium anode system (Scheme 7) *(20)*. *N*-Boc indole was found to be a suitable substrate with a reduction potential of −2.76 V vs SCE and initial reactions afforded the mono-carboxylated product in 60% yield along with 18% of the dicarboxylated reduced indole. Further optimization of the reaction conditions enabled a switch in major product, utilizing a different electrolyte and MeCN as solvent afforded the dicarboxylic acid as the major product in 80% yield along with 10% of the mono-carboxylate.

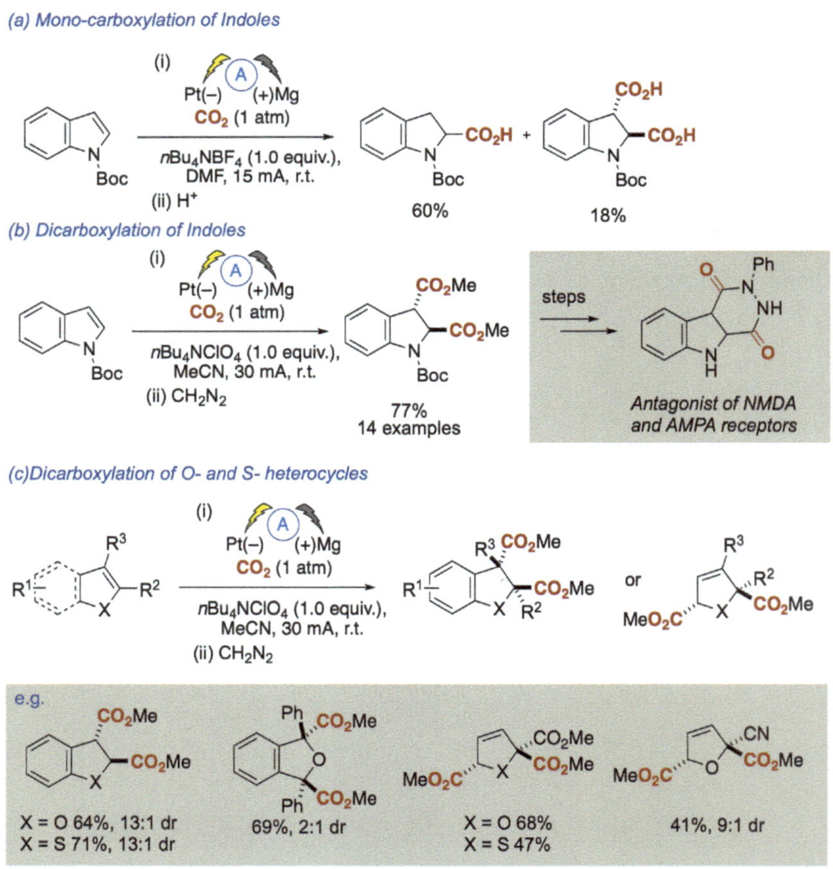

Scheme 7 Maeda and Mita's route toward the carboxylation of C=C bonds embedded in heterocyclic systems.

A range of *N*-Boc protected indoles were then screened and afforded the corresponding dicarboxylates (after treatment with CH_2N_2) in moderate to excellent yield. Additionally a range of reduction potentials were calculated for various heterocycles and those that fell between −1.80 to −2.94 V were found to successfully undergo carboxylation. Exploitation of the reaction conditions developed led the group to develop novel routes to a range of potentially biologically important compounds, exemplified in the synthesis of an antagonist of *N*-methyl D-aspartate (NMDA) and α-amino-3-hydroxy-5-methyl-4-isoxazolepropionic acid receptor (AMPA) receptors. The reaction is believed to proceed through the CO_2 radical anion rather than reduction of the indole and subsequent trapping of CO_2 as electrophile. Preliminary evidence from a controlled voltage reaction, where the voltage was kept at −2.20 V vs SCE that is well below the −2.76 V vs SCE for *N*-Boc indole, afforded the desired carbonylated products. The mechanism was also supported by DFT calculations which showed that if the process was to proceed through indole reduction first carboxylation should preferentially be observed at C3 however only C2 and dicarboxylated reduced indoles were isolated.

4.2 Insertion of one CO_2H group

Selective mono-carboxylation of alkenes has until very recently rarely been reported using electrochemical systems even though this is prevalent in a range of recent publications in the traditional catalysis literature and is a direct competitor to the highly important hydroformylation process that relies on carbon monoxide *(21–28)*. This is due in part to the problem of controlling the regiochemical outcome of the reaction and mono vs dicarboxylation. In the electrosynthesis literature Duñach's reaction system reported for dicarboxylation above (Scheme 4) utilizing sacrificial electrodes and Ni catalysis did also produce β-carboxylated product, however, this was as a non-selective mixture with the dicarboxylated dominating *(12)*. Based on Duñach's work Tokuda and co-workers have reported the mono-carboxylation of bicyclic systems, utilizing a platinum plate electrode and a zinc plate sacrificial electrode in an undivided cell *(29)*. However, in some cases the yields were low and the product selectivity between mono- and di-carboxylation poor and unpredictable. For example, in Scheme 8 when R = Me the mono-carboxylic acid is the major product, but when the methyl ester is replaced with an ethyl ester the dicarboxylic acid product dominates.

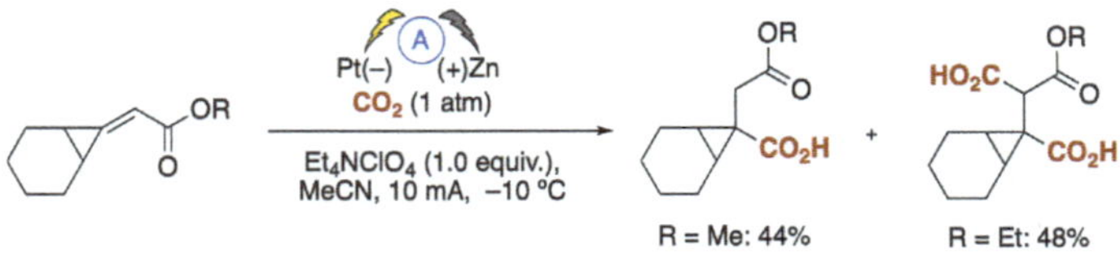

Scheme 8 Bicyclic alkenes as substrates for carboxylation.

Zhang and Lin have introduced a conceptually different approach toward the carbo-functionalization of alkenes through a radical-polar crossover reaction with alkyl bromides, utilizing a sacrificial magnesium anode and carbon cathode electrochemical cell (Scheme 9). The authors examined several possible reaction mechanisms with that shown in Scheme 10 believed to be the most plausible given the differences in reduction potential of the various components. Under the reaction conditions an alkyl halide could first be reduced to form the corresponding alkyl radical which in turn could react with an aryl alkene affording a more stabilized benzylic radical. Addition of a further electron would then produce the corresponding benzylic anion. Trapping of the anion with various electrophiles (DMF, Et_2NH as well as CO_2 were employed) provided the final products in good yield and selectivity (Scheme 10). Importantly under these conditions the α-carboxylate was isolated as the major product. The reaction conditions were then applied to precursors for the synthesis of biologically active compounds such as a glucokinase activator (Scheme 9). In addition they were also able to extend the work with the use of non-styrene alkenes, the yields were somewhat lower than the styrene equivalents but demonstrated a simple functionalization approach to challenging olefins.

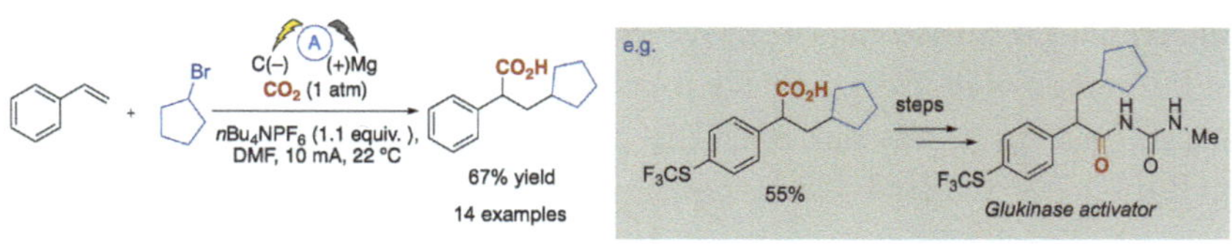

Scheme 9 Zhang and Lin's radical-polar crossover carboxylation approach.

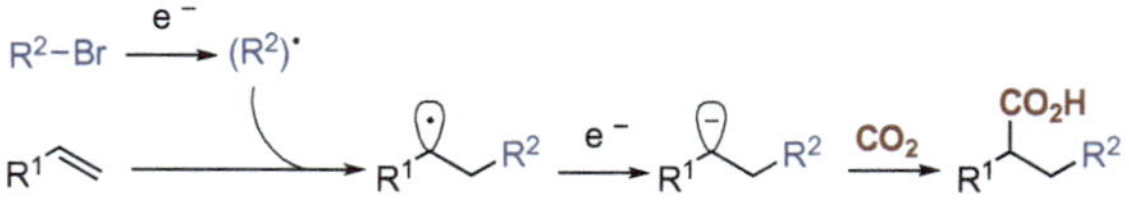

Scheme 10 Zhang and Lin's proposed radical-polar crossover mechanism.

The development of a non-sacrificial electrode approach toward olefin carboxylation has been a long-standing challenge. A recent report describes the moderately selective mono-carboxylation of four aromatic olefins. The approach utilized a divided cell and a bespoke $12CaO{\cdot}7Al_2O_3$ electrode system (Scheme 11) *(30)*. *trans*-Stilbene as well as methyl cinnamate could be selectively mono-carboxylated, however, the substrate scope was extremely limited, and the use of a bespoke electrode limits the applicability of this system. Nonetheless, this report represents one of the important first efforts in this area.

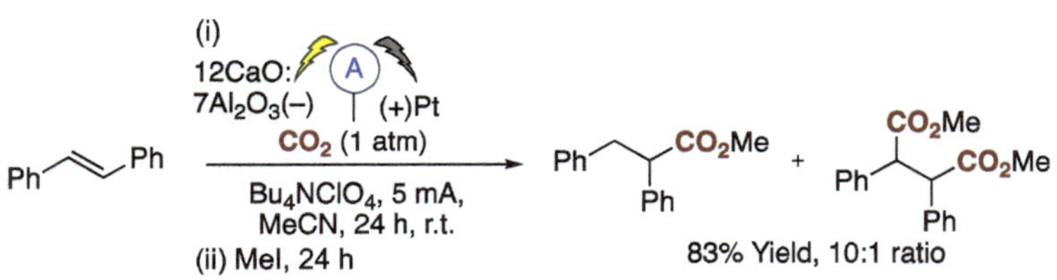

Scheme 11 A non-sacrificial anode approach to the carboxylation of *trans*-stilbene utilizing a bespoke $12CaO{\cdot}7Al_2O_3$ cathode divided cell system.

More recently, Buckley and Malkov have described an alternative undivided cell method to circumvent the use of a sacrificial metal by using widely available inert carbon electrodes *(31)*. Styrene was chosen as a test substrate, under initial conditions screened a 2:1 mixture of mono- and dicarboxylated products was achieved. However, when adding a suitable proton source (triethanol amine, TEOA) not only was full substrate conversion achieved but also the formation of dicarboxylated product was avoided and mono-carboxylation at the β-position was the primary pathway. Further optimization included the introduction of a carbon-based cathode, lowering the amounts of additives (NBu_4I, TEOA) and applied voltage. This gave the mono-carboxylated styrene with a regio-selectivity of 8:1, and the desired isomer in 70% yield. The scope of mono-carboxylated products derived from aromatic alkenes was broad allowing the electrosynthesis of a range of mono-carboxylated di-, tri- and tetra-substituted alkenes (Scheme 12).

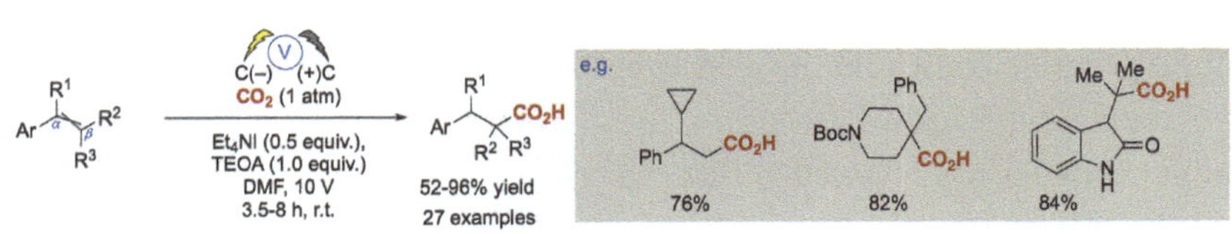

Scheme 12 A undivided cell non-sacrificial anode system for carboxylation of alkenes.

This new approach allowed for the presence of halides in the substrates and two examples of styrene substrates bearing *meta*- and *ortho*-Cl substituents could be conveniently transformed into their β-carboxylated products in moderate yields (52–54%). Notably, these kinds of substrates are typically not included in the scope of carboxylated products when metal-based methodologies are applied as likely the aryl-halides are then activated offering thus parasitic dehalogenation routes. By increasing the steric demand in the alkene precursors (i.e., using *cis*- or *trans*-stilbene or -trisubstituted alkenes), mono-carboxylation was still feasible under the electrosynthetic conditions affording the products (isolated as their Me-esters) in 53–71% yield. Such challenging -disubstituted and -trisubstituted alkenes cannot be carboxylated using photocatalysis, whereas Ni-catalyzed hydrocarboxylation of trisubstituted alkenes to date has been elusive.

Importantly, the electrosynthetic protocol proved to be effective in converting substrates that are otherwise inert to photochemical/metal-mediated procedures. Thus giving access to carboxylated compounds featuring quaternary centers in moderate to good yields (65–92%) and with excellent regioselectivity (>99:1).

The authors have postulated a mechanism for the electrosynthetic approach toward regio-selective mono-carboxylation of aromatic alkenes after conducting various control experiments, including radical clock experiments (using several cyclopropyl-substituted substrates), deuterium labeling experiments (showing predominant deuterium incorporation at the benzylic position) and performing the electrosynthesis at a saturated concentration of CO_2 (not changing the onset of the reduction process suggesting the initial formation of a surface-bound radical species). These combined mechanistic experiments clearly indicate the formation of radical intermediates. In the proposed mechanistic pathway, the aromatic alkene is first adsorbed on the cathode surface and converted to a radical anion by single-electron transfer (SET), followed by carboxylation, protonation, a second SET, dissociation from the electrode and finally a second protonation (Scheme 13). Given the applied potential, the formation of a reduced CO_2 radical anion cannot be ruled out but would lead to a similar carboxylated radical intermediate prior to protonation. In later studies the authors identified, under the standard reaction conditions, trace levels of formic acid from direct reduction of CO_2 indicating that the CO_2 radical anion pathway could indeed also be in operation *(32)*.

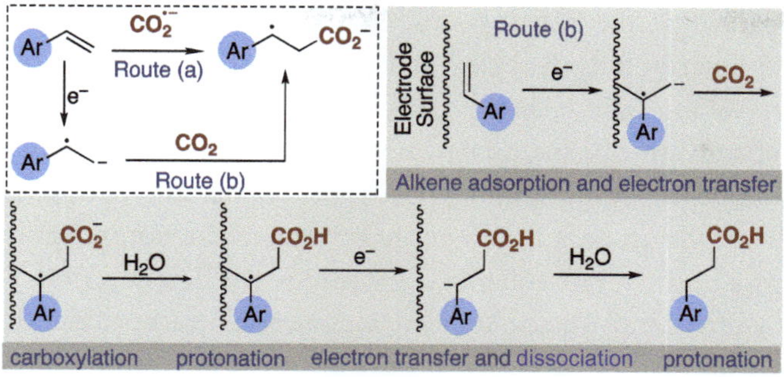

Scheme 13 Mechanistic rational for the mono-carboxylation process.

In addition, Buckley and co-workers have also reported a new electrochemical method for hydrocarboxylation of electron-deficient trisubstituted unsaturated alkenes (Scheme 14) *(27)*. Previously reported work in this area has required the presence of aryl substituents and the use of sacrificial metal electrodes such as magnesium and aluminum although successful photochemical routes have also been reported *(33)*. In contrast, this protocol employs an inexpensive carbon anode and cathode and works efficiently for alkyl substituents. The reaction is highly regioselective for the β-position and is generally high yielding. The authors also carried out a functional group tolerance/robustness screen to benchmark the electrochemical process for future researchers wanting to compare chemical, electrochemical or photochemical processes and since the electrochemical system does not rely on a sacrificial electrode they were successfully able to demonstrate the use of flow conditions to produce a precursor to ethosuximide on gram scale (Scheme 15).

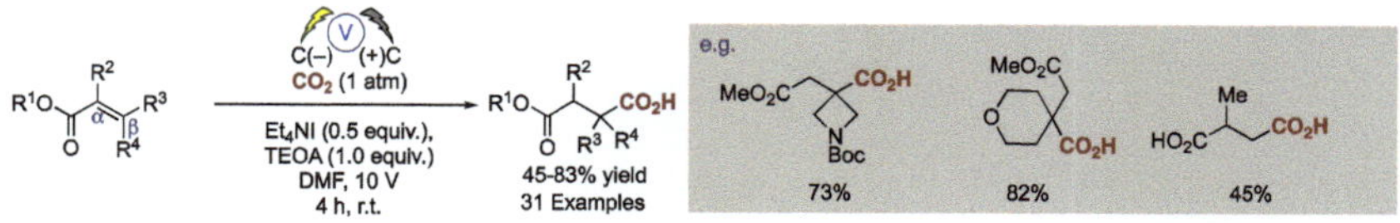

Scheme 14 Application of the non-sacrificial system to unsaturated esters.

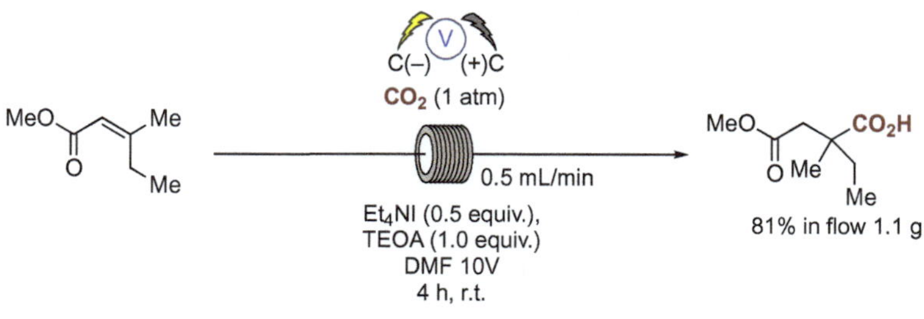

Scheme 15 Utilization of the non-sacrificial conditions applied to continuous flow synthesis.

The carboxylation of α,β-unsaturated ketones under the reaction conditions described by Buckley and co-workers above was unsuccessful. However, Wang and Zhang have reported a novel electrochemical β-carboxylation of α,β-unsaturated ketones using silica nanowires (SiNWs) as the cathode in conjunction with light irradiation (Scheme 16). This specific electrode absorbs light energy to reduce the consumption of electrical energy and overall enhances the reaction efficiency *(34)*. A number of aryl substituted α,β-unsaturated ketones were tolerated under the reaction conditions.

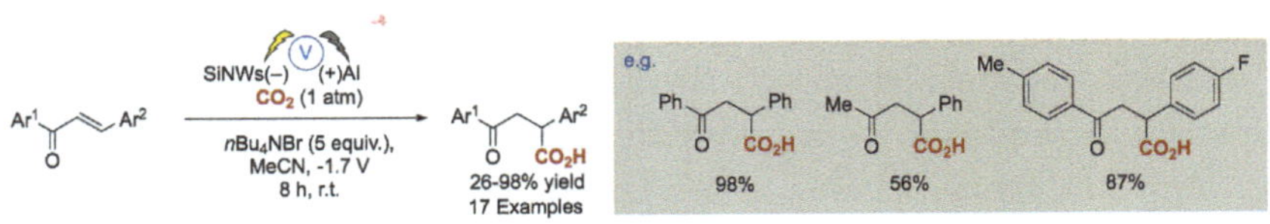

Scheme 16 Wang and Zhang's silica nanowire photoelectrochemical system for β-carboxylation of α,β-unsaturated ketones.

Finally for C=C bond carboxylation, Zhou and co-workers have developed an interesting regioselective defluorinative/carboxylation of α-trifluoromethyl olefins using a two platinum electrode system affording a wide range of gem-difluoroalkenes in good yield (Scheme 17A).

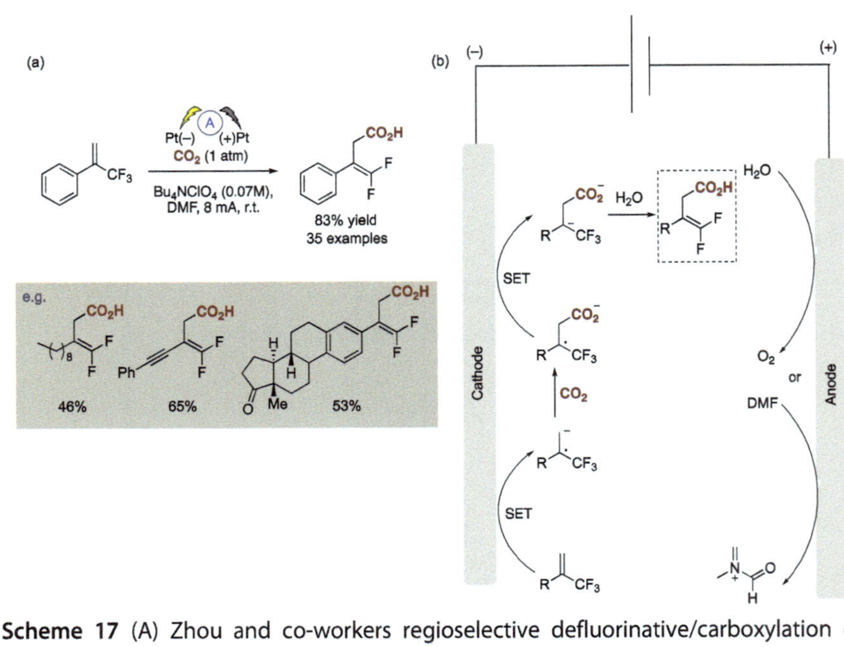

Scheme 17 (A) Zhou and co-workers regioselective defluorinative/carboxylation of α-trifluoromethyl olefins. (B) Proposed mechanism.

Mechanistically the authors propose, in conjunction with DFT studies that cathodic SET to the α-trifluoromethyl olefin and a subsequent cathodic SET reduction to form the corresponding benzylic anion, which then undergoes rapid fluoride elimination through an E1cB-type mechanism leading to the final gem-difluoroalkene products (Scheme 17B) *(35)*.

5. Carboxylation of C≡C bonds

The electrochemical carboxylation of alkynes has also been reported by several research groups, affording a variety of differently substituted products, and again these electrochemical reactions are dominated by the use of sacrificial anodes. Mono-carboxylation of a range of aliphatic and aromatic terminal alkynes through reduction of the alkyne bond has been achieved by Duñach and co-workers (Scheme 18 and 19) *(36)*. Using a carbon cathode, sacrificial magnesium anode and a nickel bipyridine ligated catalyst they were able to hydrocarboxylate terminal alkynes with the α-carboxylated terminal alkene as the major product, mono-carboxylation of disubstituted alkynes followed the same trend *(37)*. However, unsymmetrical disubstituted alkynes proved challenging and with little discrimination between α-carboxylation sites.

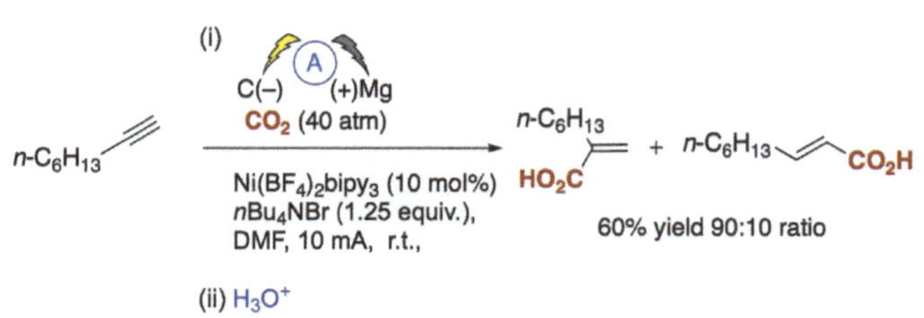

Scheme 18 α-Carboxylation of alkynes using Duñach's Ni catalysis.

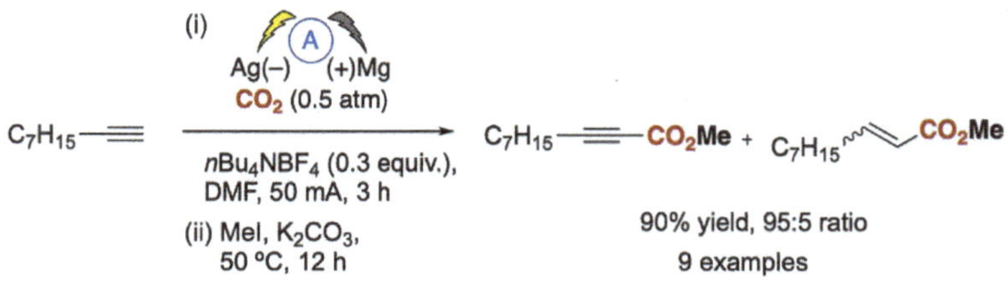

Scheme 19 Terminal alkyne carboxylation without reduction of the alkyne bond.

Duñach and co-workers later went on to report a related process in which carboxylation occurs at the terminal position of alkynes without reduction of the alkyne bond (Scheme 19) *(38)*. Sacrificial magnesium electrode conditions were employed but the reaction was found to proceed without the need for a nickel catalyst. Optimum conditions involved the use of a silver cathode. Up to 95% selectivity was observed for terminal carboxylation over a limited range of substrates.

A reductive dicarboxylation of alkynes using a nickel–aluminum electrode couple has been reported by Jiang and co-workers (Scheme 20) *(39)*. Under the reaction conditions reported a mixture of the dicarboxylic acid and the corresponding anhydride were produced from both terminal and disubstituted alkynes. However, from the limited number of substrate patterns screened the ratio of the two products appeared unpredictable.

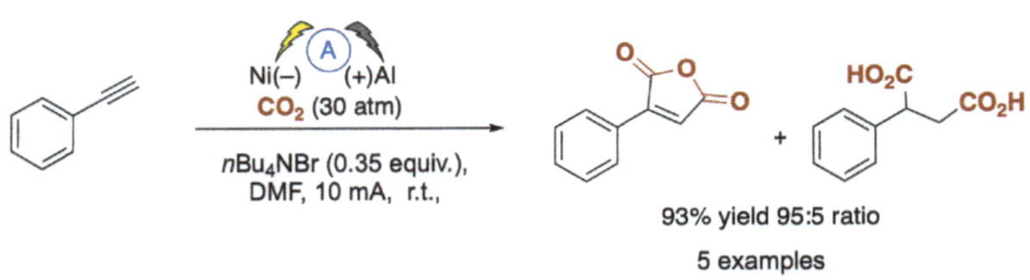

Scheme 20 Dicarboxylation of alkynes to afford anhydrides and dicarboxylic acids.

In a related report from Huanfeng and co-workers incorporation of two or three CO_2 molecules has been reported *(40)*. Addition of a copper iodide catalyst utilizing a cell set-up similar to that described by Jiang afforded the corresponding fully reduced anhydride or anhydride acid (Scheme 21). It is unclear why the addition of the copper catalyst alters the product profile of the reaction and the authors offer no reasoning as to this phenomenon.

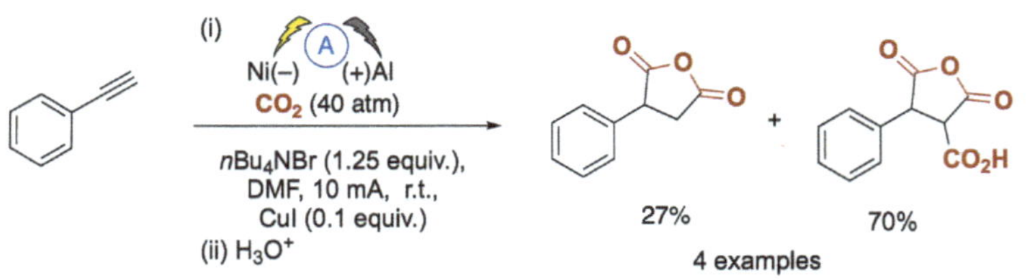

Scheme 21 Dicarboxylation of alkynes to reduced anhydrides and dicarboxylic acids.

A very nice example of the utility of electrochemical carboxylation reactions has been reported whereby an intermediate phenyl centered radical and subsequent intramolecular addition to an alkyne results in useful dicarboxylated heterocycles (Scheme 22) *(41)*.

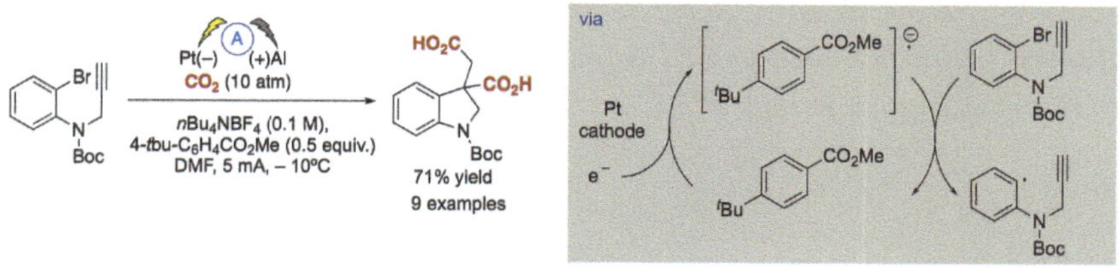

Scheme 22 Tandem electrochemical radical formation and trapping with carbon dioxide.

6. Carboxylation of conjugated C=C—C=C or C≡C—C≡C bonds

6.1 Carboxylation of 1,3-dienes

Low-value conjugated dienes have been extensively investigated for carbon dioxide utilization to prepare value-added materials such as adipic acid. Dicarboxylation of dienes has also received attention in the traditional transition metal catalysis literature *(42)*, however, as with the sacrificial electrode systems highlighted below the use of a metal stoichiometric reductant is a mandatory requirement.

In an early report Loveland described the electrolytic production of acyclic carboxylic acids from olefins, using a mercury–carbon electrochemical cell under 10 atmospheres of carbon dioxide (Scheme 23) *(43)*.

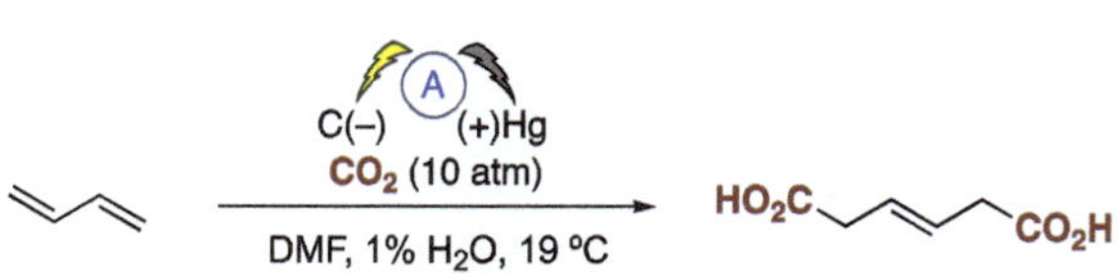

Scheme 23 Loveland's dicarboxylation of 1,3-butadienes.

Development of a more user-friendly approach through replacement of the mercury electrode was originally reported by Dinjus in the transition metal assisted electrocarboxylation of 1,3-butadiene to afford predominantly the straight chain dicarboxylic acid, utilizing a sacrificial magnesium anode

(44). Duñach and Perichon later described a sacrificial magnesium/nickel catalyzed system with exclusive dicarboxylic acid formation and good (*Z*)-olefin selectivity *(17)*. Li and co-workers then reported a related nickel-aluminum electrode system in which good yields and exclusively *trans*-substituted dicarboxylated alkenes were obtained (Scheme 24) *(45)*. They reported that an increase in CO_2 pressure allows for a higher yield, and that this is believed to be due to the solubility of CO_2 in DMF which increases at higher pressure.

Scheme 24 A mercury electrode free approach to the dicarboxylation of butadienes.

Buckley and co-workers have applied their breakthrough non-sacrificial electrode system to the carboxylation of 1,3-dienes and even extended the process to the carboxylation of trienes (Scheme 25A, one example, not shown) *(5)*. Depending on the reaction conditions the major product of the reaction could be tuned to be either the dicarboxylated or mono-carboxylated olefin. Selective mono-carboxylation occurred at the terminal

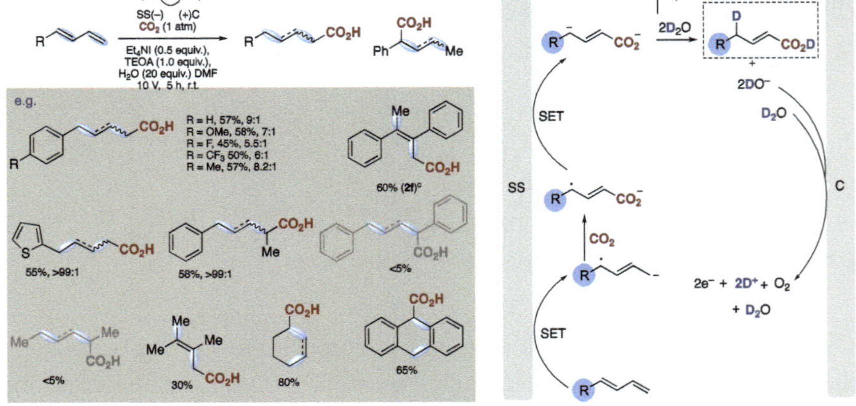

Scheme 25 (A) A non-sacrificial electrode approach toward the carboxylation of 1,3-dienes. (B) Mechanistic proposal for the reaction.

position of the diene and the resulting double bond(s) migrated to disrupt conjugation with the aryl unit. Selectivity for the major product was approximately 7 to 1 with E/Z ratios of 3:1. These promising preliminary results could prove synthetically useful if the selectivity of the process could be increased thus generating carboxylated alkenes, dienes or even polyenes ready for further functionalization. The mechanism of the reaction is believed to proceed in a similar fashion to the olefin carboxylation described in section 4.2 (Scheme 25B). Single electron transfer (SET) to the 1,3-diene proceeds to form the adsorbed radical anion of the 1,3-diene, subsequent carboxylation, SET and protonation from water affords the final α,δ-mono-carboxylated product. The reaction at the counter electrode has been described by Chang and co-workers in their study of deuteration reactions employing electrosynthesis and D_2O *(46)*. In the absence of water the dicarboxylated product dominates.

6.2 Carboxylation of 1,3-diynes

The carboxylation of 1,3-diynes is much less prominent when compared to 1,3-dienes in the literature, in part due to the lack of readily available diyne starting materials. Again Duñach and co-workers have pioneered the area utilizing their simultaneous activation of carbon dioxide and diynes by an electro-generated LNi^0 pre-catalyst (Scheme 26) *(47)*. The reactions were carried out in a single-compartment cell fitted with a sacrificial magnesium anode, and a catalytic amount of an Ni(II) complex to afford a mixture of terminal and internal mono-carboxylated products. Cyclic voltammetry showed that carbon dioxide and the diyne coordinate to a LNi^0-generated species in DMF. Under similar reaction conditions a non-conjugated diynes afforded linear or cyclic adducts depending on the ligand employed.

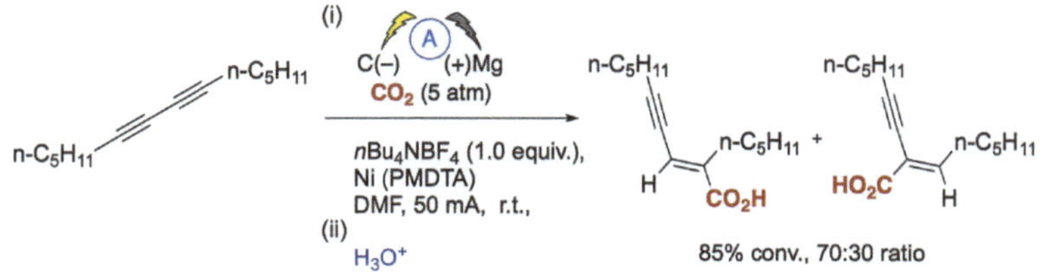

Scheme 26 Reductive carboxylation of 1,3-diynes.

Alkylidene lactones have been prepared by Jiang using an undivided sacrificial electrochemical cell (Scheme 27) *(48)*. The products were isolated from the reaction in moderate to good yields and could be controlled by the addition of copper iodide. In the absence of copper iodide the γ-keto carboxylic acids were obtained as the major product.

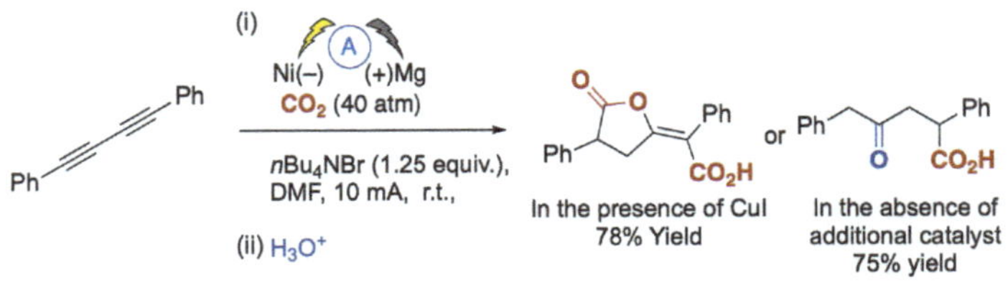

Scheme 27 Alkylidene lactones and γ-keto carboxylic acids from 1,3-diynes.

7. Carboxylation of C=O or C=N bonds

The electrocarboxylation of ketones with CO_2 has been an area of significant investigation, particularly focused on the use of a sacrificial magnesium anode and the carboxylation of acetophenone or closely related substrates *(49)*. Wawzonek and Gundersen reported the first example of electrocarboxylation of ketones *(50)*. This work then provided the basis for a number of ketone carboxylation reactions based on similar electrochemical systems and later α-aryl propionic acids were produced by Chan and co-workers using a lead-aluminum sacrificial electrochemical cell (Scheme 28) *(51)*. Two examples of carboxylation of ketones based on the naproxen and ibuprofen structures were produced. The carboxylated naproxen derived product was produced in high yield (91%) under atmospheric pressure CO_2 but to achieve high yields for *p*-isobutylacetophenone, the process had to be carried out at 5 atmospheres of CO_2. Unfortunately, this approach required the use of a lead cathode and the authors only reported two examples of the approach. However, this protocol was later converted into a flow cell reactor by Wagenknect and co-workers. *(52)* The flow system was carried out in a 1 L reactor cell which produced an

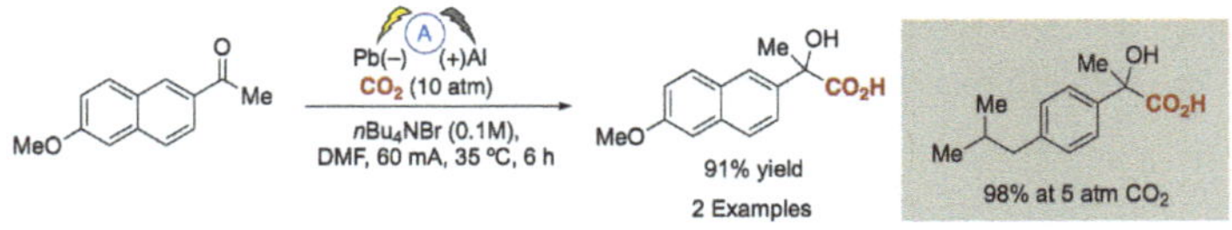

Scheme 28 Synthesis of hydroxyl-naproxen and –ibuprofen.

impressive conversion of 92.5% with 96% selectivity to the carboxylated product. However, an attempt to scale the system to 75 L was met with a significant reduction in yield to 58%.

Interestingly asymmetric electrochemical carboxylation of acetophenone has been reported by Lu and co-workers *(53)*. Utilizing a stainless-steel cathode and sacrificial magnesium anode with cinchonine up to 30% *ee* of the hydroxy acid could be obtained. Further modification of the reaction utilizing a modified chinchonine alkaloid increased the *ee* to a modest 50% *(54)*. Although the enantioselectivity observed here was low the ability to induce any chirality transfer holds promise for future modification of the reaction conditions for the development of highly enantioselective electrochemical carboxylation. The electrocarboxylation of imine substrates has also been an area of interest utilizing similar sacrificial electrochemical systems as those reported for ketones *(55)*. Of particular note Silvestri and co-workers in the 1980's showed that benzalaniline could be converted to the corresponding α-amino acid (Scheme 29) *(56)*. This process was carried out using an aluminum sacrificial anode in a semi continuous reaction vessel. In order for the continuous processing to be achieved replenishment of the sacrificial anode occurred during the reaction with recirculation of the reaction mixture. The cell was shown to be very efficient achieving a conversion of 92% and a yield of 85% for the desired product at an impressive 200 g scale. More recently a microreactor has been developed that does not require the use of a sacrificial anode, producing a limited number of diaryl amino acids from benzalaniline precursors in moderate yield *(57)*.

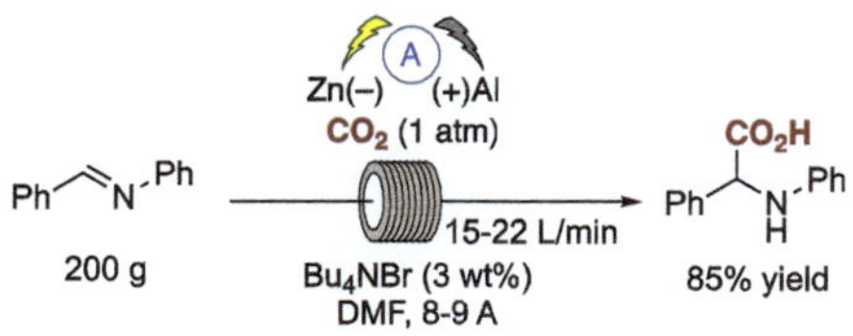

Scheme 29 Electrocarboxylation of benzalaniline to the corresponding α-amino acid.

8. Carboxylation of C—X bonds

Typically, the carboxylation of the carbon halide/pseudo halide bond has been achieved using a magnesium or aluminum sacrificial electrode although more recently several non-sacrificial electrode systems have been developed. The carboxylation of activated halides was first reported in the

1970's but provided low yields of the corresponding carboxylates *(58)*. Since this time many research groups have investigated the control of the reaction conditions to improve carboxylic acid yield and selectivity. Benzyl halides have been carboxylated by Gennaro and co-workers using a silver cathode and an aluminum anode (Scheme 30) *(59)*. The authors found that silver cathode materials lowered the reduction potential of benzyl chlorides whilst the reduction potential of CO_2 remained relatively unchanged, indicating initial reduction of the C—X bond and subsequent trapping of CO_2.

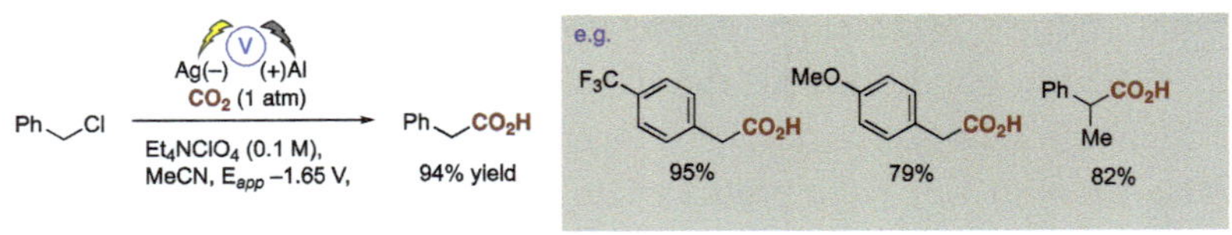

Scheme 30 Carboxylation of benzyl chlorides.

Several other studies on electrochemical carboxylation of benzyl halides using a silver cathode have since been published *(60–62)*. Ye and co-workers have developed a related method that enables desulfonylative electrocarboxylation of a range of benzylic substituted phenylsulfonyl compounds (Scheme 31) *(63)*. Importantly, Manthiram has developed a divided cell, non-sacrificial electrode approach to the carboxylation of a range alkyl/benzylic/aryl bromides/chlorides/iodides in moderate to good yield (Scheme 32).

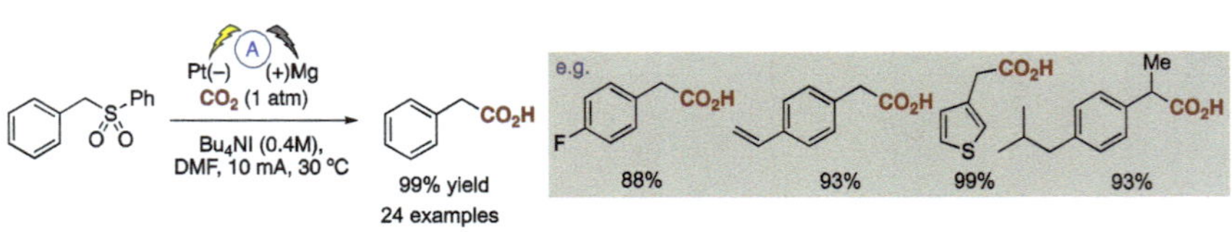

Scheme 31 Desulfonylative electrocarboxylation using a magnesium sacrificial electrode.

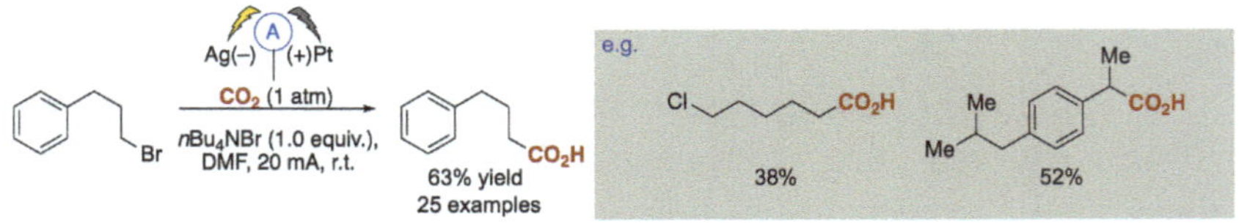

Scheme 32 Manthiram's non-sacrificial approach to alkyl/aryl halide carboxylation.

Carboxylation of benzyl fluorides has been reported to avoid using hazardous fluorinating agents for the synthesis of fluorinated analogs of non-steroidal anti-inflammatory α-fluorofenoprofen and α-fluoroloxoprofen (Scheme 33) *(64)*. A platinum cathode, magnesium sacrificial electrode system was used to carboxylate a limited number of benzylic gem-difluoro substrates.

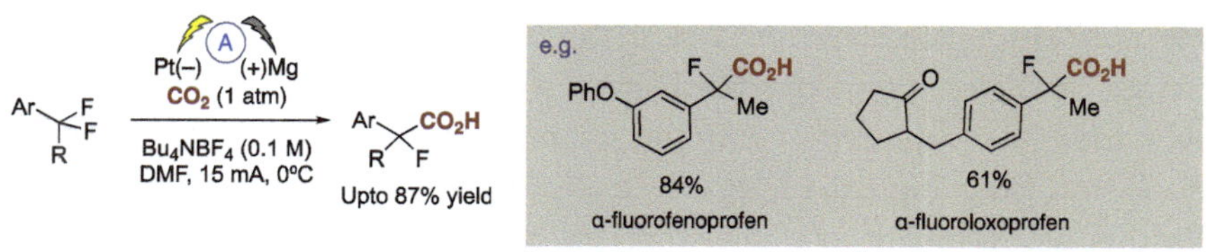

Scheme 33 Carboxylation of benzyl fluorides.

Yu and co-workers have reported a nickel-catalyzed electrochemical carboxylation of aryl and unactivated alkyl halides/pseudo halides with CO_2 using a carbon cathode and sacrificial zinc anode undivided cell system. The approach afforded the corresponding carboxylic acids in good yields ranging from 50% to 91%. Further development of the system through movement to a undivided cell with a carbon cathode and anode also afforded good yields of the corresponding carboxylic acids and looks to be a promising route to these compounds especially if they were to be prepared through a continuous approach (Scheme 34A) *(65)*. Extensive mechanistic and cyclic voltammetry studies led to the proposed catalytic cycle described in Scheme 34B. The active Ni^0 species is generated through cathodic reduction of the complex derived from $Ni^{II}(acac)_2$, di-*tert*-butyl-bipyridine

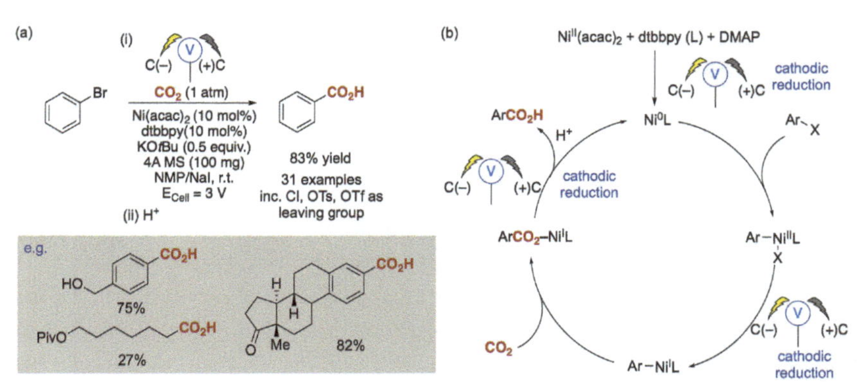

Scheme 34 Yu and co-workers non-sacrificial approach to electrochemical carboxylation of aryl halides.

(L, dtbbpy) and DMAP. After oxidative addition of the Ni^0L species to the aryl halide, the metal center is then reduced to a Ni^I species through cathodic reduction. This Ar– Ni^IL species is then reacted with CO_2 to give a Ni^I carboxylate intermediate. Following ligand exchange and reduction of the Ni^I carboxylate complex to regenerate the active Ni^0L catalyst and upon protonation affords desired aryl carboxylic acid.

The asymmetric carboxylation of related activated alkyl halides has been reported through the use of a chiral auxiliary *(66)*. It was found that the Evans auxiliary only provided low to moderate diastereoselectivity, switching to Oppolzer's camphor sultam resulted in good yield (up to 80%) and very high diastereoselectivity (up to >98:2 *d.r.*, Scheme 35). Further development to a non-sacrificial electrode and a catalytic asymmetric route would be of significant interest although this has not yet been followed up.

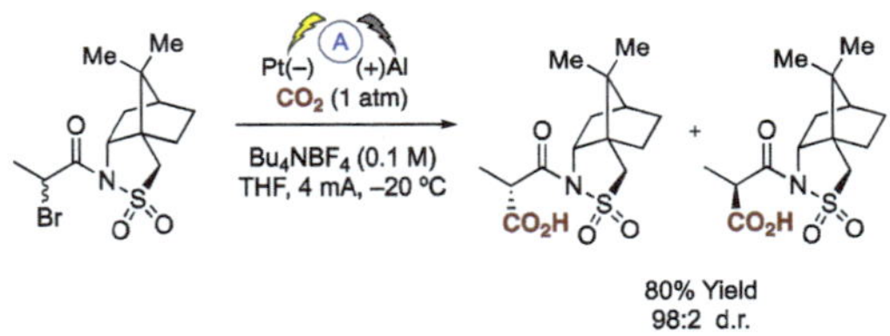

Scheme 35 Asymmetric auxiliary approach to cathodic carboxylation of alkyl halides.

The carboxylation of allylic halides/pseudo halides and acetates has been reported *(67)*. In order to attempt to control the regioselectivity of the reaction the use of a metal catalyst has been examined *(68)*. Originally reported by Ackermann utilizing cobalt catalysis *(69)*, others have reinvestigated the process examining the electro-reductive transition metal carboxylation of allyl chlorides with a sacrificial electrode system providing products with up to 9:1 regioselectivity being observed, although the major isomer varied with the substitution pattern of the aryl ring (Scheme 36A) *(70)*. A range of mechanistic, cyclic voltammetry and DFT studies were carried out leading to the proposed mechanism described in Scheme 36B. Formation of the active cobalt(I) catalyst is achieved on addition of PPh_3 followed by coordination of the allylic alkene. The adjacent allylic C—H bond is then cleaved, resulting in oxidative addition of the allylic alkene to form a η^3-allyl-cobalt(III) intermediate. At this point, the intermediate then undergoes rearrangement to form the two regio-isomeric η^1-allyl-cobalt(III) intermediates. Both then undergo cathodic reduction to afford

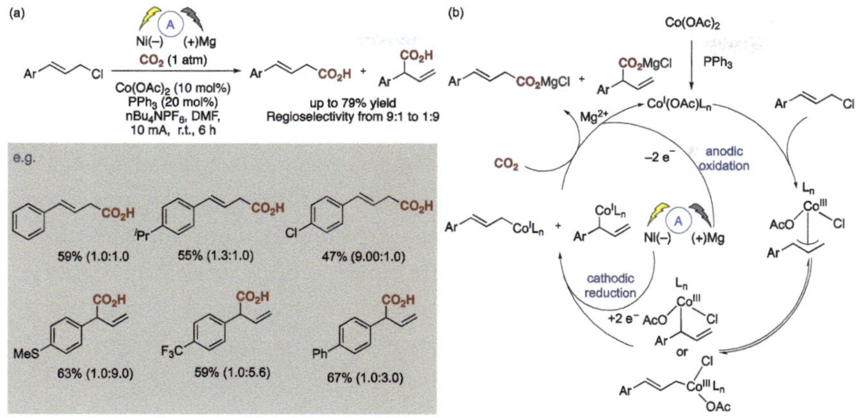

Scheme 36 (A) A cobalt catalyzed approach to allyl chloride carboxylation. (B) Postulated mechanism.

the corresponding low-valent η^1-allyl-cobalt(I) species. This determines the regioselectivity of the product which is highly dependent on the ligand employed. Finally the products are generated through Co—Mg *trans*-metalation, C—C bond formation with CO_2 and subsequent hydrolysis. Utilization of palladium mediated catalysis reported by Mei and co-workers has enabled the development of an asymmetric electrocarboxylation of allyl esters with good yields, regioselectivity and *ee*'s of up to 67% *(71)*.

9. Concluding remarks and future outlook

From the reports described in this review it is clear to see that the electrocarboxylation has been an area of significant interest for quite some time. The development of electrochemical carboxylation reactions has mainly relied on the use of a metal cathode and a sacrificial metal anode in an undivided cell. This general approach has enabled the carboxylation of a wide variety of substrate types across all of the areas covered in this review. Addition of a transition metal catalyst has been shown to promote and to influence the product distribution of the reactions and has even proven useful for preliminary efforts to render some of the carboxylation reactions asymmetric. However, continued development and uptake of the reactions in an academic as well as in an industrial setting has been hampered by the lack of understanding of electrochemical equipment required for the synthetic chemist. Over recent years the availability of standardized electrochemical equipment, the requirement for new methods to sustainably activate/utilize carbon dioxide and students/post-doctoral workers being trained in these areas has seen a renewed academic and industrial interest in electrosynthesis and hence electrocarboxylation reactions. The development of undivided or divided cell systems that do not require a sacrificial electrode have been a major step forward in the area and these processes now offer the opportunity of application within a continuous process setting, since the anode no longer requires replacement and precipitated electrode material no longer requires removal. Coupled with the significant developments in related transition-metal catalyzed and photocatalyzed reactions the area of electrocarboxylation looks particularly bright.

References

1. (a) Leech, M. C.; Lam, K. A practical guide to electrosynthesis. *Nat. Rev. Chem.* **2022**, *6*, 275; (b) Leech, M. C.; Garcia, A. D.; Petti, A.; Dobbs, A. P.; Lam, K. A. *React. Chem. Eng.* **2020**, *5*. 997–990; (c) Horn, E. J.; Rosen, B. R.; Baran, P. S. Synthetic organic electrochemistry: an enabling and innately sustainable method. *ACS Cent. Sci.* **2016**, *2*, 302–308; (d) Yan, M.; Kawamata, Y.; Baran, P. S. Synthetic organic electrochemical methods since 2000: on the verge of a renaissance. *Chem. Rev.* **2017**, *117*, 13230–13319.
2. Matthessen, R.; Fransaer, J.; Binnemans, K.; De Vos, D. E. Electrocarboxylation: towards sustainable and efficient synthesis of valuable carboxylic acids. *Beilstein J. Org. Chem.* **2014**, *10*, 2484–2500.
3. Martín, C.; Fiorani, G.; Kleij, A. W. Recent advances in the catalytic preparation of cyclic organic carbonates. *ACS Catal.* **2015**, *5*, 1353–1370.
4. Ion, A.; Parvulescu, V.; Jacobs, P.; De Vos, D. Synthesis of symmetrical or asymmetrical urea compounds from CO_2 via base catalysis. *Green Chem.* **2007**, *9*, 158–161.
5. Liu, Q.; Wu, L.; Jackstell, R.; Beller, M. Using carbon dioxide as a building block in organic synthesis. *Nat. Commun.* **2018**, *6*, 5933.
6. Gomes-Jelonek, J. *Metathesis: The Green Method of Chemical Synthesis*; Royal Society of Chemistry June: Chemistry World, 2018.
7. Gennaro, A.; Isse, A. A.; Severin, M.-G.; Vianello, E.; Bhugun, I.; Savéant, J.-M. Mechanism of the electrochemical reduction of carbon dioxide at inert electrodes in media of low proton availability. *J. Chem. Soc. Faraday Trans.* **1996**, *92*, 3963–3968.
8. Gennaro, A.; Isse, A. A.; Savéant, J.-M.; Severin, M.-G.; Vianello, E. Homogeneous electron transfer catalysis of the electrochemical reduction of carbon dioxide. Do aromatic anion radicals react in an outer-sphere manner? *J. Am. Chem. Soc.* **1996**, *118*, 7190–7196.
9. Goodridge, F.; Presland, G. The electrolytic reduction of carbon dioxide and monoxide for the production of carboxylic acids. *J. Appl. Electrochem.* **1984**, *14*, 791–796.
10. Rudolph, M.; Dautz, S.; Jäger, E.-G. Macrocyclic N_4^{2-} coordinated nickel complexes as catalysts for the formation of oxalate by electrochemical reduction of carbon dioxide. *J. Am. Chem. Soc.* **2000**, *122*, 10821–10830.
11. Sun, L.; Ramesha, G. K.; Kamat, P. V.; Brennecke, J. F. Switching the reaction course of electrochemical CO_2 reduction with ionic liquids. *Langmuir* **2014**, *30*, 6302–6308.
12. Reissig, H.-U.; Zimmer, R. Donor – acceptor-substituted cyclopropane derivatives and their application in organic synthesis. *Chem. Rev.* **2003**, *103*, 1151–1196.
13. For recent electrochemical examples, see: (a) Kolb, S.; Petzold, M.; Brandt, F.; Jones, P. G.; Jacob, C. R.; Werz, D. B. Electrocatalytic activation of donor-acceptor cyclopropanes and cyclobutanes: an alternative $C(sp^3)$–$C(sp^3)$ cleavage mode. Angew.

Chem. Int. Ed. 2021, 60, 15928 – 15934; (b) Peng, P.; Yan, X.; Zhang, K.; Liu, Z.; Zeng, L.; Chen, Y.; Zhang, H.; Lei, A. Electrochemical C–C bond cleavage of cyclopropanes towards the synthesis of 1,3-difunctionalized molecules. *Nat. Commun.* **2021**, *12*, 3075.

14. For recent photochemical examples, see: (a) Pitts, C. R.; Ling, B.; Snyder, J. A.; Bragg, A. E.; Lectka, T. Aminofluorination of cyclopropanes: a multifold approach through a common, catalytically generated intermediate. J. Am. Chem. Soc. **2016,** 138, 6598 – 6609; (b) Petzold, D.; Singh, P.; Almqvist, F.; König, B. Visible-light-mediated synthesis of β-chloro ketones from aryl cyclopropanes. *Angew. Chem. Int. Ed.* **2019**, *58*, 8577–8580; (c) Ge, L.; Wang, D. X.; Xing, R.; Ma, D.; Walsh, P. J.; Feng, C. Photoredox-catalyzed oxo-amination of aryl cyclopropanes. *Nat. Commun.* **2019**, *10*, 4367; (d) Zuo, Z.; Daniliuc, C. G.; Studer, A. Cooperative NHC/photoredox catalyzed ring-opening of aryl cyclopropanes to 1-aroyloxylated-3-acylated alkanes. *Angew. Chem. Int. Ed.* **2021**, *60*, 25252–25257.
15. Liao, L. L.; Wang, Z.-H.; Cao, K.-G.; Sun, G.-Q.; Zhang, W.; Ran, C.-K.; Li, Y.; Chen, L.; Cao, G.-M.; Yu, D.-G. Electrochemical ring-opening dicarboxylation of strained carbon–carbon single bonds with CO_2: facile synthesis of diacids and derivatization into polyesters. *J. Am. Chem. Soc.* **2022**, *144*, 2062–2068.
16. Shul'zhenko, G. I.; Vasil'ev, Y. B. Electrochemical synthesis of dicarboxylic acids by means of the electroreduction of carbon dioxide in the presence of ethylene and its derivatives. *Russ. Chem. Bull.* **1991**, *40*, 1217–1220.
17. Dérien, S.; Clinet, J.-C.; Duñach, E.; Périchon, J. Electrochemical incorporation of carbon dioxide into alkenes by nickel complexes. *Tetrahedron* **1992**, *48*, 5235–5248.
18. Senboku, H.; Komatsu, H.; Fujimura, Y.; Tokuda, M. Efficient electrochemical dicarboxylation of phenyl-substituted alkenes: synthesis of 1-phenylalkane-1,-2-dicarboxylic acids. *Synlett* **2001**, 418–420.
19. Ohkoshi, M.; Michinishi, J.-Y.; Hara, S.; Senboku, H. *Tetrahedron* **2010**, *66*, 7732–7737.
20. You, Y.; Kanna, W.; Takano, H.; Hayashi, H.; Maeda, S.; Mita, T. Electrochemical dearomative dicarboxylation of heterocycles with highly negative reduction potentials. *J. Am. Chem. Soc.* **2022**, *144*, 3685–3695.
21. Hoberg, H.; Peres, Y.; Krüger, C.; Tsay, Y. H. A 1-oxa-2-nickela-5-cyclopentanone from ethene and carbon dioxide: preparation, structure, and reactivity. *Angew. Chem. Int. Ed.* **1987**, *26*, 771–773.
22. Williams, C. M.; Johnson, J. B.; Rovis, T. Nickel-catalyzed reductive carboxylation of styrenes using CO_2. *J. Am. Chem. Soc.* **2008**, *130*, 14936–14937.
23. Greenhalgh, M. D.; Thomas, S. P. Iron-catalyzed, highly regioselective synthesis of α-aryl carboxylic acids from styrene derivatives and CO_2. *J. Am. Chem. Soc.* **2012**, *134*, 11900–11903.
24. Gaydou, M.; Moragas, T.; Juliá-Hernández, F.; Martin, R. Site-selective catalytic carboxylation of unsaturated hydrocarbons with CO_2 and water. *J. Am. Chem. Soc.* **2017**, *139*, 12161–12164.
25. Toki, S.; Hida, S.; Chemischer, S. T. Photochemical reaction of styrenes with triethylamine—photofixation of carbon dioxide by radical anion. *Nippon Kagaku Kaishi* **1984**, 152–157.
26. Seo, H.; Liu, A.; Jamison, T. F. Direct β-selective hydrocarboxylation of styrenes with CO_2 enabled by continuous flow photoredox catalysis. *J. Am. Chem. Soc.* **2017**, *139*, 13969–13972.
27. Murata, K.; Numasawa, N.; Shimomaki, K.; Takaya, J.; Iwasawa, N. Construction of a visible light-driven hydrocarboxylation cycle of alkenes by the combined use of Rh(I) and photoredox catalysts. *Chem. Commun.* **2017**, *53*, 3098–3101.
28. Meng, Q.-Y.; Wang, S.; Huff, G. S.; König, B. Ligand-controlled regioselective hydrocarboxylation of styrenes with CO_2 by combining visible light and nickel catalysis. *J. Am. Chem. Soc.* **2018**, *140*, 3198–3201.

29. Chowdhury, M. A.; Şenboku, H.; Tokuda, M. Electrochemical carboxylation of bicyclon. 1.0 alkylidene derivatives. *Tetrahedron* **2004**, *60*, 475–481.
30. Li, J.; Inagi, S.; Fuchigami, T.; Hosono, H.; Ito, S. Selective mono-carboxylation of olefins at $12CaO{\cdot}7Al_2O_3$ electride cathode. *Electrochem. Commun.* **2014**, *44*, 45–48.
31. Alkayal, A.; Tabas, V.; Montanaro, S.; Wright, I. A.; Malkov, A. V.; Buckley, B. R. Harnessing applied potential: selective β-hydrocarboxylation of substituted olefins. *J. Am. Chem. Soc.* **2019**, *142*, 1780–1785.
32. Sheta, A. M.; Alkayal, A.; Mashaly, M. A.; Said, S. B.; Elmorsy, S. S.; Malkov, A. V.; Buckley, B. R. Selective electrosynthetic hydrocarboxylation of α,β-unsaturated esters with carbon dioxide. *Angew. Chem. Int. Ed.* **2021**, *60*, 21832–21837.
33. (a) Huang, H.; Ye, J.-H.; Zhu, L.; Ran, C.-K.; Miao, M.; Wang, W.; Chen, H.; Zhou, W.-J.; Lan, Y.; Yu, B.; Yu, D.-G. Visible-light-driven anti-Markovnikov hydrocarboxylation of acrylates and styrenes with CO_2. *CCS Chem.* **2020**, 1746–1756; (b) Kang, G.; Catalysis, D. R. A. Photocatalyzed, β-selective hydrocarboxylation of α,β-unsaturated esters with co_2 under flow for B-lactone synthesis. *ACS Catal.* **2021**, *2021* (11), 1309–1315.
34. Chen, R.; Tian, K.; He, D.; Gao, T.; Yang, G.; Xu, J.; Chen, H.; Wang, D.; Zhang, Y. Carboxylation of α,β-unsaturated ketones by CO2 fixation through photoelectro-chemistry. *ACS Appl. Energy Mater.* **2020**, *3*, 5813–5818.
35. Gao, X.-T.; Zhang, Z.; Wang, X.; Tian, J.-S.; Xie, S. L.; Zhou, F.; Zhou, J. Direct electrochemical defluorinative carboxylation of A-CF_3 alkenes with carbon dioxide. *Chem. Sci.* **2020**, *11*, 10414–10420.
36. Duñach, E.; Périchon, J. Electrochemical carboxylation of terminal alkynes catalyzed by nickel complexes: unusual regioselectivity. *J. Organomet. Chem.* **1988**, *352*, 239–246.
37. Duñach, E.; Dérien, S.; Périchon, J. Nickel-catalyzed reductive electrocarboxylation of disubstituted alkynes. *J. Organomet. Chem.* **1989**, *364*, C33–C36.
38. Köster, F.; Dinjus, E.; Duñach, E. Electrochemical selective incorporation of CO_2 into terminal alkynes and diynes. *Eur. J. Org. Chem.* **2001**, 2507–2511.
39. Yuan, G.-Q.; Jiang, H.-F.; Lin, C. Efficient electrochemical dicarboxylations of arylacetylenes with carbon dioxide using nickel as the cathode. *Tetrahedron* **2008**, *64*, 5866–5872.
40. Li, C.; Yuan, G.; Huanfeng, J. Electrocarboxylation of alkynes with carbon dioxide in the presence of metal salt catalysts. *Chin. J. Chem.* **2010**, *28*, 1685–1689.
41. Katayama, A.; Senboku, H.; Hara, S. Aryl radical cyclization with alkyne followed by tandem carboxylation in methyl 4-tert-butylbenzoate-mediated electrochemical reduction of 2-(2-propynyloxy)bromobenzenes in the presence of carbon dioxide. *Tetrahedron* **2016**, *72*, 4626–4636.
42. (a) Takimoto, M.; Mori, M. Cross-coupling reaction of oxo-π-allylnickel complex generated from 1,3-diene under an atmosphere of carbon dioxide. *J. Am. Chem. Soc.* **2001**, *123*, 2895–2896; (b) Tortajada, A.; Ninokata, R.; Martin, R. Ni-catalyzed site-selective dicarboxylation of 1,3-dienes with CO_2. *J. Am. Chem. Soc.* **2018**, *140*, 2050–2053.
43. Loveland, J. W. *Electrolytic Production of Acyclic Carboxylic Acids From Hydrocarbons*; U.S. Patent 3,032,489, May 1, 1962.
44. Bringmann, J.; Dinjus, E. Electrochemical synthesis of carboxylic acids from alkenes using various nickel-organic mediators: CO_2 as C1-synthon. *Appl. Organometal. Chem.* **2001**, *15*, 135–140.
45. Jiang, H.; Li, C.; Yuan, G.; Qi, C.; Ji, X. Method for synthesizing 3-alkene(cycloalkene)-1,6-dicarboxylic acid by using electrochemical reaction. *Faming Zhuanli Shenqing CN* **2009**, 10192927.
46. Liu, X.; Liu, R.; Qiu, J.; Cheng, X.; Li, G. Chemical-reductant-free electrochemical deuteration reaction using deuterium oxide. *Angew. Chem. Int. Ed.* **2020**, *59*, 13962–13967.

47. Dérien, S.; Clinet, J.-C.; Duñach, E.; Périchon, J. Activation of carbon dioxide: nickel-catalyzed electrochemical carboxylation of diynes. *J. Org. Chem.* **2002**, *58*, 2578–2588.
48. Li, C.-H.; Yuan, G.-Q.; Qi, C.-R.; Jiang, H.-F. Copper-catalyzed electrochemical synthesis of alkylidene lactones from carbon dioxide and 1,4-diarylbuta-1,3-diynes. *Tetrahedron* **2013**, *69*, 3135–3140.
49. (a) Lateef, S. K.; Raju, R. R.; Mohan, S. K.; Reddy, S. J. Electrochemical synthesis of α-hydroxycarboxylic acids from acetophenones. *Synth. Commun.* **2006**, *36*, 31–36; (b) Singh, K.; Sohal, H. S.; Singh, B. Synthesis of α-hydroxycarboxylic acids from various aldehydes and ketones by direct electrocarboxylation: a facile, efficient and atom economy protocol. *Asian J. Chem.* **2021**, *33*, 839–845; (c) Yuan, G.; Li, Z.; Jiang, H. Electrosyntheses of α-hydroxy carboxylic acids from carbon dioxide and aromatic ketones using nickel as the cathode. *Chin. J. Chem.* **2009**, *27*, 1464–1470; (d) Stalcup, M. A.; Nilles, C. K.; Lee, H.-J.; Subramaniam, B.; Blakemore, J. D.; Leonard, K. C. Organic electrosynthesis in CO_2-expanded electrolytes: enabling selective acetophenone carboxylation to atrolatic acid. *ACS Sustainable Chem. Eng.* **2021**, *9*, 10431–10436; (e) Zhao, S.-F.; Horne, M.; Bond, A. M.; Zhang, J. Electrocarboxylation of acetophenone in ionic liquids: the influence of proton availability on product distribution. *Green Chem.* **2014**, *16*, 2242–2251; (f) Damodar, J.; Raju, R. R.; Reddy, S. J. Electrochemical synthesis of precursors of non-steroidal antiinflammatory agents. *Ind. J. Chem. B.* **2002**, *41B*, 2655–2658; (g) Liu, R.; Yuan, G.; Joe, C. L.; Lightburn, T. E.; Tan, K. L.; Wang, D. Silicon nanowires as photoelectrodes for carbon dioxide fixation. *Angew. Chem. Int. Ed.* **2012**, *51*, 6709–6712; (h) Feng, Q.; Huang, K.; Liu, S.; Yu, J.; Liu, F. *Electrochim. Acta* **2011**, *56*, 5137–5141; (i) Ikeda, Y.; Manda, E. Syntheses of benzilic acids through electrochemical reductive carboxylation of benzophenones in the presence of carbon dioxide. *Bull. Chem. Soc. Jpn.* **1985**, *58*, 1723–1726; (j) Zhao, S.-F.; Horne, M.; Bond, A. M.; Zhang, J. Electrochemical reduction of aromatic ketones in 1-butyl-3-methylimidazolium-based ionic liquids in the presence of carbon dioxide: the influence of the ketone substituent and the ionic liquid anion on bulk electrolysis product distribution. *Phys. Chem. Chem. Phys.* **2015**, *17*, 19247–19254; (k) Tian, K.; Chen, R.; Xu, J.; Yang, G.; Xu, X.; Zhang, Y. Understanding the photo- and electro-carboxylation of o-methylbenzophenone with carbon dioxide. *Catalysts* **2020**, *10*, 664; (l) Zhang, K.; Wang, H.; Wu, L.; Zhang, J.; Lu, J. Efficient electrocarboxylation of *p*-methylpropiophenone in the presence of carbon dioxide. *Chin. J. Chem.* **2010**, *28*, 509–513.
50. Wawzonek, S.; Runner, M. E. Polarographic studies in acetonitrile. *J. Electrochem. Soc.* **1952**, *99*, 457–459.
51. Chan, A. S. C.; Huang, T. T.; Wagenknecht, J. H.; Miller, R. E. A novel synthesis of 2-aryllactic acids via electrocarboxylation of methyl aryl ketones. *J. Org. Chem.* **1995**, *60*, 742–744.
52. Datta, A. K.; Marron, P. A.; King, C. J. H.; Wagenknecht, J. H. Process development for electrocarboxylationof 2-acetyl-6-methoxynaphthalene. *J. Appl. Electrochem.* **1998**, *28*, 569–577.
53. (a) Zhang, K.; Wang, H.; Zhao, S.-F.; Niu, D.-F.; Lu, J.-X. Asymmetric electrochemical carboxylation of prochiral acetophenone: an efficient route to optically active atrolactic acid via selective fixation of carbon dioxide. *J. Electroanal. Chem.* **2009**, *630*, 35–41; (b) Zhao, S.-F.; Zhu, M.-X.; Zhang, K.; Wang, H.; Lu, J.-X. Alkaloid induced asymmetric electrocarboxylation of 4-methylpropiophenone. *Tetrahedron Lett.* **2011**, *52*, 2702–2705.
54. Chen, B.-L.; Tu, Z.-Y.; Zhu, H.-W.; Sun, W.-W.; Wang, H.; Lu, J.-X. CO_2 as a C1-organic building block: enantioselective electrocarboxylation of aromatic ketones with CO_2 catalyzed by cinchona alkaloids under mild conditions. *Electrochim. Acta* **2014**, *116*, 475–483.

55. (a) Sathe, A. A.; Hartline, D. R.; Radosevich, A. T. A synthesis of α-amino acids via direct reductive carboxylation of imines with carbon dioxide. *Chem. Commun.* **2013**, *49*, 5040–5042; (b) Li, C.-H.; Song, X.-Z.; Tao, L.-M.; Li, Q.-G.; Xie, J.-Q.; Peng, M.-N.; Pan, L.; Jiang, C.; Peng, Z.-Y.; Xu, M.-F. Electrogenerated-bases promoted electrochemical synthesis of *N*-bromoamino acids from imines and carbon dioxide. *Tetrahedron* **2010**, *70*, 1855–1860; (c) Koshechko, V. G.; Titov, V. E.; Bondarenko, V. N.; Pokhodenko, V. D. Electrochemical carboxylation of fluorocontaining imines with preparation of fluorinated N-phenylphenylglycines. *J. Fluorine Chem.* **2008**, *129*, 701–706.
56. Silvestri, G.; Gambino, S.; Filardo, G.; Tedeschi, F. A filter-press electrolytic cell with semi-continuous renewal of sacrifical electrodes. *J. Appl. Electrochem.* **1989**, *19*, 946–948.
57. (a) Qu, Y.; Tsuneishi, C.; Tateno, H.; Matsumura, Y.; Atobe, M. Green synthesis of α-amino acids by electrochemical carboxylation of imines in a flow microreactor. *React. Chem. Eng.* **2017**, *2*, 871–875; (b) Naito, Y.; Nakamura, Y.; Shida, N.; Senboku, H.; Tanaka, K.; Atobe, M. Integrated flow synthesis of α-amino acids by in situ generation of aldimines and subsequent electrochemical carboxylation. *J. Org. Chem.* **2021**, *86*, 15953–15960.
58. (a) Baizer, M. M.; Chruma, J. L. Electrolytic reductive coupling. XXI.1 Reduction of organic halides in the presence of electrophiles. *J. Org. Chem.* **1972**, *37*, 1951–1960; (b) Tyssee, D. A.; Baizer, M. M. Electrocarboxylation, I. Mono- and dicarboxylation of activated olefins. *J. Org. Chem.* **1974**, *39*, 2819–2823.
59. Isse, A. A.; Gennaro, A. Electrocatalytic carboxylation of benzyl chlorides at silver cathodes in acetonitrile. *Chem. Commun.* **2002**, *2798–2799.*
60. Scialdone, O.; Galia, A.; Errante, G.; Isse, A. A.; Gennaro, A.; Filardo, G. Electrocarboxylation of benzyl chlorides at silver cathode at the preparative scale level. *Electrochim. Acta* **2008**, *53*, 2514–2528.
61. Niu, D.-F.; Xiao, L.-P.; Zhang, A.-J.; Zhang, G.-R.; Tan, Q.-Y.; Lu, J.-X. Electrocatalytic carboxylation of aliphatic halides at silver cathode in acetonitrile. *Tetrahedron* **2008**, *64*, 10517–10520.
62. Niu, D.; Zhang, J.; Zhang, K.; Xue, T. Electrocatalytic carboxylation of benzyl chloride at silver cathode in ionic liquid $BMImBF_4$. *Chin. J. Chem.* **2009**, *27*, 1041–1044.
63. Zhong, J.-S.; Yang, Z.-X.; Ding, C.-L.; Huang, Y.-F.; Zhao, Y.; Yan, H.; Ye, K.-Y. Desulfonylative electrocarboxylation with carbon dioxide. *J. Org. Chem.* **2021**, *86*, 16162–16170.
64. Yamauchi, Y.; Fukuhara, T.; Hara, S.; Senboku, H. Electrochemical carboxylation of α,α-difluorotoluene derivatives and its application to the synthesis of α-fluorinated nonsteroidal anti-inflammatory drugs. *Synlett* **2008**, 438–442.
65. Sun, G.-Q.; Zhang, W.; Liao, L. L.; Li, L.; Nie, Z.-H.; Wu, J.-G.; Zhang, Z.; Yu, D.-G. Nickel-catalyzed electrochemical carboxylation of unactivated aryl and alkyl halides with CO_2. *Nat. Commun.* **2021**, *12*, 7086.
66. Feroci, M.; Orsini, M.; Palombi, L.; Sotgiu, G.; Colapietro, M.; Inesi, A. Diastereoselective electrochemical carboxylation of chiral α-bromocarboxylic acid derivatives: an easy access to unsymmetrical alkylmalonic ester derivatives. *J. Org. Chem.* **2004**, *69*, 487–494.
67. Sock, O.; Troupel, M.; Perichon, J. Electrosynthesis of carboxylic acids from organic halides and carbon dioxide. *Tetrahedron Lett.* **1985**, *26*, 1509–1512.
68. (a) Torii, S.; Tanaka, H.; Hamatani, T.; Morisaki, K.; Jutand, A.; Pfluger, F.; Fauvarque, J.-F. Pd(0)-catalyzed electroreductive carboxylation of aryl halides, β-bromostyrene, and allyl acetates with CO_2. *Chem. Lett.* **1986**, *15*, 169–172; (b) Senboku, H.; Kanaya, H.; Fujimura, Y.; Tokuda, M. Stereochemical study on electrochemical carboxylation of vinyl triflates. *J. Electroanal. Chem.* **2001**, *507*, 82–88; (c) Medeiros, M. J.; Pintaric, C.; Olivero, S.; Duñach, E. Nickel-catalysed

electrochemical carboxylation of allylic acetates and carbonates. *Electrochim. Acta* **2011**, *56*, 4384–4389; (d) Wu, L.-X.; Zhao, Y.-G.; Guan, Y.-B.; Wang, H.; Lan, Y.-C.; Wang, H.; Lu, J.-X. Silver encapsulated copper salen complex: efficient catalyst for electrocarboxylation of cinnamyl chloride with CO_2. *RSC Adv.* **2019**, *9*, 32628–32633.
69. Folest, J.-C.; Duprilot, J.-M.; Perichon, J.; Robin, Y.; Devynck, J. Electrocatalyzed carboxylation of organic halides by a cobalt-salen complex. *Tetrahedron Lett.* **1985**, *26*, 2633–2636.
70. (a) Ang, N. W. J.; de Oliveira, J. C. A.; Ackermann, L. Electroreductive cobalt-catalyzed carboxylation: cross-electrophile electrocoupling with atmospheric CO_2. *Angew. Chem. Int. Ed.* **2020**, *59*, 12842–12847.
71. Jiao, K.-J.; Li, Z.-M.; Xu, X.-T.; Zhang, L.-P.; Li, Y.-Q.; Zhang, K.; Mei, T.-S. Palladium-catalyzed reductive electrocarboxylation of allyl esters with carbon dioxide. *Org. Chem. Front.* **2018**, *5*, 2244–2248.

About the author

Benjamin Buckley is currently a Reader at Loughborough University (UK). In 2003 he received his PhD with Prof. Philip C. Bulman Page on organocatalytic asymmetric epoxidation. He then took up a post-doctoral position in the Page group for a further 5 years working on a range of projects with industrial and academic partners. In late 2007 he was appointed to a Research Councils UK Fellowship at Loughborough and initiated independent research in the areas of catalysis, electrosynthesis, and carbon dioxide utilization. He was promoted to a Senior Lectureship in 2014 and also moved to the University of Utah, USA on sabbatical leave for a year with Professor Matthew S. Sigman's group. In February 2020 he was appointed to a Readership in Chemistry at Loughborough. His group are active in the areas of carbon dioxide utilization and electrosynthesis.

CHAPTER FOUR

Photocatalytic carboxylation with CO_2

Hong Zhao†, Wei Wang†, Hai-Peng Zhang, Yi He, Si-Shun Yan*, and Da-Gang Yu*

Key Laboratory of Green Chemistry & Technology of Ministry of Education, College of Chemistry, Sichuan University, Chengdu, People's Republic of China

*Corresponding author: e-mail address: yssluck@163.com; dgyu@scu.edu.cn

Contents

Abstract

Due to the structural characteristics of carbon dioxide (CO_2) molecule, which has high thermodynamic stability and kinetic inertness, it is challenging to achieve efficient activation and selective conversion of CO_2 under mild conditions. Since carboxylic acids do not only exist widely in natural products and bioactive drugs, but also serve as important bulk chemicals in industry, chemists have made great efforts to develop novel light-induced carboxylation employing CO_2 in recent years. In this chapter, we will introduce the photocatalytic carboxylation of various substrates with CO_2 to construct diverse carboxylic acids. This article is arranged based on the different types of reaction substrates with CO_2, including carboxylation of C—H bonds, C—(pseudo)halide bonds, unsaturated substrates and miscellaneous. We highlight the mechanisms of these

† Equal contribution.

Advances in Catalysis, Volume 70
ISSN 0360-0564
https://doi.org/10.1016/bs.acat.2022.06.003

reactions, which can proceed in the presence of a photoredox catalyst or a photoredox/transition metal dual catalytic system. We also discuss a perspective on their prospects, hopefully stimulating the further research and development of this appealing field.

Fossil fuels have served as the industrial "blood" for our civilizations. However, accompanied with the utilization of fossil fuels, the over-emission of carbon dioxide (CO_2) brings serious problems, like the global warming. Delightedly, chemists are ongoingly pursuing the high-value transformation of readily-available feedstocks, which may provide an opportunity in the reduction of CO_2 emission in the atmosphere. Actually, long-term efforts have been devoted to the conversion of CO_2 to various value-added chemicals, such as urea, carbonates and salicylic acid etc. *(1)*. Among these products, carboxylic acids and derivatives are common motifs in pharmaceuticals, agrochemicals and bioactive compounds *(2)*, which attracts persistent attention both in academia and industry. The well-developed strategies to construct carboxylic acids, including the hydrolysis of nitriles, the oxidation of alcohols and aldehydes, and the carbonylation of organic (pseudo) halides with carbon monoxide, can't meet the need for the atomic economy or environmental friendliness. On account for its abundance and low cost, CO_2 is employed as an ideal feedstock to synthesize carboxylic acids via the carboxylation of nucleophilic reagents. The use of stoichiometric amounts of reactive organometallic reagents, such as organo lithium and Grignard reagents, is a direct way to realize this transformation, which might result in the poor tolerance of functional groups. Later, less reactive organometallic reagents, such as organo boronic acids and stannanes, as well as C—H bonds have been widely investigated to furnish the carboxylic acids by employing transition-metal complexes or catalysts *(3)*. In addition, electrochemistry provides a promising tool to access the reductive carboxylation. However, the utilization of sacrificial anodes in many cases has hampered their wide applications. Therefore, the chemists are appealing for environment-friendly and much more efficient methodologies to furnish the carboxylic acids with CO_2. With the renaissance of photochemistry, chemists have developed versatile visible light-induced organic synthesis via single-electron transfer (SET) or energy transfer (ET) process *(4)*. Given their mild reaction conditions, visible light-driven reactions possess broad functional group compatibility. Notably, the cooperation of photocatalysts and transition metal catalysts provide a revolutionary strategy to enrich the organic transformations *(5)*. Driven by the above-mentioned advantages, increasing numbers of groups are trying to employ photochemistry to realize the CO_2 fixation. For example, Murakami, Jamison, Ishida, and others developed excellent works in UV light-promoted carboxylation with CO_2 *(6–9)*. Iwasawa, He, Martin, Wu, König, Li, Xi, Yu and others achieved remarkable visible-light photoredox-catalytic carboxylic functionalizations of alkenes, alkynes or C—H bonds with CO_2 *(10–13)*. In this chapter, we summarize the photocatalytic carboxylation of CO_2 into four parts, including the carboxylation of C—H bonds, C−(pseudo)halide bonds, unsaturated substrates and others. We emphasized discussions on the investigation of the mechanism in each work. Hopefully, this summary can point out the directions to future researches in the field of CO_2 fixation.

1. Carboxylation of C—H bonds

As we know, the C—H bonds are one of the most abundant chemical bonds in organic compounds. Therefore, the direct carboxylation of the C—H bonds with CO_2 is an attractive way to obtain high-valued carboxylic acid straightly. For the past few years, scientists have already discovered many ways to achieve this goal by applying transition metal catalysis *(14)*. For example, Iwasawa's group has already reported the Ni- and Ru-catalyzed C—H bonds carboxylation of ethylene with CO_2 *(15)*. In 2010, Hou's group also reported the Cu-mediated carboxylation of C—H bonds with CO_2 *(16)*. Besides transition metal catalysis, due to the advantages such as being environmental-friendly and the mild reaction condition, the photocatalysis way was also developed recently.

1.1 Carboxylation of sp^3 C—H bonds

1.1.1 UV light

In 2015, Murakami's group pioneered the carboxylation of sp^3 C—H bond in *o*-alkylphenyl ketones through the Norrish Type II photoreaction and a [4+2] cycloaddition reaction (Scheme 1) *(6)*. The standard substrate *o*-methylbenzophenone **1a** could transform into the corresponding carboxylic acid **1b** in 89% yield under simple reaction conditions: CO_2 (1 atm), DMSO, UV light (365 nm), room temperature and 2 h. What's more, this reaction could also conduct under 7 h' solar irradiation with a 72% yield. Through the mechanistic investigation and DFT calculation, the author supposed that under UV irradiation, the substrate could undergo the Norrish Type II photoreaction to form the *o*-quinodimethane **1c**. The formed *o*-quinodimethane was a highly reactive 1,3-diene, which could undergo a [4+2] cycloaddition reaction with CO_2 and then the

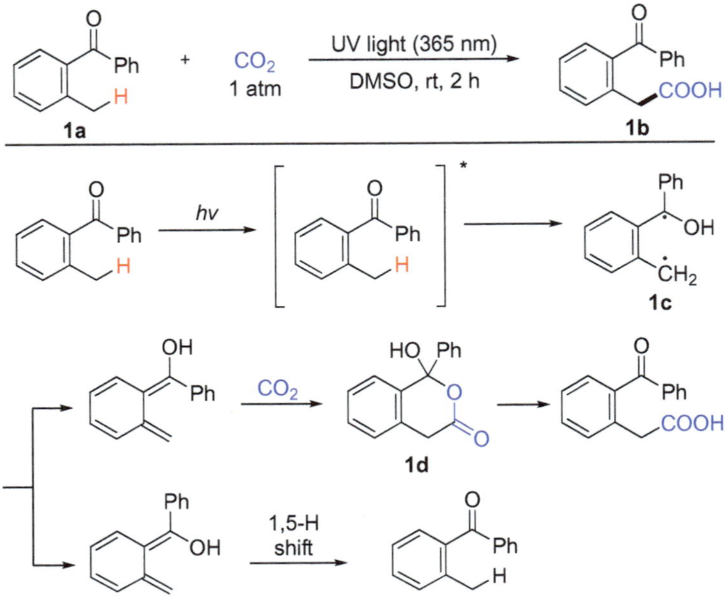

Scheme 1 UV light-induced carboxylation of sp^3 C—H bonds in *o*-alkylphenyl ketones.

six-membered cycloadduct **1d** product transformed into the desired carboxylic acid through a ring-opening reaction. It's worth noting that only (*E*)-*o*-quinodimethane could react with CO_2 because of its long lifetime. The (*Z*)-*o*-quinodimethane rapidly underwent a 1,5-hydrogen atom shift to revert to the starting ketone. This was also the first report that CO_2 could react as a dienophile in the [4+2] cycloaddition reaction.

After this work, Murakami's group continued to investigate the use of CO_2 in the carboxylation of C—H bonds with CO_2 and reported another UV light-induced carboxylation of allylic C—H bonds of alkenes with a light/ketone/copper system in 2016 *(7)*. In the presence of ketone, copper complex, ^{t}BuOK, benzene, UV light (365 nm), and high temperature (110 °C), the standard substrate cyclohexene **2a** could react with CO_2 (1 atm) and formed the β,γ-unsaturated carboxylic acid **2b** with 45 TON (based on the copper complex) after 12 h. After the investigation of the mechanism and DFT calculation, the author suggested that the reaction was conducted as follows (Scheme 2). First, the ketone was excited by the UV light, and the carbonyl oxygen abstracted hydrogen of the allylic C—H bond of the substrate. The formed radicals coupled with each other to get the alcohol intermediate. After that, the copper tert-butoxide **2c** which was produced from the copper complex **2d** and ^{t}BuOK abstracted

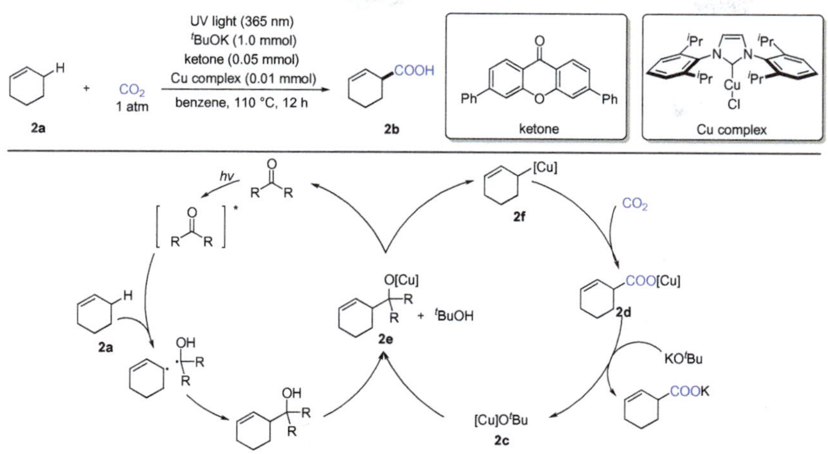

Scheme 2 UV light-induced carboxylation of allylic sp^3 C—H bonds of alkenes.

proton from the alcohol intermediate. The formed copper alkoxide **2e** underwent a *β*-carbon elimination to furnish the allylcopper species **2 f**. The allylcopper species could attack CO_2 and subsequently follow with a ligand exchange process to get the desired carboxylate salt. This work showed the potential industry ability of photocatalysis in the field of CO_2 utilization.

Based on the work published in 2016, Murakami's group further reported the carboxylation of benzylic and aliphatic C—H bonds induced by UV light in 2019 *(8)*. In this work, the author found that nickel and xanthone were the appropriate dual catalysts. The standard substrate *p*-methoxytoluene could react with CO_2 to get the corresponding carboxylic acid in 84% yield which was much higher than the result in previous work. The mechanism of this reaction was similar to the copper-mediated one in the initiation step (Scheme 3). However, the formed two radicals didn't couple with each other. In this reaction, the ketyl radical **3a** was deprotonated by ^{t}BuOK and gave a ketyl radical anion **3b** which could transfer its electron to the nickel (II) salt. And a nickel(0) species might be generated and further combine with the benzylic radical **3c**. The formed benzylnickel(I) species **3d** reacted with CO_2 to give a nickel(I) carboxylate **3e** which could be further reduced by ketyl radical anion **3f** to release the desired carboxylate anion and regenerated the nickel(0). Notably, this reaction could also tolerate aliphatic substrates such as cyclopentane and pentane, which also showed the industrial application potential of this reaction.

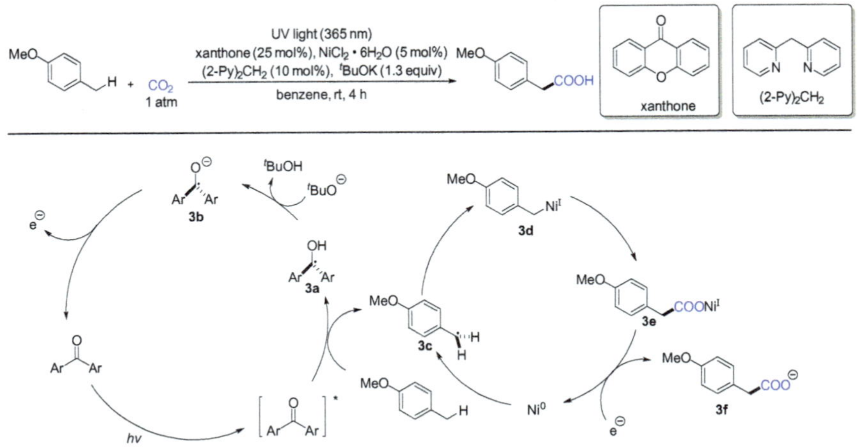

Scheme 3 UV light-induced carboxylation of benzylic and aliphatic sp^3 C—H bonds.

The works discussed above mainly focused on the photo-activation of the substrate and following two-electron activation of CO_2. Different from these works, in 2017, Jamison's group found that CO_2 could undergo a single-electron transfer (SET) reduction through photocatalysis and the formed carbon dioxide radical anion ($CO_2^{\bullet-}$) participated in the synthesis of α-amino acids (Scheme 4) *(9)*. After the comparison of the reduction potential of CO_2 and many kinds of photocatalysts, the authors chose *p*-terphenyl to achieve the SET reduction of CO_2. Through the investigation of the standard reaction condition, the standard substrate *N*-benzylpiperidine could form the desired carboxylic acid in continuous flow with 92% yield in the presence of CO_2, *p*-terphenyl, $KOCOCF_3$, DMF, and UV light (>280 nm). Through the investigation of the mechanism, the author proposed that the irradiation of the *p*-terphenyl with UV light produced the excited state of *p*-terphenyl, which underwent SET with the amine **4a** to provide the strongly reducing *p*-terphenyl radical anion and the corresponding amine radical cation. CO_2 was reduced to form the $CO_2^{\bullet-}$ by the strong reductant. Concurrently, deprotonation of **4b** affords the neutral α-amino radical **4c**. Then, the $CO_2^{\bullet-}$ bonded with the neutral α-amino radical **4c** to produce the desired α-amino acid.

1.1.2 Visible light

Compared to UV light, the visible-light-induced carboxylation of the sp^3 C—H bonds was rarely reported, which might arise from lower energy

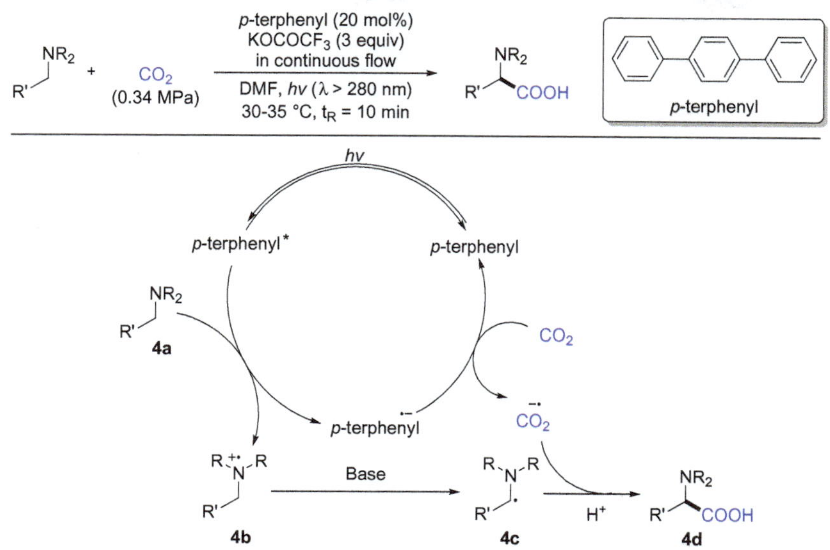

Scheme 4 Photoredox activation of CO_2 for amino acid synthesis in continuous flow.

of visible light. In 2019, König's group firstly reported a visible-light photoredox-catalyzed carboxylation of benzylic C—H bonds with CO_2 under metal-free conditions via a hydrogen atom transfer (HAT) strategy (Scheme 5) *(11)*. The investigation of standard reaction conditions showed that the standard substrate ethylbenzene could form the desired carboxylic acid with a 53% yield in the presence of CO_2 (4 atm), 2,4,5,6-tetra(carbazol-9-yl)isophthalonitrile (4CzIPN), triisopropylsilanethiol, DMF and blue light. During the investigation of the mechanism, the author detected the formation of 2,3,4,6-tetra(9Hcarbazol-9-yl)-5-(1-phenylethyl)benzonitrile (4CzPEBN), which was proposed as the main active catalyst for carbanion generation in previous work *(12)*. Therefore, the author proposed that in this reaction, 4CzIPN was first transformed into 4CzPEBN. 4CzPEBN was excited by blue light irradiation and then underwent a single electron transfer with triisopropylsilanethiol (R–SH) to form the $4CzPEBN^{\bullet -}$ and $R–SH^{\bullet +}$. Subsequently, $R–SH^{\bullet +}$ was deprotonated to yield $R–S^{\bullet}$ which further abstracted a hydrogen atom from the benzylic position of ethylbenzene and produced a benzylic radical. The previously formed $4CzPEBN^{\bullet -}$ was able to reduce the benzylic radical to afford the carbanion of ethylbenzene which was readily captured by CO_2 to generate the final product. Compared to previous works, this work proceeded

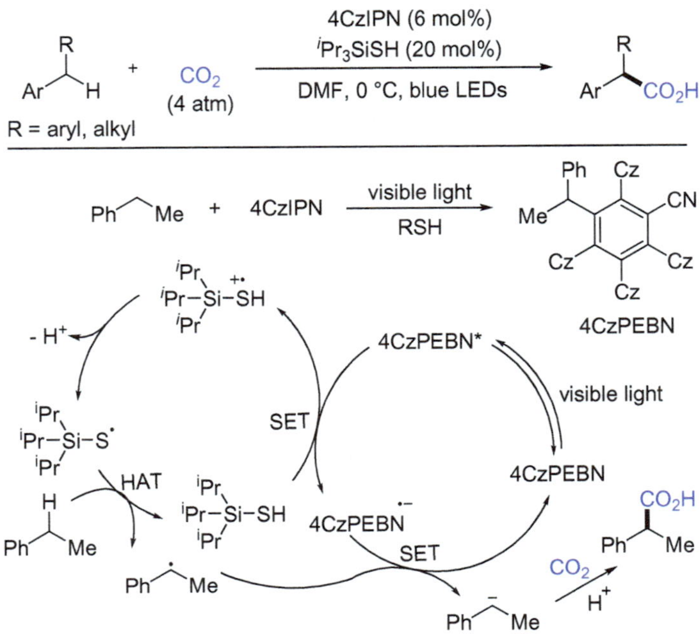

Scheme 5 Visible-light-mediated carboxylation of benzylic sp^3 C—H bonds.

smoothly without adding any metal reagent, sacrificial electron donors, electron acceptors or stoichiometric additives, thus showing great advantages in synthesis of aryl acetic acids.

Utilizing the SET and HAT strategy, Yu's group reported a visible-light photoredox-catalyzed remote difunctionalizing carboxylation of unactivated alkenes with CO_2 in 2020 (Scheme 6) *(17)*. The author chose Langlois reagent (CF_3SO_2Na) as a radical precursor and successfully achieved the transformation of the standard substrate into the corresponding product with a 77% yield in the presence of CO_2 (1 atm), Ir(ppy)$_2$(dtbbpy) PF$_6$, Cs_2CO_3, KOPiv, DMAc and blue light. This reaction is suitable for several kinds of radical precursors and alkene substrates. After the mechanistic study and the DFT calculation, the authors gave a plausible mechanism. First, the Ir(III) photocatalyst was photoexcited and reductively quenched by CF_3SO_2Na to deliver a CF_3 radical and an Ir(II) species. The CF_3 radical then attacked the substrate **6a** and produced a new radical **6d**. After that, the rate-determining 1,5-HAT process occurred to afford the more stable benzylic radical **6e**, which could be reduced by the formed Ir(II) species to give the benzylic anionic species **6 f**. The benzylic anionic species were captured by CO_2 and following protonation could provide

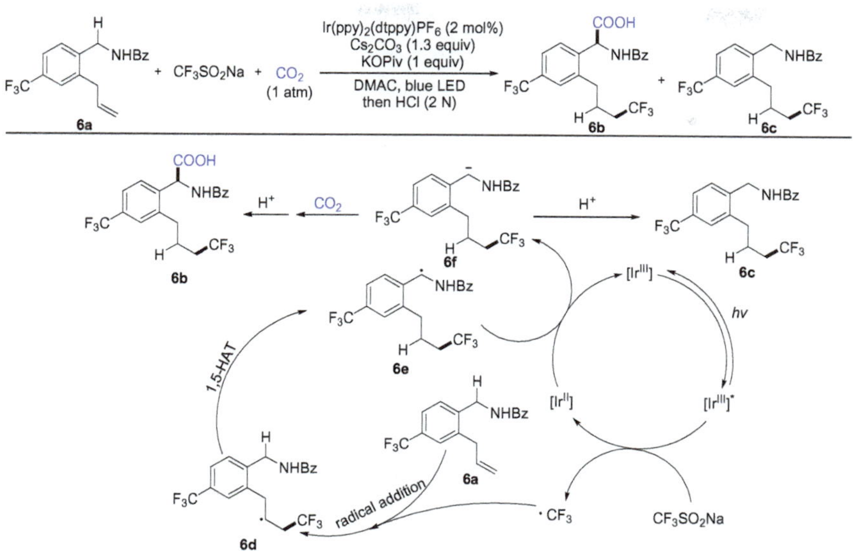

Scheme 6 Visible-light photoredox-catalyzed remote difunctionalizing carboxylation of unactivated alkenes.

the desired carboxylic acids **6b**. Meanwhile, the direct protonation of **6f** would lead to the formation of the byproduct **6c**. This work successfully overcame many challenges, including the low reactivity of unactivated alkenes as well as chemo- and regioselectivity issues arising from competitive monofunctionalization and other side reactions.

1.2 Carboxylation of sp^2 C—H bonds

1.2.1 UV light

In 2018, Chang's group reported the first direct photochemical C—H carboxylation of aromatic diamines with CO_2 under electron donor- and base-free conditions *(18)*. Based on previous work *(19)*, the authors found that under UV light irradiation, the in situ formed Fe(II) complex $[Fe^{II}(opda)_3][ClO_4]_2$ (opda = *o*-phenylenediamine) could transformed into 2,3-diaminobenzoic acid (DBA) with a 58% yield. Notably, opda could also be transformed into DBA with 27.5% yield in the absence of Fe(II) which indicated that Fe(II) was not the essential factor. After the investigation of the mechanism, the authors proposed a plausible mechanism (Scheme 7). First, the photo-irradiation induced the generations of hydrogen and aminyl radical intermediates. The aminyl radical then formed a C—C bond with CO_2 via the delocalization of the unpaired electron, thus forming the

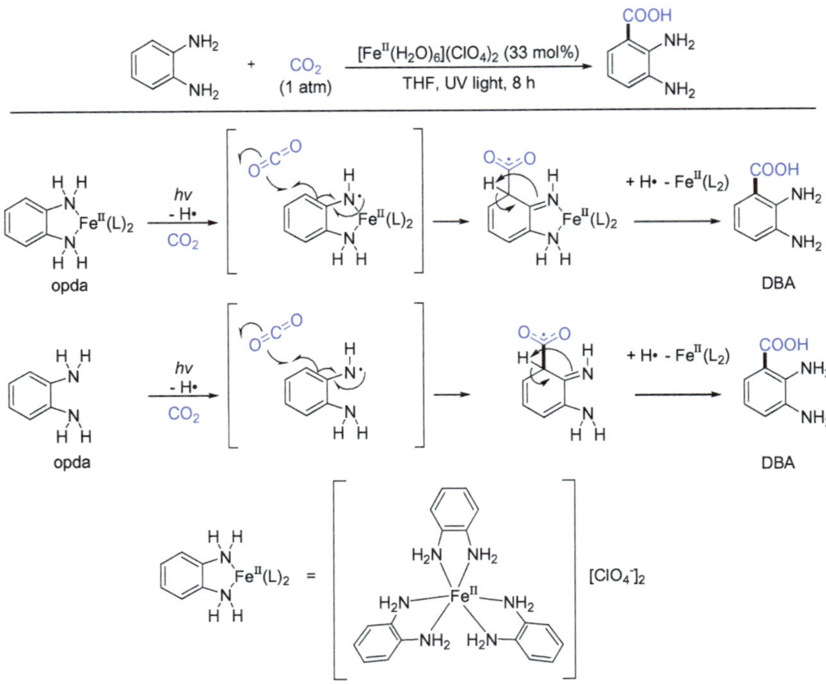

Scheme 7 UV light-induced carboxylation of sp^2 C—H bonds in aromatic diamines.

carboxyl radical intermediate. Subsequently, the methine proton transferred to the imino nitrogen and formed a 2,3-diaminobenzoic radical species. The 2,3-diaminobenzoic radical species finally reacted with hydrogen radical to form the desired product.

1.2.2 Visible light

In 2020, König's group presented a novel redox-neutral C—H carboxylation of arenes and styrenes using a photocatalytic approach (Scheme 8) *(20)*. Different from Jamison's work *(9)*, upon blue-light excitation, the anthrolate anion photocatalyst was able to reduce many aromatic compounds to their corresponding radical anions, which react with CO_2 to afford carboxylic acids. In this work, the authors chose acenaphthene as the standard substrate, 2,3,6,7-tetramethoxyanthracen-9(10H)-one (TMAH) as the photocatalyst, Cs_2CO_3 as the base, DMSO as the solvent, and successfully obtain desired carboxylic acid with a 68% yield under blue light irradiation. It's worth noticing that this reaction had a wide substrate scope such as naphthalenes, thiophenes, benzothiophenes, furans,

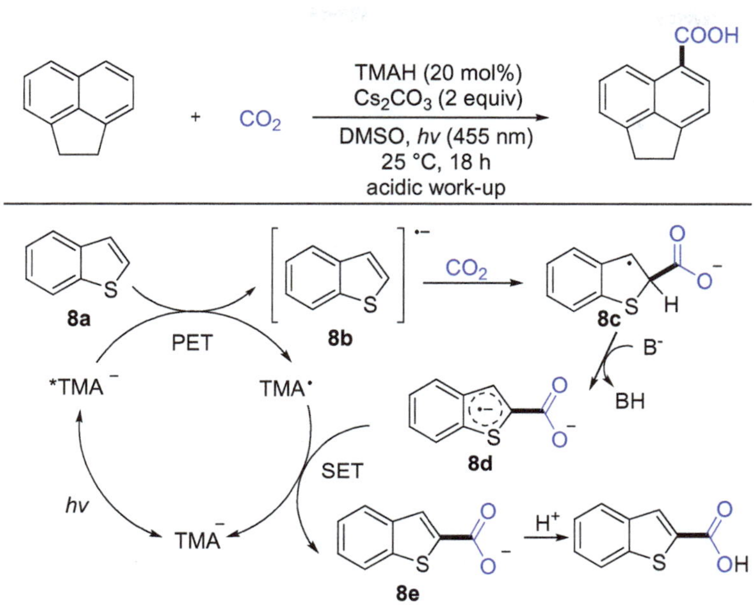

Scheme 8 Visible-light photoredox-catalyzed carboxylation of sp^2 C—H bonds in arenes and styrenes.

benzofuran, boc-protected indoles, thiocarbazole and styrenes. Based on the mechanistic study, the authors proposed the following reaction mechanism. First, in the presence of a base, the TMAH was deprotonated to form an anionic species TMA$^-$ which could be excited by the irradiation of blue light. The excited anion TMA-* then underwent a photoinduced electron transfer (PET) which reduced the substrate **8a** to form the radical anion **8b**. After that, CO_2 was attacked by **8b** and the formed radical carboxylate **8c** was deprotonated immediately in the presence of the base. The resulting radical dianion **8d** was oxidized by the formed radical TMA$^\bullet$ via SET and afforded the desired product **8e**.

2. Carboxyation of C−(pseudo)halide bonds

Since Osakada's pioneering work *(13)*, the carboxylation of organic (pseudo)halides with CO_2 has witnessed rapid development and serves as a powerful alternative to provide carboxylic acids *(21)*. Organo(pseudo) halides are important synthetic raw materials and widely investigated in organic transformations. In the beginning, organic halides were widely used to generate reactive Grignard reagents or lithium reagents, which could directly react with CO_2. Over the past few decades, transition-metal-promoted or -catalyzed carboxylation of organic halides, such as aryl, alkenyl, and alkyl halides has been widely reported. During this period, huge efforts have been devoted, and great process using Ni, Pd and Cu catalysts has been realized recently by Martin *(22)*, Yamamoto *(23)*, Tsuji *(24)* and others *(25)*. Although these reactions can be achieved at higher yields, equivalent metal reductants and expensive transition metals are unavoidable. Therefore, many chemists turned attention to visible light photoredox catalysis by using mild electron donors.

2.1 Carboxylation of C(sp^2)—(pseudo)halide bonds

In 2017, Iwasawa and coworkers reported a pioneering visible-light-driven carboxylation of aryl halides by the combined use of palladium and photoredox catalysts (Scheme 9) *(26)*. In this original report, this carboxylation

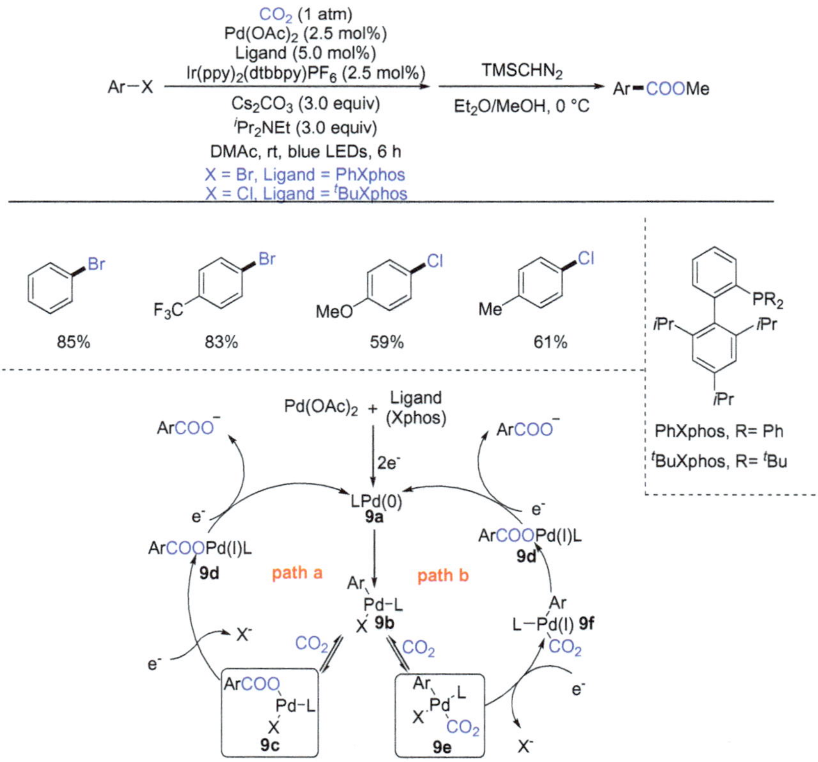

Scheme 9 Visible light photoredox/Pd dual catalysis for carboxylation of aryl halides with CO_2.

reaction proceeded in high yields under 1 atm of CO_2 with a variety of functionalized aryl bromides and chlorides without the necessity of using stoichiometric metallic reductants. And it was found that a better result was obtained by carrying out the reaction in the presence of Cs_2CO_3 as a base. A wide range of aryl bromides bearing alkyl, alkoxy, halide, alkyne, alkene, ester, thiophene and indole reacted well with good yields. In addition to aryl bromides, a wide variety of aryl chlorides were also well reactive. For the mechanism of this reaction, the author found that the first reduction potential of the Pd complex was −2.28 V (vs Fc/Fc^+), which was much lower than the reduction potential of the reductant Ir(II). More importantly, the CV measurement of this complex under a CO_2 atmosphere showed a new peak at about −1.4 V. Based on the above mechanism experiments, the authors speculated two mechanisms for the actual palladium species. The first one is (ArCOO)PdX−(Xphos) **9c** generated in a small amount in equilibrium with ArPdX(Xphos) **9b** under CO_2 (path a), and a SET reduction of **9c** gave (ArCOO)Pd(I)(Xphos) **9d** and X^- by the photocatalysis, and then the generated **9d** could undergo further SET reduction to give Pd(0)(Xphos) and $ArCOO^-$. For this case, the reduction of the carboxylate complex **9c** could shift the equilibrium and the carboxylation reaction would proceed catalytically. Another one is a CO_2-coordinated ArPd(*II*)X(Xphos) species **9e**, which is also generated in a small amount. This species could be able to undergo SET reduction to give a CO_2-coordinated ArPd(I)(Xphos) species **9f**, which would have greater ability to undergo addition to CO_2 to give (ArCOO)Pd(I)(Xphos) **9d** (path b), followed by the similar SET reduction from the photocatalysis. It is noteworthy that such a process is different from the two-electron and two-proton transfer processes proposed in the hydrocarboxylation reaction of alkenes via photoredox/Rh dual catalysis *(27)*.

Soon after this report, König and coworkers reported carboxylation of aromatic and aliphatic bromides/triflates with CO_2 via visible-light photoredox/nickel dual catalysis (Scheme 10) *(28)*. The donor-acceptor dye 4CzIPN and HEH were chosen as the photosensitizer and stoichiometric reductant, along with $NiBr_2$·glyme combined with neocuproine as a cocatalyst. Interestingly, they found that K_2CO_3 could be used as a source of CO_2. A broad range of para-, meta- and challenging ortho-substituted aryl bromides reacted with CO_2 smoothly to afford carboxylic acids in moderate to excellent yields. In addition, primary alkyl bromides and aryl triflates were also compatible. They also realized the synthesis of MCPB, which exhibits potent biological activity. For the reaction mechanism,

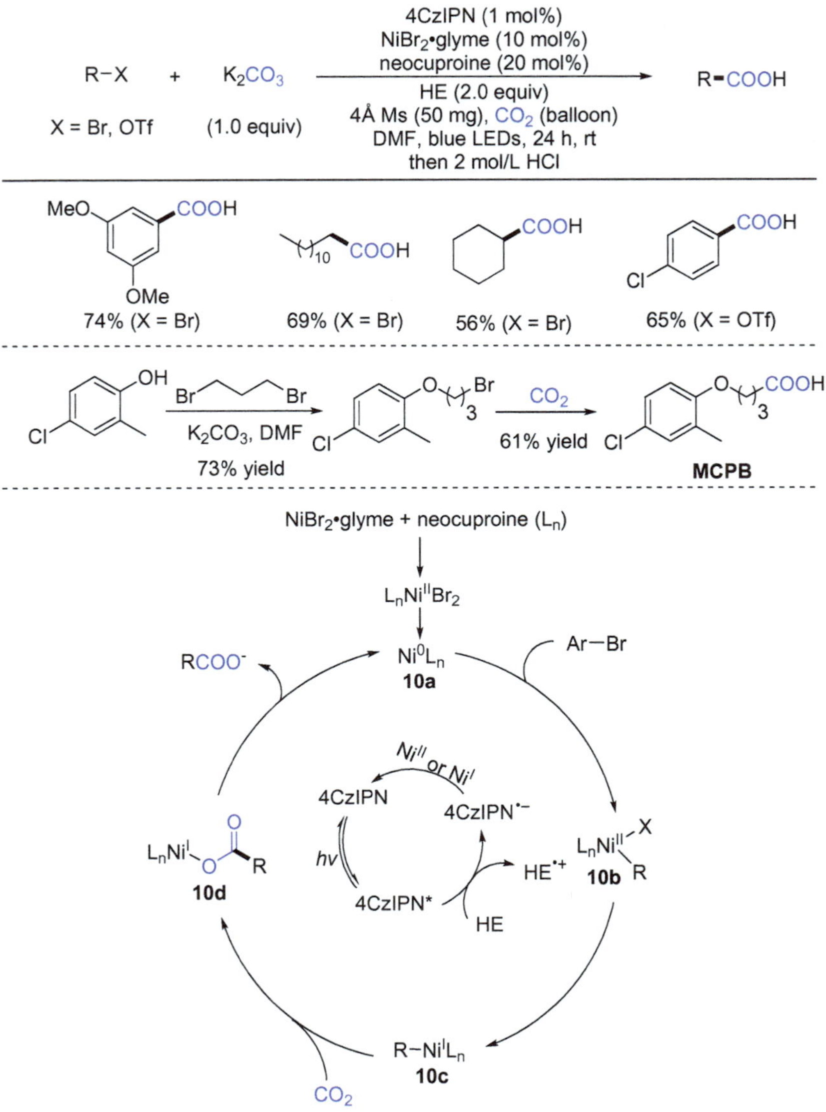

Scheme 10 Visible light photoredox/Ni dual catalysis for carboxylation of aryl bromides/triflates and alkyl bromides with CO_2.

the author suggested that $NiBr_2$·glyme and neocuproine form the initial catalyst, as when the amount of the ligand neocuproine was decreased to be equimolar with the amount of $NiBr_2$·glyme, the final product was still obtained in 61% yield. To prove their hypothesis, neocuproine with $NiBr_2$·glyme in DMF were mixed, and characteristic absorptions of

$LNiBr_2$ (L = neocuproine) were observed, which supported the generation of $LNiBr_2$ in situ during the initial process. Moreover, when $LNiBr_2$ was used as the catalyst instead of $NiBr_2$·glyme and neocuproine, benzoic acid was produced in 77% yield. Based on the previous mechanism experiment, a possible catalytic cycle was proposed. Initially, $LNiBr_2$ was obtained by the ligand exchange between $NiBr_2$·glyme and neocuproine, which can be further reduced to Ni(0) intermediate **10a** via the photocatalysis, followed by oxidative addition of the aryl bromide to yield the Ni(II) species **10b**. Then Ni(I) species **10c** was obtained by single-electron reduction of Ni(II) species **10b**. Subsequently, intermediate **10c** reacted with CO_2 to give the nickel carboxylate intermediate **10d**, which could undergo further reduction to get the target product and regenerate catalyst **10a**.

Using a similar photoredox/Pd dual-catalytic system, in 2019, Iwasawa and coworkers reported palladium-catalyzed visible-light-driven carboxylation of aryl and alkenyl triflates with CO_2 (Scheme 11) *(29a)*. Both $C(sp^2)$—O bonds of aryl and alkenyl triflates underwent carboxylation under mild conditions with good functional group tolerance and yields. For the reaction mechanism, initial oxidative addition of ROTf to Pd(0)(Xphos) **11a** generated electron-deficient [RPd(Xphos)]OTf **11b**, followed by SET reduction to give nucleophilic RPd(I)(Xphos) **11c**. Intermediates **11c** reacted with CO_2 to give the corresponding (RCOO)Pd(I)(Xphos) **11d**, which could be further reduced to Pd(0)(Xphos) **11a** with the release of $RCOO^-$ by the photocatalysis. Compared to previously reported articles from König's group, in which only substrates equipped with three aryl triflates were feasible, a range of aryl and alkenyl triflates with different substitutions could be tolerated in this work. Meanwhile, Jana group also reported the carboxylation of aryl triflates using the similar photoredox/Pd dual-catalytic system and the mechanism was similar to the previously reported work *(29b)*.

In the above reports, CO_2 is fixed into aryl halides to furnish aromatic carboxylic acids. Such CO_2 fixation reactions proceed, in most cases, via the way of nucleophilic attack of arylmetal intermediates to CO_2. In 2019, Murakami and coworkers report a new pathway to fix CO_2 into aryl bromides by the photo-assisted three-step procedure (Scheme 12) *(30)*. CO_2 was temporarily captured by benzophenone with the assistance of light, and subsequently, was transferred to aryl bromides in the presence of a palladium catalyst. A series of aryl bromides were subjected to the cross coupling reaction with the ester **12c** and a wide variety of substituents were tolerated on the aromatic rings. The reaction mechanism was going

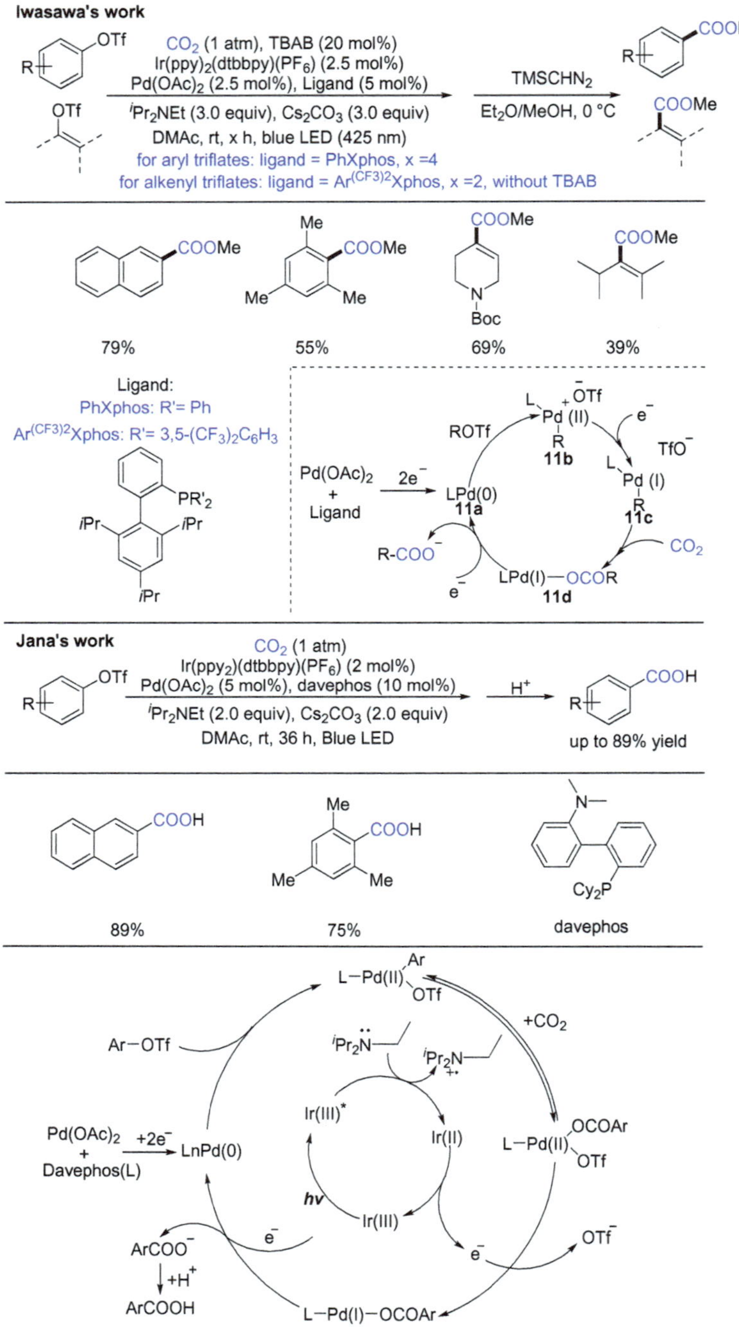

Scheme 11 Visible light photoredox/Pd dual catalysis for carboxylation of aryl and alkenyl triflates with CO_2.

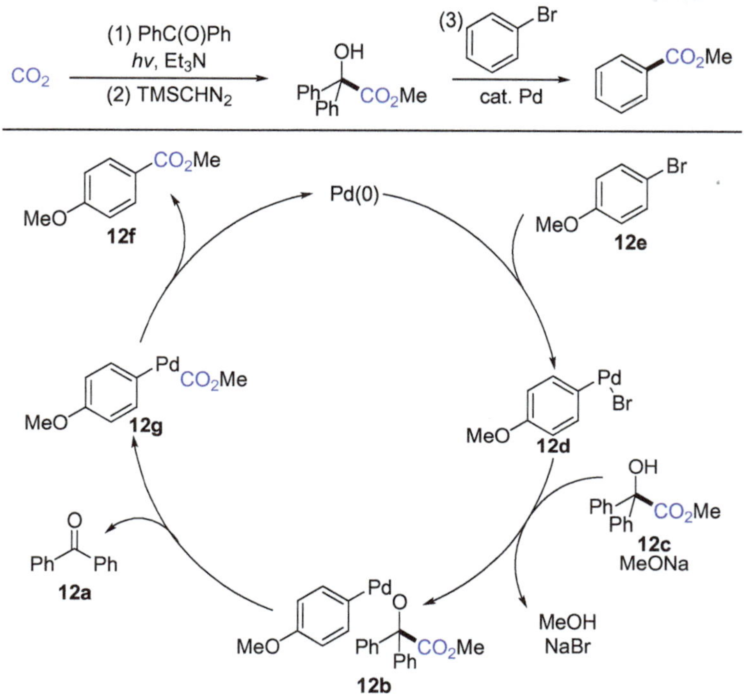

Scheme 12 Photo-assisted fixation of CO_2 with aryl bromides producing aromatic esters.

as follows. First, **12a** underwent light-induced reductive dimerization. A sterically strained C—C bond was formed between two carbonyl carbons to form adjacent diols. The hydroxyl group of benzopinacol was deprotonated and the resulting alkanol underwent *β*-carbon elimination to form **12a** and the carbon anion which could attack CO_2 to give **12c**. The obtained product **12a** could be easily regenerated into benzopinacol by photo-induced reductive dimerization. After that, aryl palladium bromide **12d** was formed by the oxidative addition of **12e** with palladium(0) species, followed by the ligand exchange with **12c** to give the palladium alkoxide **12b**. *β*-Carbon elimination of **12b** will provide benzophenone **12a** and intermediate **12g**. Finally, intermediate **12g** underwent reductive elimination to give target product **12f** with the regeneration of the palladium catalyst. On the whole, the present method exemplifies a unique way to incorporate CO_2 into aryl bromides using light as the source of energy.

Various catalytic carboxylation methodologies with the C—X bond, such as I, Br, Cl and OTf, have been well established, which enable

CO_2 incorporation into important chemical architectures. However, catalytic C—F bond carboxylation has been rarely reported due to the lack of efficient protocol to promote C—F bond scission. The development of selective C—F bond carboxylation is very important due to the growing demand for fluorinated carboxylic acids and derivatives, which are widely applicated in medicinal chemistry and materials science. Fortunately, in 2019, Feng and coworkers reported selective C—F bond carboxylation of *gem*-difluoroalkenes with CO_2 via photoredox/palladium dual catalysis (Scheme 13) *(31)*. This novel strategy can realize the carboxylation of

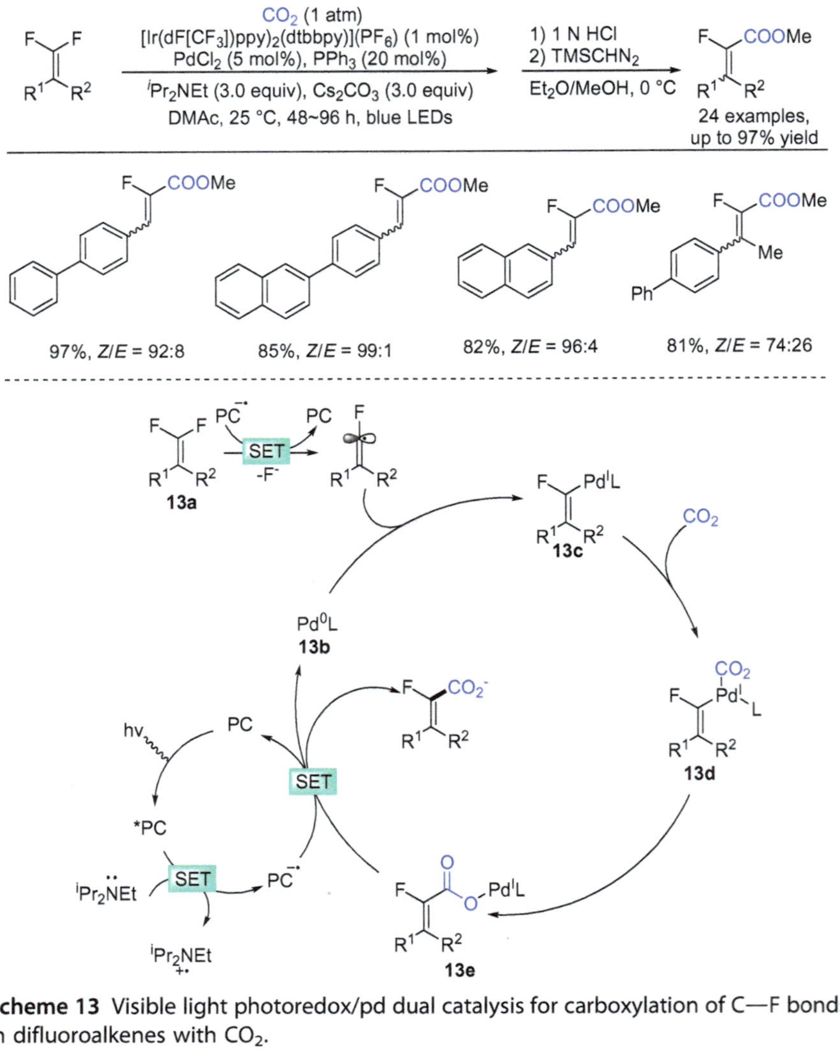

Scheme 13 Visible light photoredox/pd dual catalysis for carboxylation of C—F bonds in difluoroalkenes with CO_2.

various *gem*-difluoroalkenes with good yields and regioselectivity. Based on the control experiments and previous studies, a possible mechanism was proposed. Initially, fluoroalkenyl radical **13a** was obtained via SET reduction with reduced PC, which was generated via reductive quenching of excited PC* by $^{i}Pr_2NEt$. Then the fluoroalkenyl radical **13a** was captured by Pd(0) complex **13b** to provide a putative Pd(I) intermediate **13c**, followed by coordination of CO_2 and migratory insertion, giving the Pd(I) carboxylate intermediate **13e**. The desired product was obtained via SET reduction of **13e** with another reduced PC, accompanied with the regeneration of Pd(0) catalyst.

2.2 Carboxylation of C(sp^3)—(pseudo)halide bonds

As the above mentioned, carboxylation of C(sp^2)—(pseudo) halides with CO_2 offers new prospects for the preparation of industrially relevant carboxylic acids without stoichiometric organometallic reagents. However, the carboxylation of C(sp^3)—(pseudo) halides bond, which is more difficult to be activated, has also attracted extensive attention from chemists in recent years.

In 2019, Crespi, König, Martin, and coworkers reported a remote carboxylation of sp^3 C—H bonds (Scheme 14) *(32)*. A variety of benzylic and primary C(sp^3)—H bonds could undergo highly selective carboxylation under mild conditions with good functional group tolerance via a chain walking process. Interestingly, when using **L1** as the ligand, the best yield with excellent linear/branched selectivity (66%, linear:branched = 10:90) was given. More challenging carboxylation of remote alkyl bromides was also implemented when using **L2** as the ligand, showing that the modification of the auxiliary ligand had an important effect on the reaction performance. Based on the control experiments and theoretical studies, a possible mechanism was proposed. Initially, the oxidative addition of Ni(0) species to an alkyl C—Br bond generated an alkyl Ni complex **14b**, followed by the formation of alkene derivatives **14c** through β-H elimination. Species **14d** was formed via the migratory insertion, which could further undergo SET reduction to provide Ni(I) species **14e**. The subsequent nucleophilic attack to CO_2 would give carboxylate intermediate **14f**, which underwent a SET process with the reduced PC would finish the catalytic cycle and give the desired product. Both kinetic isotope effect and the regioselectivity were observed at remote primary C(sp^3)—H sites, indicating that the formation of Ni(I) species might be the rate-determining step in this reaction.

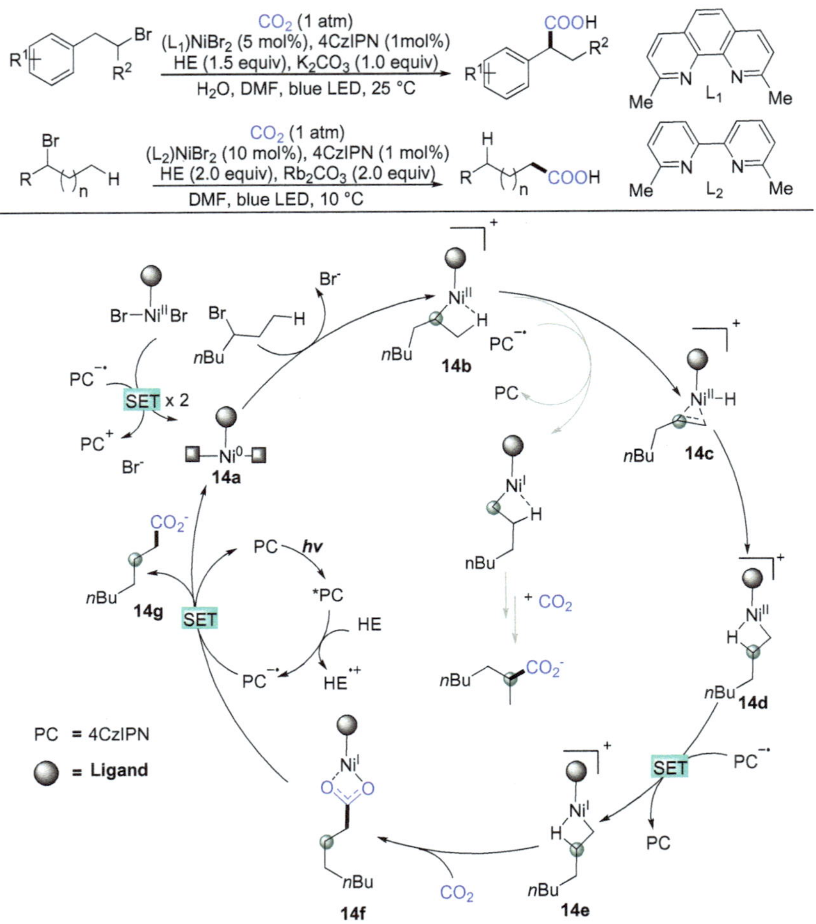

Scheme 14 Visible light photoredox/Ni dual catalysis for remote carboxylation of sp^3 C—H bonds with CO_2.

Later, a transition metal-free carboxylation of benzylic halides was developed by the Yu group and coworkers *(33)*. The donor-acceptor dye 4DPAIPN and TMEDA were chosen as the photosensitizer and stoichiometric reductant, respectively. A wide range of commercially available, inexpensive primary, secondary, and more challenging tertiary benzyl halides underwent carboxylation to give valuable aryl acetic acids, including several pharmaceutical molecules and drug precursors. Based on the control experiments and previous studies, the authors proposed two possible pathways (Scheme 15). The preferred path I involves a successive single-electron transfer (SSET) process, in which photocatalyst $PC^{\bullet -}$ was generated from

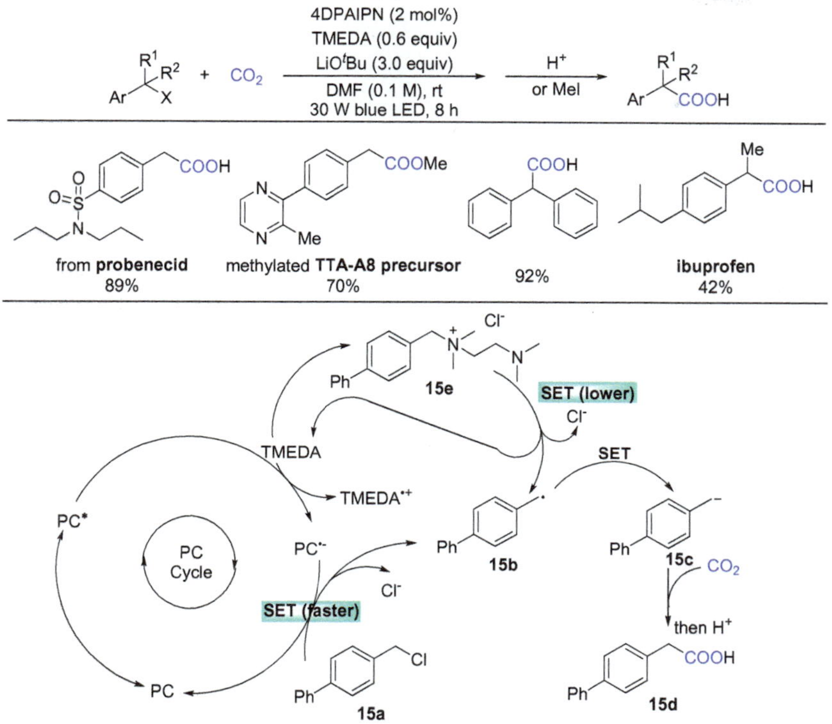

Scheme 15 Visible-light photoredox-catalyzed carboxylation of benzyl halides with CO_2.

the reductive quenching between an excited photocatalyst PC and TMEDA. Subsequently, the SET reduction of **15a** gave a radical anion, which would generate benzylic radical **15b** after the cleavage C—X bond. The benzylic radical was further reduced via SET process to deliver the key benzylic anion intermediate **15c**, which underwent subsequent nucleophilic attack to CO_2 to furnish the target carboxylic acid **15d**. However, path II, involving the in situ formed tetraalkyl ammonium salt **15e**, cannot be ruled out at this stage.

Compared with organic halides, the carboxylation of readily-available organic (pseudo)halides has also been widely reported in recent years, such as the transition metal-catalyzed cross-coupling reactions of electrophilic organoammonium salts *(34)*. These reactions tend to produce amines as byproducts, which lead to a lower atomic economy. In 2018, Yu and coworkers reported visible-light photoredox-catalyzed external reductant-free cross couplings of tetra-alkylammonium salts with CO_2 under mild conditions (Scheme 16) *(35)*. The reaction featured good substrate tolerance

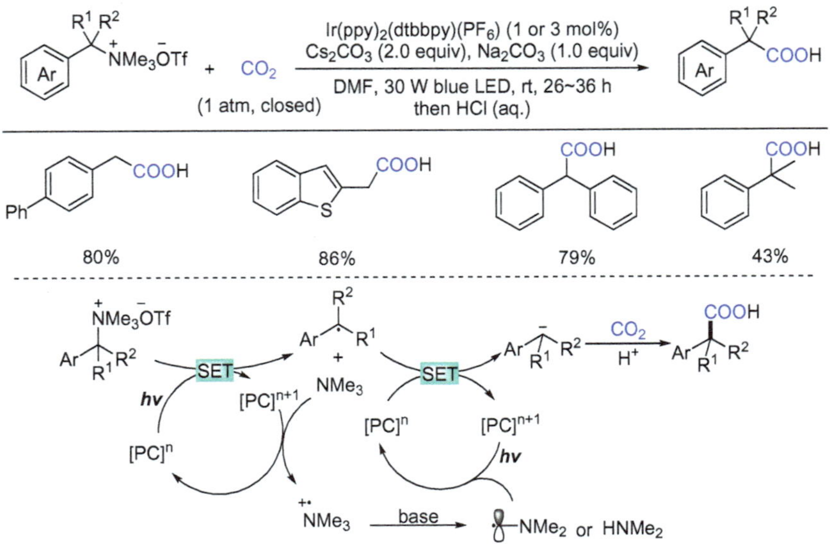

Scheme 16 Visible-light photoredox-catalyzed carboxylation of tetraalkyl ammonium salts with CO_2.

for primary, secondary and tertiary substituted quaternary ammonium salts. Interestingly, this reaction did not require an external reducing agent due to the efficient utilization of simultaneously generated trimethylamine as the electron donor. Mechanistic studies indicated that benzyl radicals and anions might be generated as the key intermediates via photocatalysis, providing a new direction for cross-electrophile couplings.

Besides tetra-ammonium salts derivatives, carboxylation of benzyl alcohol derivatives with CO_2 was reported by the Iwasawa's (Scheme 17A) and Yu's group (Scheme 17B), independently *(36)*. Based on their previous work, Iwasawa's group were curious to know the possibility to extend the catalytic system to sp^3 carbon atoms. Fortunately, a wide range of primary, secondary benzyl alcohol derivatives could be employed to provide benzylic carboxylic acids in moderate to high yields by using a similar photoredox/palladium dual catalysis. Based on control experiments, a possible mechanism was proposed. Initially, oxidative addition of the benzyl C—O bonds to Pd^0 species **17a** took place to obtain π-benzyl Pd(II) species **17b** or σ-benzyl Pd(II) species **17c**. Then the formed species **17b** or **17c** underwent a SET process with the photocatalyst to generate key intermediate benzyl—Pd(I) **17d**. Subsequent nucleophilic attack of benzyl-Pd(I) **17d** to CO_2 provided Pd(I) carboxylate **17e** and furnished the target

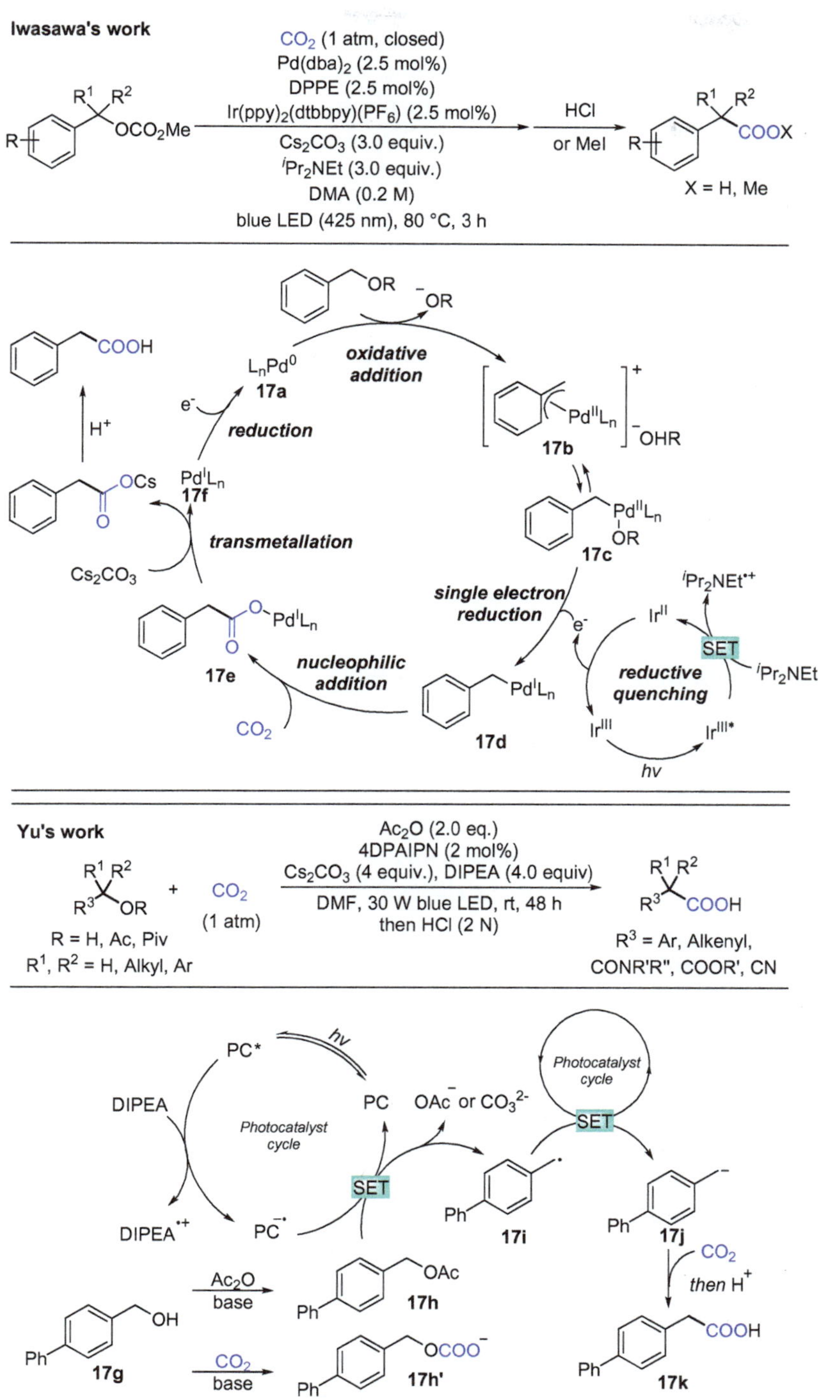

Scheme 17 Visible-light photoredox-catalyzed carboxylation of activated $C(sp^3)$—O bonds with CO_2.

product in the presence of base. Mechanistic studies indicated that oxidative addition might be the rate-determining step. At the same time, Yu group and coworkers reported visible-light photoredox-catalyzed carboxylation of activated C(sp^3)—O bonds with CO_2 in the absence of transition metals (Scheme 17B) *(37)*. This strategy can tolerate a series of primary, secondary, and more challenging tertiary benzyl alcohols and benzylic carbonates without any transition metal catalysts, generating various valuable carboxylic acids, such as ibuprofen, naproxen, fenoprofen, and flurbiprofen. Moreover, this strategy is also applicable to other carboxylates of α-hydroxyl amides, esters, and nitriles, as well as allyl alcohols, representing a practical method for the synthesis of aliphatic carboxylic acid. Based on the control experiments and previous work, a possible mechanism was proposed. Initially, the interaction between **17g** and Ac_2O could generate intermediate **17h**, which further underwent SET reduction by the reduced photocatalyst to deliver benzyl radical **17i**. Further SET reduction of benzyl radical **17i** under another photocatalytic cycle formed benzylic carbanion **17j**, which underwent nucleophilic attack to CO_2 to obtain the target carboxylic acid **17k**.

Different from Iwasawa's and Yu's work, Xi group developed the carboxylation of allylic alcohols with CO_2 by using the Photoredox/Nickel dual catalysis (Scheme 18) *(37)*. Various allylic alcohols can be tolerated with good *Z*/*E* stereoselectivity. Besides the allylic alcohols, a few cases of similar propargylic alcohols also can obtain corresponding product. Base on the control experiment, the mechanism was proposed. Initially, Ni(I) species **18a** was formed by the single-electron reduction of Ni(II) and initiated the Ni catalytic cycle in the presence of L1. The allylic alcohol was preactivated to the corresponding allylic hydrogen carbonate in the presence of water and CO_2, which significantly decrease the activation energy for oxidative addition. Then, oxidative addition of **18a** to the allylic hydrogen carbonate obtained a π-allylnickel intermediate **18b**. The ligand exchange of **18b** with LiOAc generated intermediate **18c,** which can be reduced to the intermediate **18d** by photocatalytic cycle. Nickel carboxylate **18e** was obtained by CO_2 insertion process. Finally, reduction of **18e** regenerated the catalytic Ni(0) species **18a** and gave the desired product **18f**.

Compared with most studies in the carboxylation of C—Br, C—Cl, C—OTf and C—O bonds, the carboxylation of C—F bonds with CO_2 is more challenging due to the higher bond dissociation energy of C—F bonds and resistance to oxidative addition of C—F bonds, which impeded its utilization. In 2021, Yu and coworkers reported visible-light

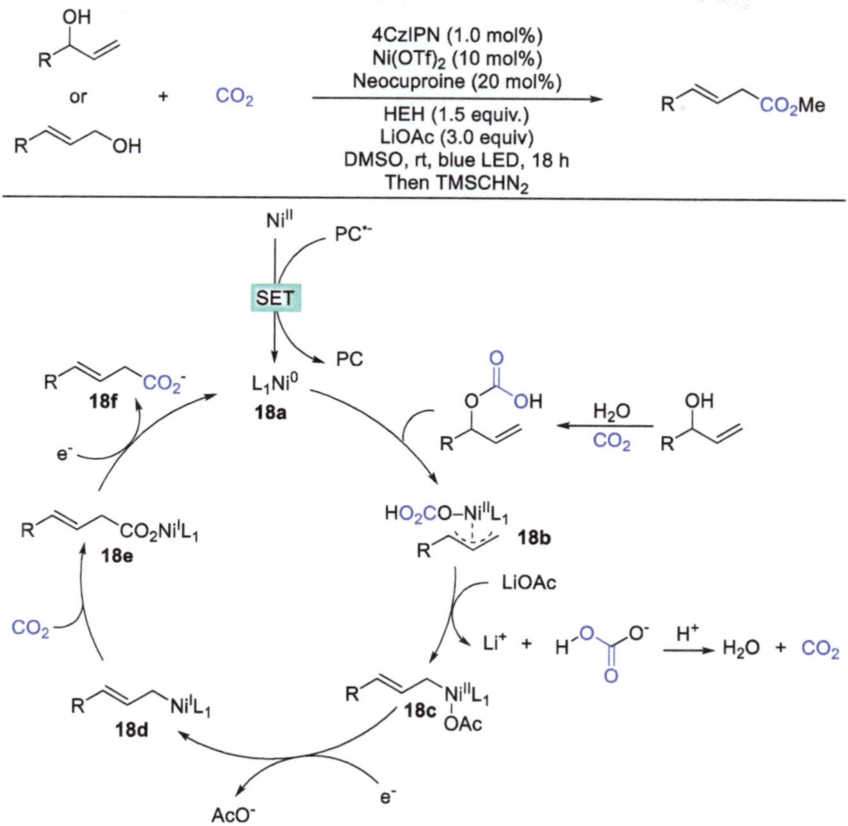

Scheme 18 Visible light photoredox/Ni dual catalysis for carboxylation of allylic alcohol with CO_2.

photoredox-catalyzed selective carboxylation of C(sp^3)—F bonds with CO_2 (Scheme 19) *(38)*. This strategy was applicable for a variety of mono-, di-, and trifluoroalkylarenes as well as α,α-difluorocarboxylic esters and amides, which can give important aryl acetic acids and α-fluorocarboxylic acids, including several drugs and analogs under mild conditions. Base on the mechanistic studies and DFT calculations, the dual roles of CO_2 as an electron carrier and electrophile during this transformation was demonstrated. Initially, the mixture of hydrosilane, Cs_2CO_3, and CO_2 could generate cesium formate **19a** and cesium silanolate **19b**, followed by a SET process with excited photocatalysts, providing the reduced photocatalyst and a siloxy-radical species **19c**. Then the siloxy-radical species **19c** underwent a HAT process with cesium formate to deliver CO_2 radical anion which might be able to reduce the fluorinated substrates via SET process.

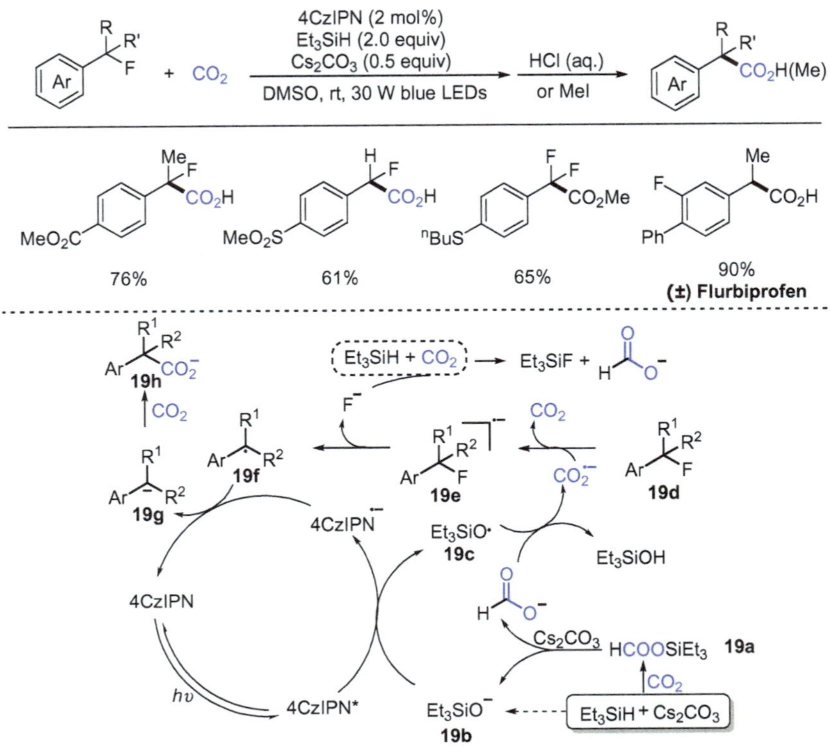

Scheme 19 Visible-light photoredox-catalyzed selective carboxylation of C(sp^3)—F bonds with CO_2.

The following cleavage of C—F bonds would provide carbon radical **19f**. The formed carbon radical **19f** further underwent a SET reduction process to generate carbanion **19g**, which then attacked CO_2 to obtain the target carboxylic acid **19h**.

3. Carboxylation of unsaturated substrates

3.1 Carboxylation of alkenes

Account for the readily-available and diversely-functionalized properties of alkenes, chemists continuously devote numerous efforts to explore various strategies to transform alkenes to valuable synthetic motifs *(39)*. Among these transformations, the synthesis of carboxylic acids in the combination of alkenes and CO_2 has attracted persistent attention. As the emergence of light-induced methodologies *(40)*, versatile carboxylation reactions of alkenes with CO_2 have been developed, highly enriching the synthetic pools of natural products, agrochemicals and pharmaceuticals.

3.1.1 Hydrocarboxylation of alkenes

Employing Pd and Rh complexes, Lapidus and coworkers reported the first catalytic synthesis of propionic acid from ethylene and CO_2 in 1978 *(41)*. In 2017, Iwasawa's group realized the first photocatalytic hydrocarboxylation of alkenes with CO_2 under visible light irradiation (Scheme 20) *(27)*. Based on the advance of the SET process and dual catalytic systems *(42)*, the authors appropriately combined the hydrocarboxylation catalyst Rh(I) complex and photoredox catalyst $[Ru(bpy)_3]^{2+}$ in the presence of $^iPr_2NEt$ as the sacrificial electron donor, avoiding the necessary use of a stoichiometric amount of organometallic reagents to regenerate the active metal hydride species from the corresponding metal carboxylates.

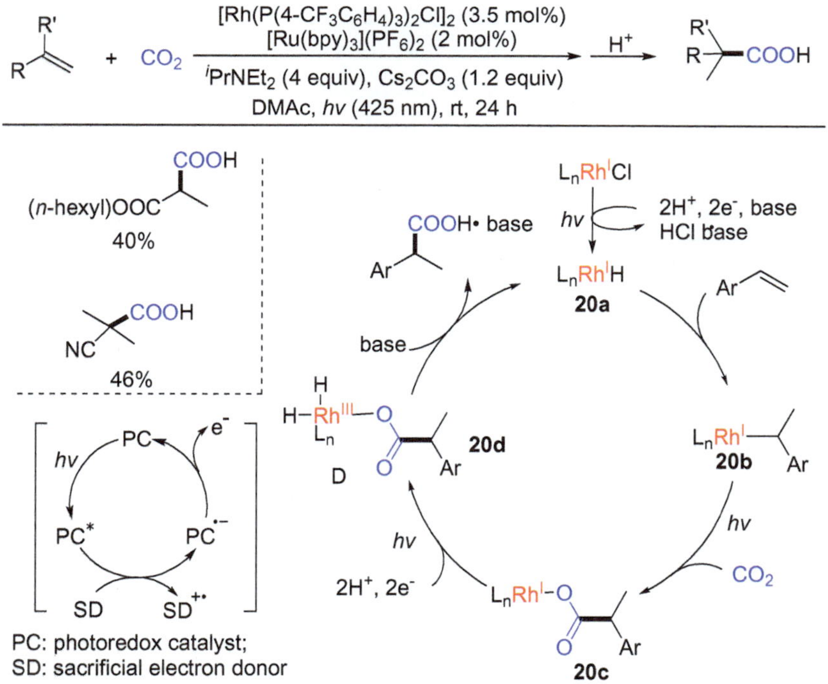

Scheme 20 Visible light-driven hydrocarboxylation of alkenes via photoredox/Rh(I) dual catalysis.

The challenges to access this protocol lie on the successive generation of Rh(I) hydride complexes from Rh(I) carboxylates and followed nucleophilic addition of Rh(I) hydride complexes to CO_2 under visible light condition. Control experiments revealed that the active Rh(I) hydride was available from the Rh(I) carboxylates under this photocatalytic condition. Besides, visible light and $[Ru(bpy)_3](PF_6)_2$ were vitally necessary for the nucleophilic addition of Rh(I) complexes to CO_2, but $^iPr_2NEt$ was dispensable. The luminescence quenching experiments demonstrated this process might go through the triplet-triplet energy transfer. In the possible catalytic mechanism, Rh(I) alkyl complex **20b** was obtained by the hydrometallation of alkene with Rh(I) hydride complex **20a**, followed by the nucleophilic addition to CO_2 to give Rh(I) carboxylate **20c**. Subsequent conversion of complex **20c** to complex **20d** was enabled by accepting two electrons and two protons from the photoredox cycle. Finally, Rh(I) monohydride **20a** was regenerated by releasing the carboxylic acid with the assistance of a base. Notably, this mechanistic process was further investigated in their following work *(43)*.

Although the above-mentioned development in the carboxylation of alkenes, the Rh(I) catalyzed strategies only produce the α-functionalized acids, due to the formation of the more stable η^3-benzylic metal species. The approach for the β-selective hydrocarboxylation of alkenes using CO_2 is limited. Envisioning the stable benzylic radical anion intermediate, Jamison's group presented an anti-Markovnikov method to access β-selective carboxylation of alkenes via a continuous flow and photoredox catalysis in 2017 *(44)*. In this methodology, the PMP (1,2,2,6,6-pentamethylpiperidine) was conducted as the sacrificial reagent and *p*-terphenyl was performed as photosensitizer. The both yield and selectivity can be further improved by the addition of 19 equivalent H_2O. Preliminary studies proved that PMP are essential for this reaction. Given the strong reduction potential of *p*-terphenyl when exposing to UV light, the reduction of CO_2 to CO_2 radical anion is feasible. The formation of deuterated product when replacing H_2O with D_2O suggested that the H_2O acted as the proton source for the protonation of benzylic anion intermediates.

In the proposed mechanism (Scheme 21), the strong reductant *p*-terphenyl radical anion was generated through the singlet state photoexcitation of *p*-terphenyl and followed SET with PMP. The *p*-terphenyl radical anion would subsequently reduce CO_2 to corresponding radical anion. After the addition of CO_2 radical anion to the β-position of alkenes, the new formed benzylic radical intermediate would be further reduced to benzylic anion by a second electron from the photoredox catalytic cycle and further protonated with H_2O to provide the hydrocarboxylation product.

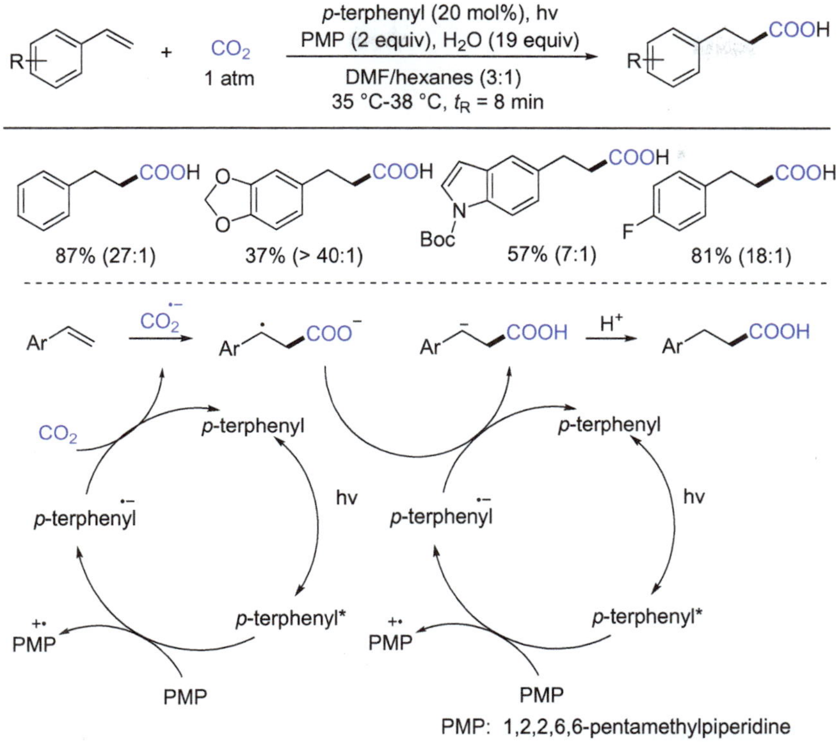

Scheme 21 Direct β-selective hydrocarboxylation of styrenes with CO_2 enabled by continuous flow photoredox catalysis.

As the aforementioned reason for the priority to Markovnikov selectivity, the controllable selectivity in the metal-catalyzed hydrocarboxylation of alkenes is much more challenging and merits further exploration. In 2018, König and coworkers addressed this issue by a ligand-controlled scenario, ensuring the regiodivergent hydrocarboxylation of styrenes using CO_2 in the combination of visible light and Ni catalyst *(45)*. In this protocol, utilizing neocuproine as the ligand lead to the Markovnikov product (b:l = 100:0; b: branching, l: linear), whereas employing 1,4-bis-(diphenylphosphino) butane (dppb) as the ligand, the construction of the anti-Markovnikov product (l:b ≥ 97:3) are dominant. Deuterium experiments indicated the hydrogen atom for the hydrocarboxylation might come from the HEH; The radical scavenging studies showed the radical process may not be feasible for this reaction. As shown in the plausible mechanism (Scheme 22), when applying **L2** (neocuproine) as the ligand in the photocatalytic reduction conditions, the Ni^0L_n species **22-B1**, which is in situ generated via two

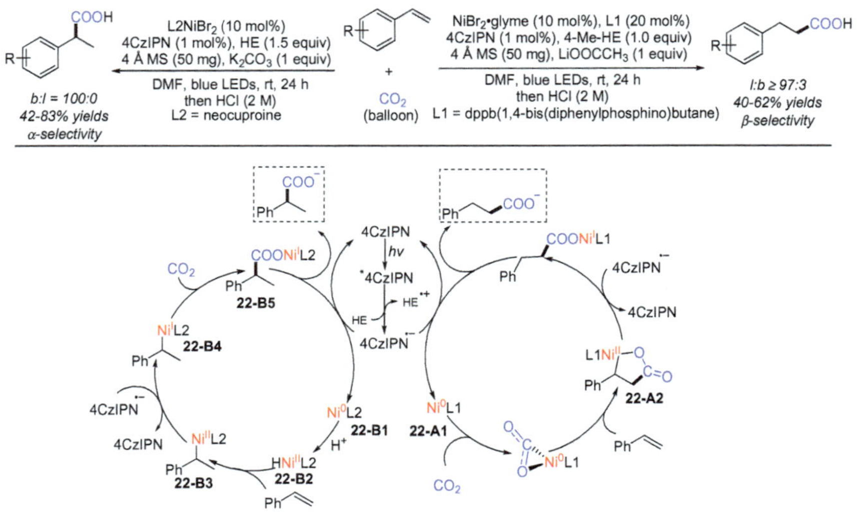

Scheme 22 Ligand-controlled regioselective hydrocarboxylation of styrenes with CO_2 by combining visible light and nickel catalysis.

SET steps, was further protonated by the oxidized HEH to produce a Ni hydride complex **22-B2**. Given the formation of a more stable organonickel complex **22-B3**, the L2-ligated Ni hydride complex **22-B2** then coordinated to the olefin with less steric hindrance. The complex **22-B4**, generated from absorbing one electron from the reduced photocatalyst, proceeded the insertion of CO_2, yielding the intermediated **22-B5**. The desired product and complex **22-B1** were released by the further SET reduction of complex **22-B5**. When employing **L12** (dppb), the Ni^0L_n species **22-A1** may coordinated with CO_2 first, and then construct a 5-membered cyclic intermediate **22-A2** by the insertion of CO_2. Further reduction of intermediate **22-A2** offers the final anti-Markovnikov product.

Driven by the development in charge-transfer complex (CTC), Yu and coworkers *(46)* disclosed the hydrocarboxylation of alkenes with CO_2 ensuring by the CTCs between thiolates and acrylates/styrenes (Scheme 23). Cyclic voltammetry (CV) test confirmed the reaction mixture created a highly reductive condition. The successful detection of both sodium formate and sodium oxalate in the absence of alkenes demonstrated the feasibility of reduction of CO_2 in the reaction condition. The UV–vis absorption spectra of the mixture of thiolates and acrylates/styrenes gave a slight bathochromic shift, indicating the construction of the CTCs in the system. The DFT calculation weakened the possibility of the formation

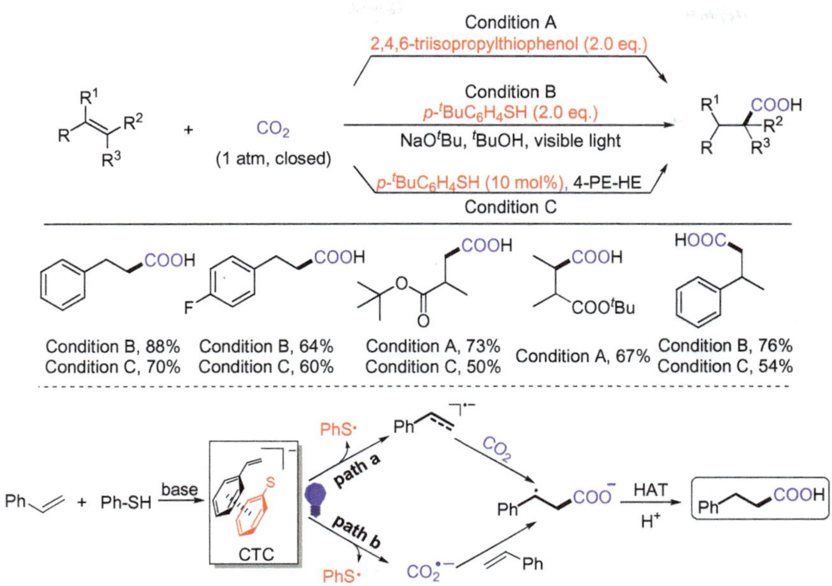

Scheme 23 Anti-markovnikov hydrocarboxylation of acrylates and styrenes with CO_2.

of CTC between thiolate and CO_2 as its higher energy gap. The higher spin density distributed in styrene and acrylate in β-position explained the high regioselectivity of this reaction. The issue in chemoselectivity was also uncovered by DFT calculation, that the hydrothiolation was thermodynamically and kinetically less favorable than hydrocarboxylation.

Very recently, Romo's group *(47)* described the photocatalyzed β-hydrocarboxylation of α,β-unsaturated esters employing CO_2 using flow chemistry (Scheme 24). In this work, the CO_2 radical anion attacked an electron-deficient alkene through the Giese-type addition. The resulting radical intermediate was quenched by hydrogen atom or protonation after further reduction. The products can be subsequently transformed to β-lactones through a halogenation-β-lactonization reaction.

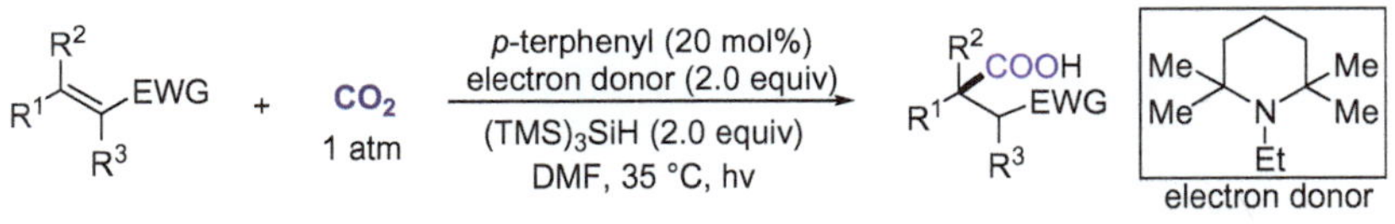

Scheme 24 β-Selective hydrocarboxylation of α,β-unsaturated esters with CO_2.

3.1.2 Difunctionalization of alkenes

The difunctionalization of alkenes was regarded as a powerful tool to deliver the highly functionalized motifs in synthetic chemistry. Compared to the hydrocarboxylation of alkenes, difunctionalization facilitates high efficiency to equip two functional groups to alkenes.

3.1.2.1 α-Carboxylation of alkenes

The general strategy (Scheme 25) to perform α-carboxylation of alkenes is going through a redox neutral process. In this methodology, the radicals, which are generated from SET oxidation of radical precursors by the excited photocatalyst, undergo radical addition to alkenes (generally including styrenes and acrylates), resulting in a much more stable radical intermediate, such as benzylic radicals, which can be further reduced to benzylic anions by the reduced state of photocatalyst. Subsequent nucleophilic attack of benzylic anions to CO_2 and protonation would furnish the desired carboxylic acids.

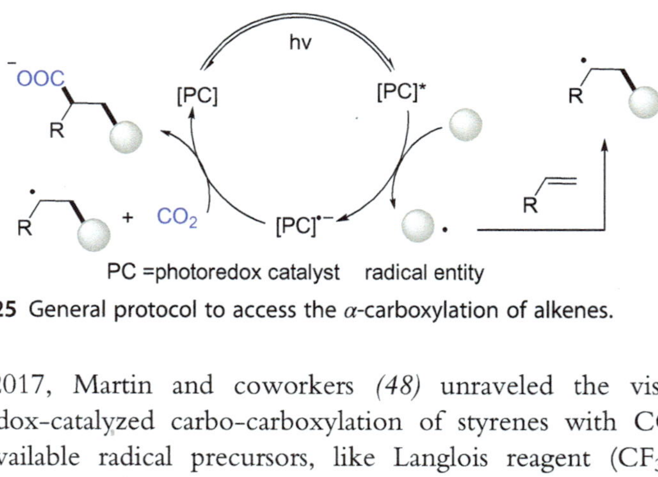

Scheme 25 General protocol to access the α-carboxylation of alkenes.

In 2017, Martin and coworkers *(48)* unraveled the visible-light photoredox-catalyzed carbo-carboxylation of styrenes with CO_2 using easily available radical precursors, like Langlois reagent (CF_3SO_2Na), CHF_2SO_2Na, oxalates and trifluoroborates, under redox-neutral conditions. This trifluoromethyl-carboxylation methodology can be widely applied to different substituted styrene derivatives. The "light-dark" experiments indicated the continuous visible-light irradiation is necessary. The Stern – Volmer luminescence studies confirmed the reductive quenching of the excited stated photocatalyst by CF_3SO_2Na was involved in this photocatalytic process.

As illustrated in the Scheme 26, the corresponding radicals, generated from the radical precursors via SET with the excited [Ir]* photocatalyst,

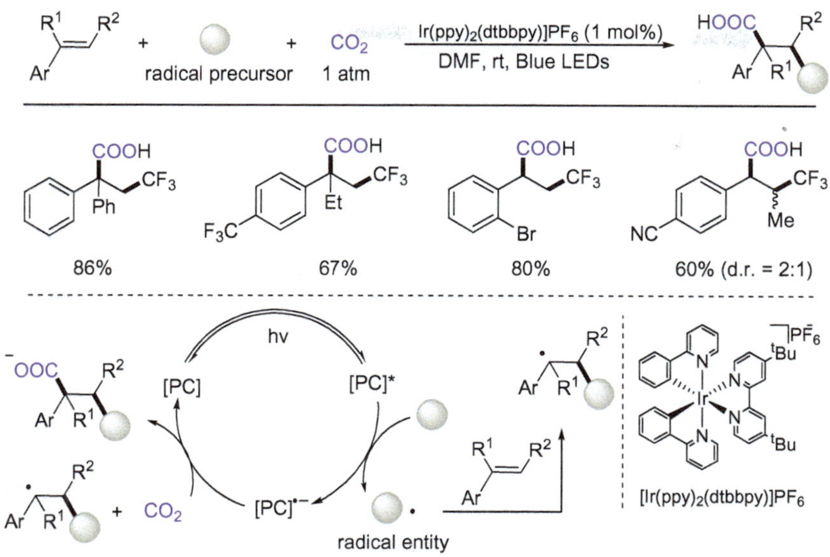

Scheme 26 Dicarbofunctionalization of styrenes with CO_2 and radical precursors.

underwent radical addition to styrenes. The delivered benzyl radical intermediate was further reduced to the benzyl anion and followed by the carboxylation with CO_2 to afford the target product.

Later, Wu and coworkers *(49)* achieved the silacarboxylation and carbocarboxylation reactions of alkenes using CO_2 as carboxylation reagent, quinuclidine-3-yl acetate as HAT catalyst and 4CzIPN as photocatalyst via $C(sp^3)$—H or Si—H activation. The isotope-labeling study confirmed the involvement of the benzylic anion intermediate in the photocatalytic cycle. Control experiments may indicate the hydrosilylation byproduct rendered a minor contribution to the construction of the desired product. Importantly, this reaction was more efficient in flow chemistry.

In the mechanism as shown in Scheme 27, the HAT reagent was oxidized by the excited state 4CzIPN* to offer the quinuclidinium radical cation intermediate, subsequent hydrogen atom scrambling from Si—H or C—H bond delivered a silyl or a carbon radical. The radical intermediate attacked alkenes to give the benzyl radical intermediates, which can transform to corresponding benzyl anion species via further reduction. The desired products were generated by followed nucleophilic addition to CO_2 and protonation.

Given that carboxylic acids containing phosphorus are of great importance in the natural products, functional materials and biological

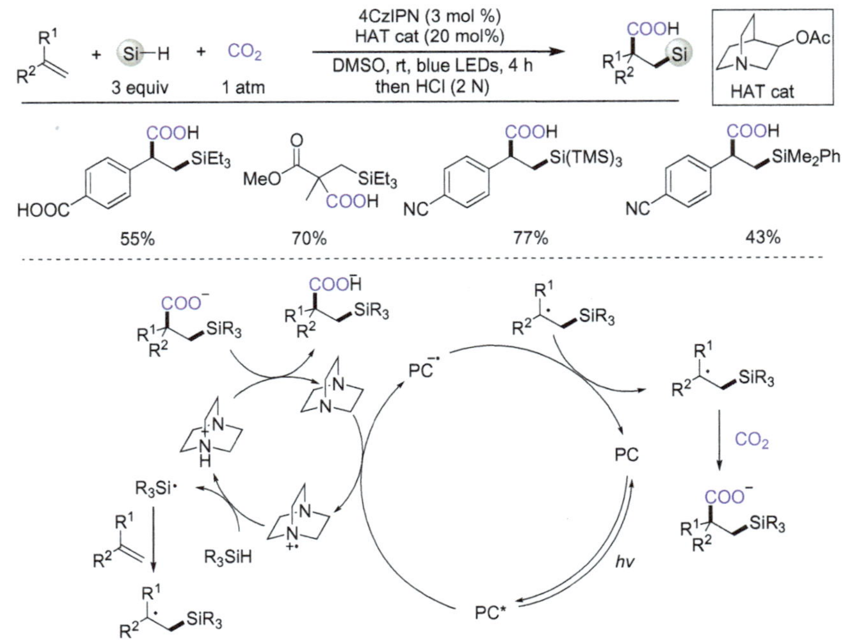

Scheme 27 Difunctionalization of alkenes with CO_2 and silanes.

pharmaceuticals, Yu and coworkers *(50)* disclosed a simultaneously photocatalytic phosphonocarboxylation of various alkenes (including enamides, styrenes, enolsilanes and acrylates) with the incorporation of CO_2 and H—P(O) compounds. The radical clock test confirmed this transformation went through a radical pathway. Besides, isotope-labeling experiments enhanced the possibility in the construction of α-amino benzylic anionic species in this reaction. Furthermore, reductive quenching process may be involved in the photocatalytic cycle.

As illustrated in the Scheme 28, after deprotonation in the presence of a base, H—P(O) compounds was further oxidized through SET with the excited photocatalyst to generate the phosphorus radicals **28a**. Then **28a** underwent radical addition to the alkenes to give the *α*-amino radicals **28b**, which can subsequently be reduced to α-amino carbanions **28c** by the reduced photocatalyst and attack CO_2 to produce the target *β*-phosphono *α*-amino acids.

Employing the common method that the tertiary amine can be transformed to α-aminoalkyl radicals via reductive quenching of photoredox catalyst and the further deprotonation by a base, Xi and coworkers *(51)* successfully afforded highly valuable γ-amino acids regio-selectively

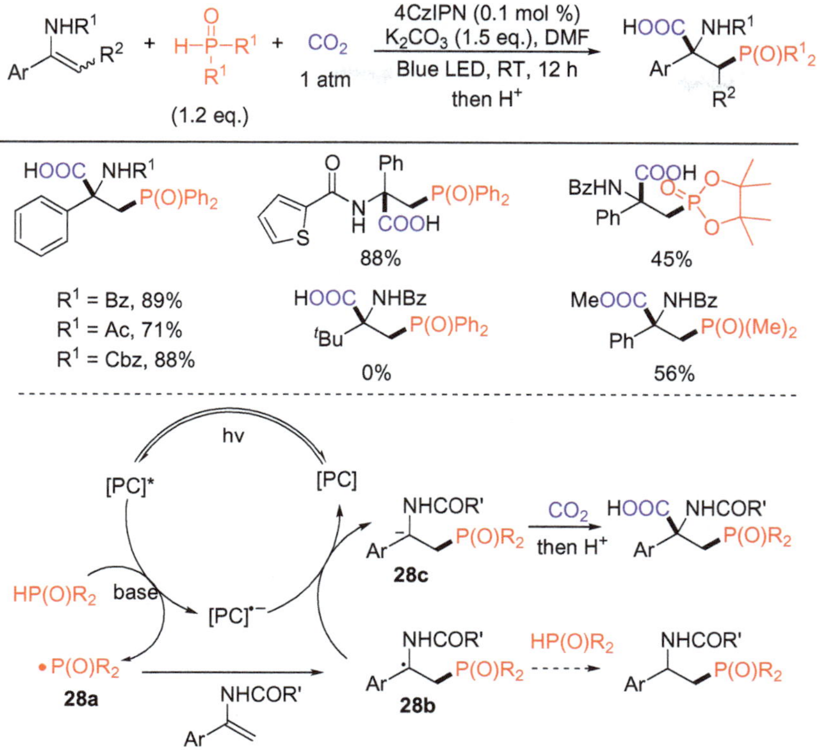

Scheme 28 Phosphonocarboxylation of alkenes.

with the incorporation of CO_2, tertiary amines and styrenes under photoredox conditions. The additive LiCl is essential to this reaction, but a HAT reagent like quinuclidine is dispensable (Scheme 29).

In 2020, Yu's group *(52)* reported the dearomative arylcarboxylation of indoles to construct complex skeletons with CO_2 via SSET strategy under visible light condition, avoiding common side reactions in transition-metal catalysis like *β*-hydride elimination (Scheme 30). The isotope-labeling studies excluded the possibility of hydrogen-atom transfer with solvent (DMSO), and proved the existence of benzylic anion intermediate in this reaction. According to the proposed mechanism, the aryl radical, which was generated from the aryl bromide through the SET reduction by reduced 4CzIPN, underwent intramolecular addition to the C2—C3 double bond of indole to produce the benzylic radical. The benzylic radical was further reduced to afford the benzylic anion, which subsequently attacked CO_2 to deliver the desired product.

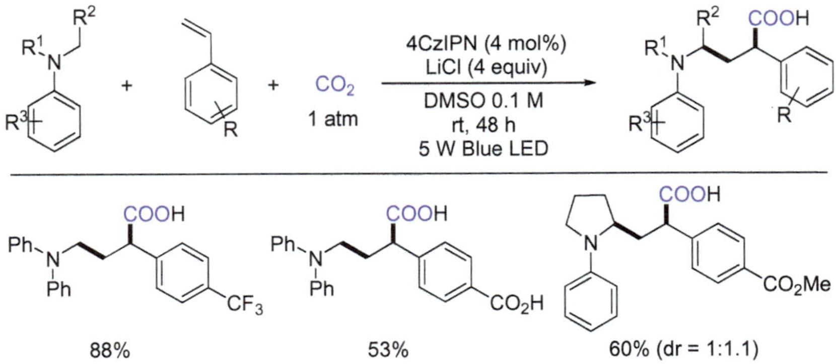

Scheme 29 Synthesis of γ-amino acids through dicarbofunctionalization of styrenes with amines and CO_2.

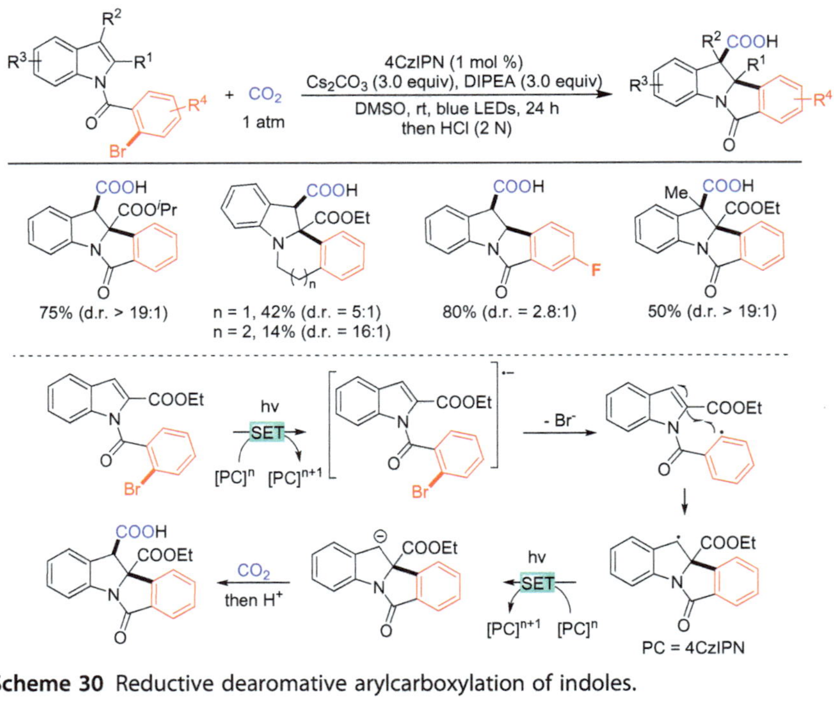

Scheme 30 Reductive dearomative arylcarboxylation of indoles.

Previously, high pressure and/or high excess of CO_2 was/were involved in the carboxylations. Besides, generally stoichiometric CO_2 was released as a waste byproduct in decarboxylation. Therefore, it is highly important to reutilize the CO_2, which is generated from decarboxylation, to undergo the further transformation. To address this issue, Yu and coworkers *(53)* disclosed a novel method to synthesize γ-amino acids by the reutilization of the in situ generated CO_2 from decarboxylation of carboxylic acids in photoredox conditions. This strategy can be widely applied to the conversion of various alkenes and carboxylic acids with good tolerance in functional groups. Furthermore, this reaction also can be carried under sunlight irradiation in 56% yield. The control experiments indicated that CsF was of great importance in this transformation, functionalizing as the base to deprotonate carboxylic acids to carboxylates, which then underwent decarboxylation. $^{13}CO_2$ isotope-labeling studies provided a strong evidence that the carboxyl group in the target product came from the carboxylic acids in the substrates.

The plausible mechanism was illustrated in the Scheme 31. After deprotonation by base, the starting material was oxidized by the excited [Ir]* to release an α-amino radical and CO_2. This radical intermediate then underwent radical addition to the activated alkenes to generate the more stable benzylic radical, which was further reduced by reduced [Ir^{II}] photocatalyst to deliver benzylic carbanion. The desired product

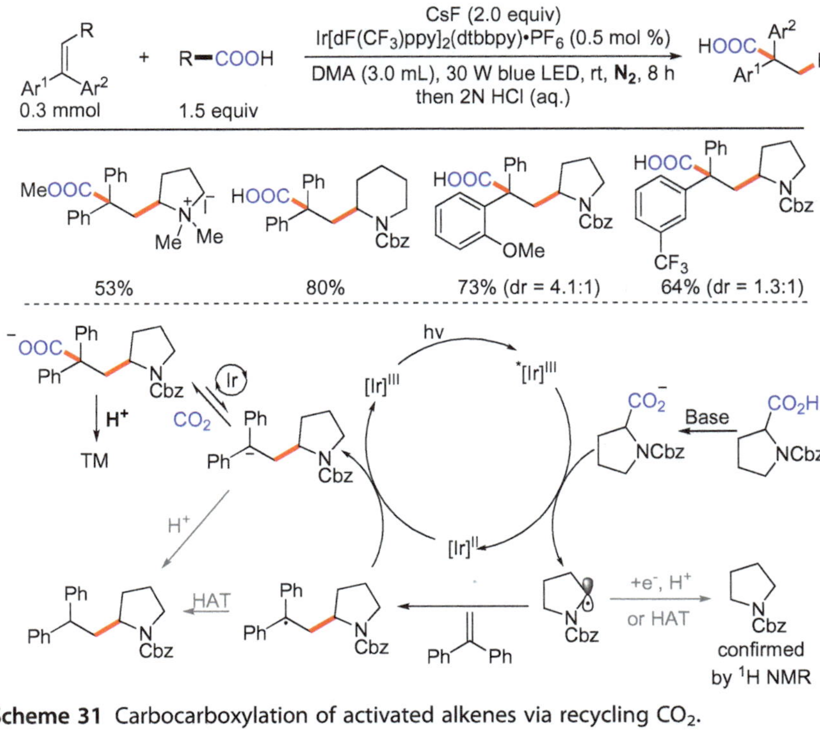

Scheme 31 Carbocarboxylation of activated alkenes via recycling CO_2.

was constructed by the nucleophilic attack of benzylic carbanion to CO_2 and followed protonation during workup. Soon after this work, Sun and coworkers *(54)* reported the similar transformation but under atmospheric CO_2.

Due to the high reducing potential of aryl halides in visible-light induced reaction, the Meerwein-arylation-type carboarylation of alkenes with CO_2 via visible-light condition is challenging. In 2020, inspired by the previous work *(55)* on the CO_2 radical anion, which possessed strong reducing potential, Li and coworkers realized the first carboarylation of styrenes involving CO_2 and aryl halides under visible light to construct valuable hydrocinnamic acid derivatives in high regioselectivity (Scheme 32). Control experiments indicated that HCO_2K was indispensable for this reaction, acting as CO_2 radical anion sources. The generation of mono- or dicarboxylation products of alkenes in the absence of aryl halides further confirmed the existence of CO_2 radical anion in this system. The $^{13}CO_2$ isotope-labeling experiments indicated the carboxyl source of the products mainly came from CO_2.

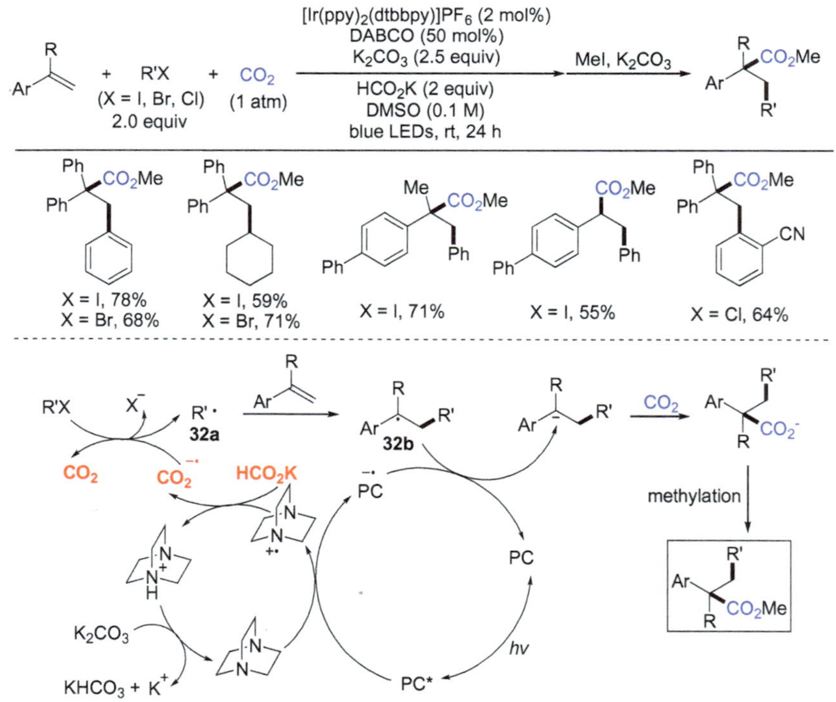

Scheme 32 Reductive carboarylation of styrenes with CO_2 and aryl halides.

Based on the mechanistic studies, a tentative mechanism was proposed in the Scheme 32. The CO_2 radical anion was generated through HAT process between HCO_2K and the radical cation of 1,4-diazabicyclo[2.2.2]octane (DABCO), which was produced via oxidation of DABCO by the excited state PC*. Subsequently aryl halides were reduced by CO_2 radical anion to yield aryl radical **32a**, which then attacked the alkenes to afford a benzyl radical **32b** regioselectively. The intermediate **32b** was further reduced to the benzyl anion through SET by the reduced photocatalyst, which then underwent nucleophilic addition to CO_2 to deliver the desired product.

3.1.2.2 β-Carboxylation of alkenes

Mimicking the protocol to access the α-carboxylation of alkenes, a simple approach to realize the β-carboxylation is to generate the CO_2 radical anion first, and then undergo radical addition to alkenes. The nascent radical intermediate is trapped by the other radical species in the system. However, due to the high reducing potential of CO_2 radical anion, cooperation of CO_2 radical anion with other reagents in one system possesses high challenges. Therefore, compared to the well development of photocatalytic α-carboxylation of alkenes involving CO_2, the β-carboxylation in the difunctionalization of alkenes has been rarely reported.

In 2017, Yu and coworkers *(56)* unraveled the first thiocarboxylation of styrenes and acrylates with CO_2, promoting by the catalytic iron/sulfur system, which is well known as the redox catalysts in the electron transfer and CO_2 fixation in photosynthesis (Scheme 33). The control experiments confirmed that visible light is essential for this reaction. Furthermore, the byproduct disulfide was excluded as the reactive intermediate. In the consideration of the high regioselectivity, the difunctionalization of alkenes is

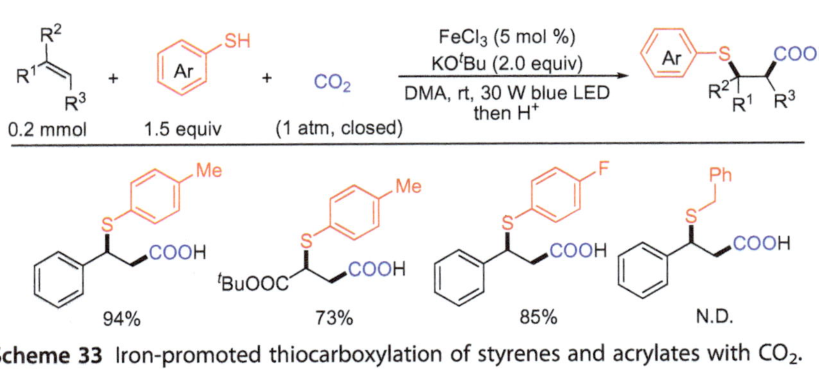

Scheme 33 Iron-promoted thiocarboxylation of styrenes and acrylates with CO_2.

believed to start with C—C bond formation with CO_2. Due to the higher reactivity of an electron-rich styrene than an electron-poor one, other pathways, such as the formation of a benzylic cation intermediate, followed by trapping with thiolate, seems impossible. The author supposed that a CO_2 radical anion and an oxidized form of Fe/S complex were generated via direct SET reduction of CO_2 by the excited Fe/S complex under visible-light irradiation. A stabilized benzylic radical was afforded after the addition of CO_2 radical anion to alkenes. Subsequently C—S bond formation between the benzylic radical and the oxidized form of Fe/S complex afforded the desired product.

In 2018, Núñez and coworkers *(57)* found that terminal olefins, equipped with an electron withdrawing *Z*-substituent, performed the addition reaction of iodine and CO_2 to the C═C double bond under UV irradiation in the acetonitrile solution of LiI (Scheme 34). The control experiments revealed that the reactive intermediate varied with the reaction condition. In the solution of acetonitrile, the CO_2 radical anion may act as a key intermediate in this reaction, which can add to the olefin reversibly. However, water is not a suitable electron accepter for CO_2, which probably account for the alternative pathway of this reaction that involving H atoms as intermediate.

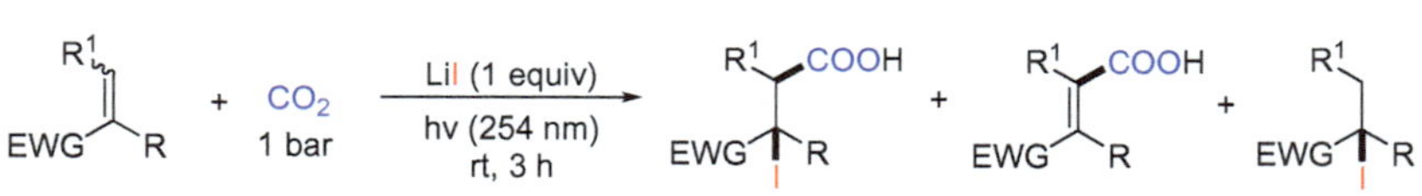

Scheme 34 Activated C—C double bonds with CO_2 and lithium iodide

3.1.2.3 Dicarboxylation of alkenes

Dicarboxylic acids widely existed in the natural products, pharmaceutical agents and functional polymers. It is important to realize dicarboxylation of unsaturated bonds with CO_2 to generate dicarboxylic acids. In 1993, Kubiak and coworkers *(58)* found the formation of *cis*- and *trans*-1,2-cyclohexanedicarboxylate with CO_2 by the direct photochemical reduction of $[Ni_3(\mu_3\text{-}I)_2(dppm)_3]$ under UV light (Scheme 35). However, this protocol was limited to the level of instrumental analysis and detection of desired product. The detection of CO and carbonate probed for the intermediacy of the CO_2 radical anion in the system, due to the disproportionation of CO_2 radical anion to CO and carbonate. After this work, Jamison's group *(9)* disclosed the *β*-selective hydrocarboxylation of styrenes with

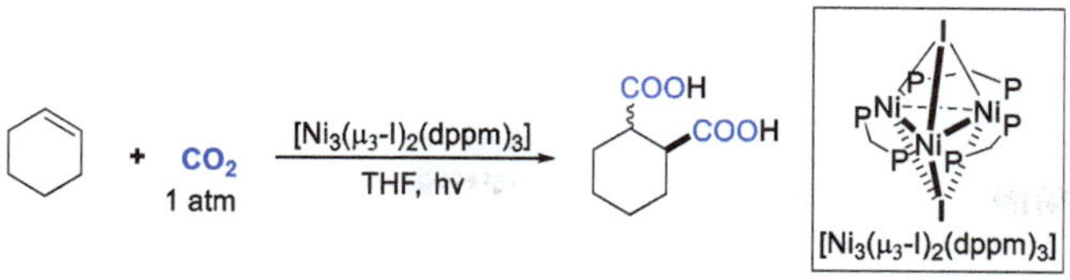

Scheme 35 Formation of two carbon–carbon bonds via addition of CO_2 radical anion to cyclohexene.

CO_2 in 2017. In this methodology, dicarboxylic acids were produced as byproducts, which yielded up to 29%.

The first systematic synthesis of dicarboxylic acids came from Yu's group *(59)* in 2021, using 4CzIPN as the photocatalyst, $^{i}Pr_2NEt$ as the reductant and Cs_2CO_3 as the base under visible-light irradiation. This reaction possessed a broad substrate scope, including styrenes, acrylates, allenes, 1,3-dienes and polycyclic aromatic hydrocarbon (Scheme 36). The product can undergo facial transformation to afford corresponding derivatives, which proved to be a great potential for the application in organic synthesis, late-stage modification of pharmaceutical chemistry.

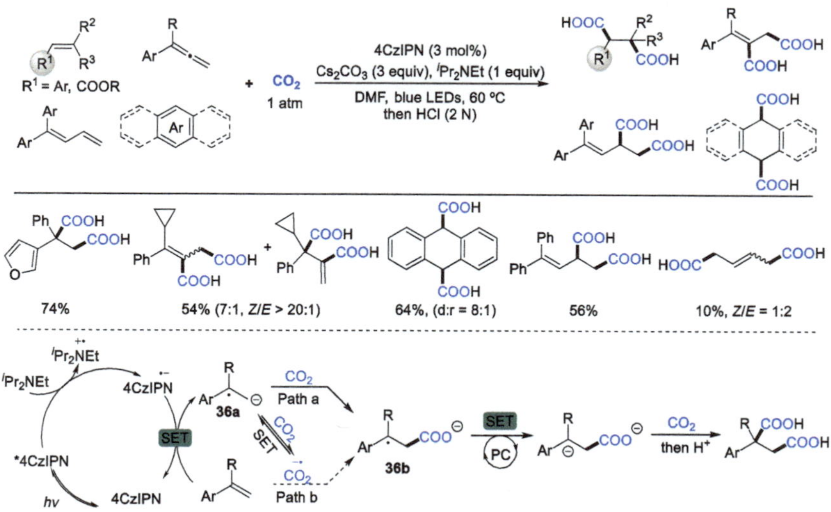

Scheme 36 Dicarboxylation of alkenes, allenes and (hetero)arenes with CO_2.

Based on the mechanistic investigations and literature reports, the author supposed the alkenes were first reduced by the reduced photocatalyst to its radical anion, which then nucleophilic attack CO_2 to give the intermediate **36b**. Further reduction of intermediate **36b** through a second SET with reduced photocatalyst delivered the benzylic anion, which can interact with a second CO_2 and go through protonation to yield the target product. Furthermore, according to the control experiments, the direct electron transfer from intermediate **36a** to CO_2 was less favorable, but possessed the possibility.

3.2 Carboxylation of alkynes

In 2018, Zhao, Wu and coworkers reported the first visible-light-driven hydrocarboxylation as well as carbocarboxylation of alkynes using CO_2 via an iridium/cobalt dual catalysis (Scheme 37) *(60)*. Their study was initiated by employing 1-phenyl-1-propyne as the model substrate to investigate the hydrocarboxylation under 1 atm of CO_2 in the presence of a photocatalyst and a cobalt catalyst with blue LED irradiation. After extensive optimization, they found that the combination of $Ir(ppy)_2(dtbbpy)PF_6$ (1 mol%), $CoBr_2$ (10 mol%), dcype (10 mol%, bis(dicyclohexylphosphino) ethane), and $ZnBr_2$ (1 equiv.) as the additive in the presence of iPr_2NEt (3 equiv.) provided the best result, affording vinyl acids in 70% combined

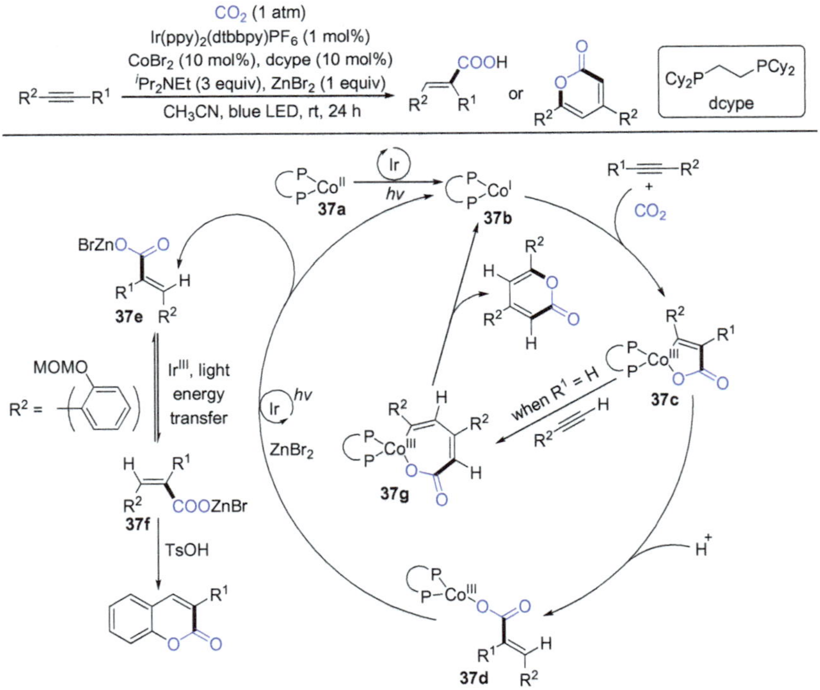

Scheme 37 Alkyne hydro-/carbocarboxylation using CO_2 via iridium/cobalt dual catalysis.

yield. Through the investigation of the mechanism, the authors gave a plausible mechanism. Tentative mechanistic pathways for alkyne hydrocarboxylation are proposed in light of all the experimental data. Reductive quenching of visible-light-excited $Ir(ppy)_2(dtbbpy)PF_6$ by $^{i}Pr_2NEt$ gives rise to an amine radical cation and a reduced Ir^{II} species, the latter ($E_{1/2}[Ir^{III}/Ir^{II}] = -1.51$ V vs SCE in MeCN) is then oxidized by Co^{II} complex **37a** ($E_{1/2} = -0.75$ V vs SCE in MeCN) to regenerate Ir^{III}. The active Co^{I} species **37b** generated in this step will react with CO_2 and alkyne substrates to produce the five-membered cobalta-cycle intermediate **37c**. Subsequent protonolysis of C—Co^{III} bond by cationic amine radicals will deliver the carboxylatocobalt intermediate **37d**. The transient complex **37d** may undergo photoredox-catalyzed reduction to regenerate Co^{I} species **37b**, and $ZnBr_2$ may facilitate this step through transmetallation to furnish carboxylate **37e**. Finally, the aryl substituted *E*-vinyl-carboxylate **37e** can undergo a reversible isomerization to its *Z* isomer **37f** through an Ir-mediated triplet – triplet energy transfer process under visible-light irradiation, following which an acid-mediated intramolecular cyclization delivers product. As for terminal alkynes, insertion of another alkyne molecule into the five-membered cobaltacycle **37c** will occur to give the seven-membered cobaltacycle **37g**. Subsequent reductive elimination will afford 2-pyrone and regenerate Co^{I} **37b**.

3.3 Hydrocarboxylation of ketone, imine and hydrazone derivatives

3.3.1 Visible light

Based on previous reports on visible light catalysis, benzophenone-based ketimines undergo SET reductions to generate carbanions through abstracting protons from water (or deuterium from D_2O) via an umpolung reactivity. In 2018, Walsh's group *(61)* reported a visible light- or sunlight-mediated photoredox hydrocarboxylation of imines and the synthesis of α,α-disubstituted α-amino acids (Scheme 38). The investigation of standard reaction conditions showed that the standard substrate benzophenone imine could form the desired carboxylic acid with a 92% yield using Cy_2NMe (*N,N*-dicy-clohexylmethylamine) as sacrificial electron donor, $[Ir(ppy)_2(4,4'\text{-}t\text{Bu}_2\text{-bpy})]PF_6$ as photoredox catalysts under blue LED irradiation. What's more, the authors also conducted preliminary investigations to probe the mechanism. UV–vis spectra indicated that only $[Ir]^{3+}$ catalyst was capable of absorbing visible light, so the reaction was likely initiated by

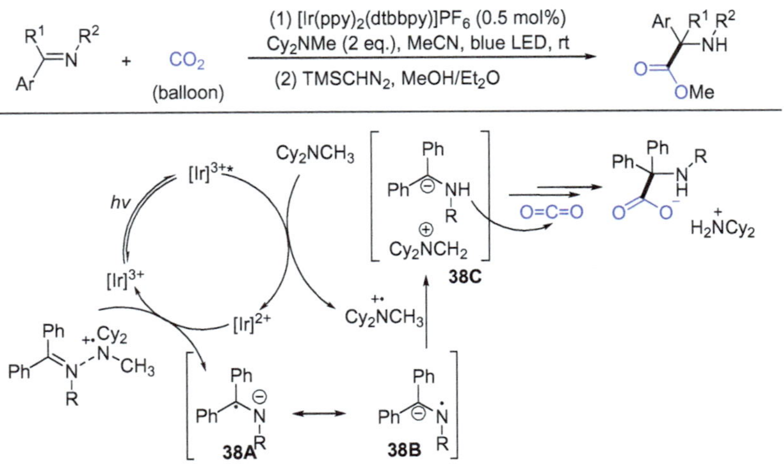

Scheme 38 CO_2 fixation with imines.

irradiation of $[Ir]^{3+}$ by light to give its excited state [Ir]3 + *. Stern–Volmer fluorescence quenching experiments indicated that Cy_2NMe acted as electron donor to reduce [Ir]3 + * to $[Ir]^{2+}$, generating the radical cation $[Cy_2NMe]^+$ at the same time. The resulting amine radical cation coordinates with imines to form a 2-center-3-electron bond, facilitating the reduction by $[Ir]^{2+}$ to form the radical anion intermediate **38a/b** and regenerating $[Ir]^{3+}$. Due to the high reactivity of the N-radical in resonance form **38b**, it is quickly quenched by the amine radical cation via HAT, to give the α-amino carbanion intermediate **38c**. **38c** acted as a strong nucleophile and attacked CO_2 to give the product. The iminium ion $[Cy_2N{=}CH_2]^+$ reacted with advantageous water to generate Cy_2NH, which forms insoluble salts with α-amino acid. The key factor on controlling reactivity in this system came from the significant contribution of resonance form **38b**. The author previously presented computational evidence that the radical anion intermediate has greater spin density on nitrogen (0.37) than that on the carbon labeled C3 (0.18), indicating that the C3 carbon carries more anion character. In addition, previous experimental results on electrochemical reduction of ketimines by Reed et al., as well as photochemical reduction of ketimines also indicated that C3 carbon carried more anionic character. The conceptual advance of this work is that stabilizing groups at the carbonyl carbon can invert the reactivity of the ketiminyl radical anion, enabling nucleophilic addition to CO_2. Such additions are usually observed with reactive organometallic reagents possessing Lewis acidic metal centers capable of activating CO_2 toward addition (i.e., Grignard and organolithium reagents).

Meanwhile, Yu's group reported the first catalytic hydrocarboxylation of enamides and imines with CO_2 to generate α,α-disubstituted α-amino acids (Scheme 39) *(62)*. After systematic investigation, the authors successfully achieved the transformation of the standard substrate to get the desired product in 83% isolated yield when using 2 mol% of 4CzIPN as photocatalyst, iPr_2NEt as reductant, and Cs_2CO_3 as the base. After the investigation of the mechanism, the authors proposed that an α-amino carbanion **39a** might be generated in this reaction, which could be trapped by CO_2 to give carboxylate form **39b** of the desired α,α-disubstituted α-amino acid. Furthermore, the reactions of N-(diphenylmethylene) benzamide with other electrophiles, such as carbon disulfide or isothiocyanate reagents, also demonstrated the possibility of the existence of carbanion **39a**.

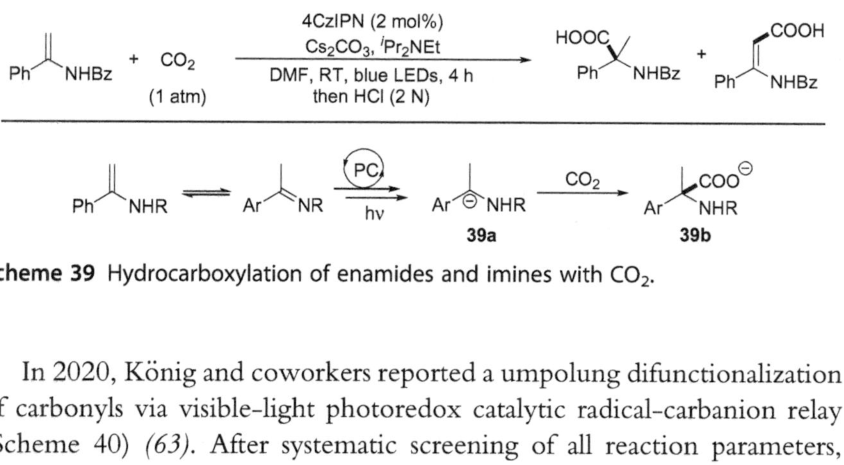

Scheme 39 Hydrocarboxylation of enamides and imines with CO_2.

In 2020, König and coworkers reported a umpolung difunctionalization of carbonyls via visible-light photoredox catalytic radical-carbanion relay (Scheme 40) *(63)*. After systematic screening of all reaction parameters, they obtain the desired functionalized carboxylic acid in 81% yield using $[Ir(dFCF_3ppy)_2dtbbpy]PF_6$ (1 mol%) under 3 atm of CO_2 in DMSO. Moreover, they successfully converted *p*-tolualdehyde into the desired product in one pot by means of a condensation and photocatalytic sequence with similar efficiency. Based on the experimental evidence and mechanistic pathways reported in the previous literatures, they propose a plausible mechanism as depicted in Scheme 40 for the reported photocatalytic generation of functionalized carbanions. Initially, the photoexcited state of $[Ir^{III}(dFCF_3ppy)_2dtbbpy]^+$ ($E_{1/2}[{}^*Ir^{III}/^{II}] = +1.21$ V vs SCE) was reductively quenched by sodium triflinate ($E_{ox} = +1.05$ V vs SCE) or thiophenolate ($E_{ox} = \sim 0.75$ V vs SCE), which was formed through the deprotonation of thiophenol by base, affording a sulfur-centered radical

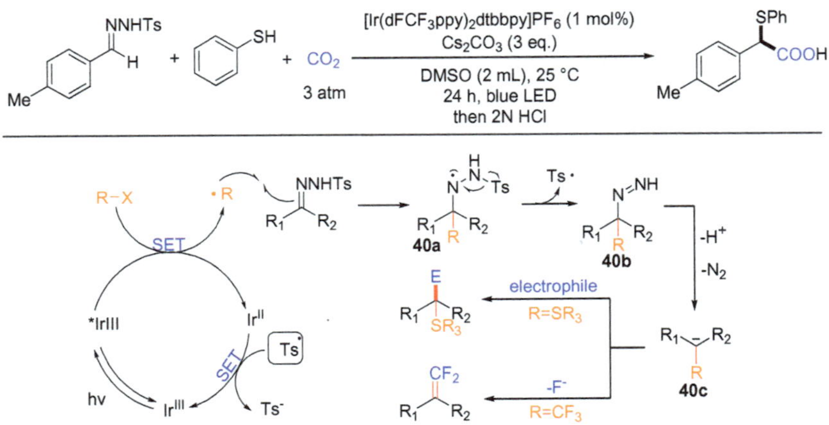

Scheme 40 Umpolung difunctionalization of carbonyls via visible-light photoredox catalytic radical-carbanion relay

and a CF_3 radical, respectively. Subsequent radical addition to the C=N bond of *N*-tosylhydrazone generates the aminyl radical species **40a**. Fragmentation of the arenesulfonyl radical from intermediate **40a** leads to a functionalized diazene intermediate **40b**, and the following Wolff–Kishner type N_2 extrusion process proceeds to give α-CF_3 or sulfur carbanion **40c** for further reactions. In the case of the α-sulfenyl carbanion, subsequent nucleophilic attack to CO_2 or aliphatic aldehydes gave carboxylic acids or alcohols. When α-CF_3 carbanions were produced, β-fluoride elimination occurred to furnish the gem-difluoroalkenes. Finally, SET between the reduced photoredox catalyst Ir^{II} ($E_{1/2}[Ir_{III/II}] = -1.37$ V vs SCE) and the arenesulfonyl radical ($E_{red} = +0.50$ V vs SCE) yielded a sulfinate anion and regenerated the photocatalyst. This strategy greatly expands the synthetic potential of Wolff-Kishner reaction.

In 2021, Yu's group reported a visible-light photoredox-catalyzed umpolung carboxylation of carbonyl compounds with CO_2 (Scheme 41) *(64)*. They found that the combination of trimethylsilylchloride (TMSCl) as additive, KO^tBu as base, and *N,N*-dimethylacetamide (DMA) as solvent was the best choice to give the desired product in 80% isolated yield. Based on the previous mechanism experiment, they proposed that the reaction proceeds with reductive quenching to give reduced Ir^{II}-catalyst ($E_{1/2}^{Red}[Ir^{III}/Ir^{II}] = -1.51$ V vs SCE). Moreover, the CV test indicated that the presence of TMSCl promoted reduction of the carbonyl by lowering the reductive potential. Thus, they believed that the SET reduction of the TMSCl-activated ketone was favored to give the corresponding

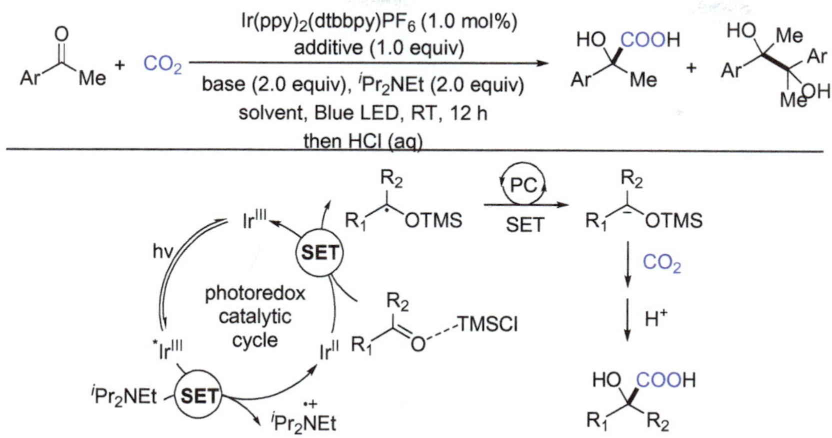

Scheme 41 Umpolung carboxylation of carbonyl compounds with CO_2.

α-silyloxyl benzyl radical, which might be further reduced by the Ir^{II}-catalyst to generate a benzylic carbon anion. The following nucleophilic attack to CO_2 would give the corresponding products. Notably, this reaction features high selectivity, broad substrate scope, good functional group tolerance, mild reaction conditions and facile derivations of products to bioactive compounds, including oxypheonium, mepenzolate bromide, benactyzine, and tiotropium.

Soon after this report, König and coworkers reported a photocatalytic umpolung reaction of carbonyl compounds to generate anionic carbinol synthons (Scheme 42) *(65)*. Through the extensive screening of the conditions for photocatalytic reduction of 2-(trifluoromethyl)benzaldehyde under CO_2 as an electrophilic trapping agent, they found that $Ir(ppy)_2(dtbbpy)PF_6$ (ppy, 2-phenylpyridinato; dtbbpy, 4,4′-di-tert-butyl-2,2′-bipyridine) as a photocatalyst and 1,3-dimethyl-2-phenyl-2,3-dihydro-1H-benzo[*d*]imidazole (DMBI) as a reductant formed an effective catalyst system. An acetonitrile solution of aldehyde, $Ir(ppy)_2(dtbbpy)PF_6$ (2 mol%), and DMBI (1.5 equiv.) was irradiated with blue light ($\lambda = 448$ nm) at 20 °C under CO_2 at atmospheric pressure. After 1 h, the reaction was quenched by HCl(aq) to give 2-(triflfluoromethyl)mandelic acid in 61% NMR yield, together with the dimeric 1,2-diol as a minor product. The formation of diol indicates that the ketyl radicals underwent radical-radical dimerization. Based on the mechanistic study, the author proposed the following reaction mechanism. The excited Ir(III) photocatalyst is reduced by DMBI to give an Ir(II) species and the DMBI radical cation **42c**. Stern – Volmer fluorescence

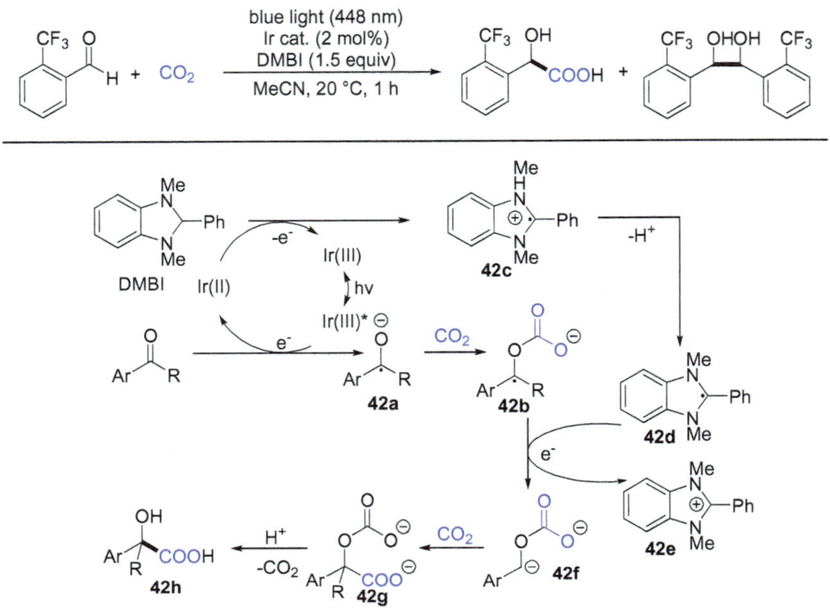

Scheme 42 Direct addition of aromatic aldehydes and ketones to CO_2.

quenching experiments have shown that the excited iridium photocatalyst is effectively quenched by DMBI. The Ir(II) species reduces an aromatic carbonyl compound to give ketyl radical **42a**, with regeneration of the Ir(III) photocatalyst. Ketyl radical **42a** is rapidly transformed into the carbinol anionic species **42f** through a successive SET reduction. The oxygen anion of **42a** reacts with CO_2 and then receives one electron from the DMBI radical **42d**, produced by deprotonation of the DMBI radical cation **42c**, to generate the carbinol anion species **42 f**. DMBI radical **42d** is known to be a strong one-electron reductant, and its redox potential is −1.68 V (vs SCE, MeCN) which is greater than that of the excited $Ir(ppy)_2$(dtbbpy) PF_6 (−1.51 V vs SCE, MeCN). Consequently, the ketyl radicals are reduced by the DMBI radical faster than they undergo radical-radical coupling. Attack by the carbinol anion species **42f** on CO_2 is followed by protonation and decarboxylation to afford the final product **42 h**.

4. Miscellaneous

The aforementioned examples mainly include the key intermediate of radical anions, carbanions, carbon dioxide radical anions and low valence active metal species, which are the key to achieve catalytic transformations of CO_2. However, in the initial studies some carboxylation via pre-formed organometallic reagent were also developed under light irradiation.

In 1977, Inoue reported the first photo-fixation of CO_2 via a metallic aluminum complex-tetraphenylporphinatoaluminium ethyl (TPPAlEt) with a low yield of carboxylates (Scheme 43) *(66)*. In the absence of 1-methylimidazole, the strategy did not proceed well because the coordination of 1-methylimidazole with aluminum increased the nucleophilicity of the ethyl group.

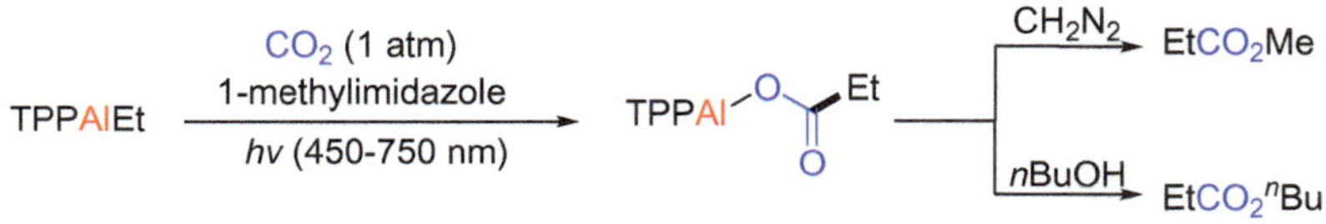

Scheme 43 Reaction of CO_2 with tetraphenylporphinatoaluminium ethyl.

Later, they use a similar photofixation method to synthesize *β*-ketocarboxylic acid with ketone as the substrate *(67)*. The difference was that the ethyl group in the metal aluminum complex was substituted by diethylamine (Scheme 44).

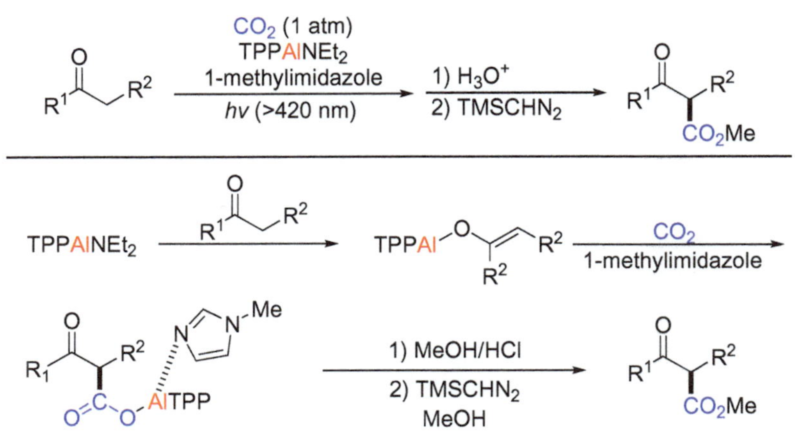

Scheme 44 Photosynthesis of *β*-ketocarboxylic acids via enolate complexes of aluminum porphyrin.

Notably, a series of malonic acid derivatives were obtained when α,β-unsaturated esters were added to the reaction *(68)*. Interestingly, only 1 mol% of alkylaluminum complex was required for this reaction, followed by the addition of equivalent zinc dialkyl as a transmetallation reagent. Based on the mechanism experiments, they ruled out the possibility that CO_2 was directly inserted into the Al—R bond. A possible mechanism was proposed, involving 1,4-addition, CO_2 insertion and transmetallation. This was the first attempt of carrying out metal/light dually catalyzed carboxylation with CO_2 (Scheme 45).

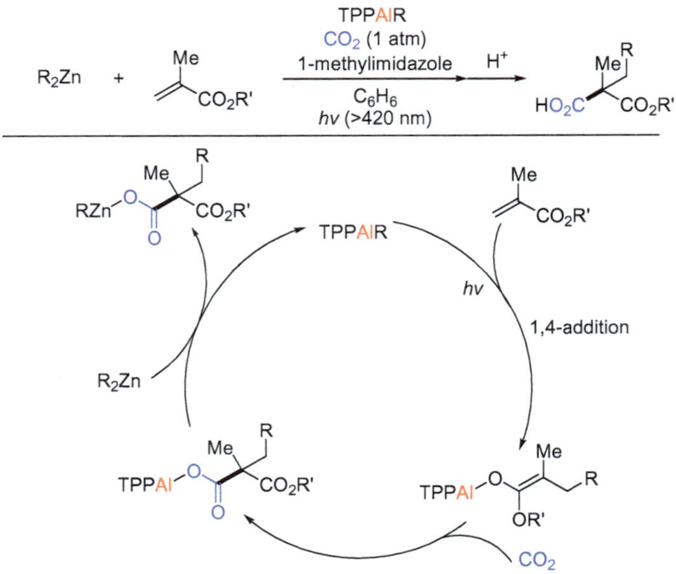

Scheme 45 Synthesis of malonic acid derivatives from CO_2 catalyzed by aluminum porphyrins.

In addition to organic aluminum reagent could react with CO_2, other organometallic reagents could also carry out the reaction smoothly. Guilard and coworkers reported a new strategy that CO_2 was inserted into the C—Indium bond *(69)*. Cooper and coworkers reported a reaction of titanocene with CO_2 to afford Ti—carboxylate under 300–750 nm irradiation *(70)*.

5. Summary

Relying on such a powerful tool as light-induced methodologies, chemists now can realize the carboxylation of inexpensive raw materials with CO_2, which highly enriches the diversity of chemical synthetic pools. However, the aforementioned protocols to fix CO_2 into various substrates to furnish valued carboxylic acids possess the limitations on the substrates and strategies. The reason for which the transformations mainly employed the activated substrates, including the benzylic or allylic C—H, styrenes, and electron-deficiency alkenes, is that a stable radical intermediate usually generating through HAT process or radical additions, is needed for the further reduction to the carbon anion which can easily perform the nucleophilic addition to CO_2. In this regard, the principal approach to the CO_2 fixation is going via a two-electron process. Furthermore, the aliphatic radicals can't be further reduced to corresponding aliphatic anions through this method due to the highly reduced potentials, which limits the substrate scope. Therefore, novel strategies are vitally necessary for the activation of CO_2 and merit persistent exploration.

References

1. (a) Hou, S.-L.; Dong, J.; Zhao, B. *Adv. Mater.* **2020**, *32*, 1806163; (b) Song, Q.-W.; Zhou, Z.-H.; He, L.-N. *Green Chem.* **2017**, *19*, 3707–3728; (c) Chen, Y.; Mu, T.-C. *Green Chem.* **2019**, *21*, 2544–2574; (d) Wang, S.; Xi, C.-J. *Chem. Soc. Rev.* **2019**, *48*, 382–404; (e) Grignard, B.; Gennen, S.; Jérôme, C.; Kleij, A. W.; Detrembleur, C. *Chem. Soc. Rev.* **2019**, *48*, 4466–4514; (f) Hu, J.-Y.; Liu, H.-Z.; Han, B.-X. *Sci. China: Chem.* **2018**, *61*, 1486–1493.
2. Hong, J.-T.; Li, M.; Zhang, J.-N.; Sun, B.-Q.; Mo, F.-Y. *ChemSusChem* **2019**, *12*, 6–39.
3. (a) Fujihara, T.; Tsuji, Y. *Front. Chem.* **2019**, *7*, 430; (b) Correa, A.; Martin, R. *Angew. Chem. Int. Ed.* **2009**, *48*, 6201–6204; (c) Ackermann, L. *Angew. Chem. Int. Ed.* **2011**, *50*, 3842–3844; (d) Boogaerts, I. I.; Nolan, S. P. *Chem. Commun.* **2011**, *47*, 3021–3024.
4. (a) Narayanam, J. M. R.; Stephenson, C. R. J. *Chem. Soc. Rev.* **2011**, *40*, 102–113; (b) Xuan, J.; Xiao, W.-J. *Angew. Chem. Int. Ed.* **2012**, *51*, 6828–6838; (c) Meggers, E. *Chem. Commun.* **2015**, *51*, 3290–3301.
5. (a) Prier, C. K.; Rankic, D. A.; MacMillan, D. W. C. *Chem. Rev.* **2013**, *113*, 5322–5363; (b) Fabry, D. C.; Rueping, M. *Acc. Chem. Res.* **2016**, *49*, 1969–1979; (c) Tellis, J. C.; Kelly, C. B.; Primer, D. N.; Jouffroy, M.; Patel, N. R.; Molander, G. A. *Acc. Chem. Res.* **2016**, *49*, 1429–1439.
6. Masuda, Y.; Ishida, N.; Murakami, M. *J. Am. Chem. Soc.* **2015**, *137*, 14063–14066.
7. Ishida, N.; Masuda, Y.; Uemoto, S.; Murakami, M. *Chem. A Eur. J.* **2016**, *22*, 6524–6527.
8. Ishida, N.; Masuda, Y.; Imamura, Y.; Yamazaki, K.; Murakami, M. *J. Am. Chem. Soc.* **2019**, *141*, 19611–19615.
9. Seo, H.; Katcher, M. H.; Jamison, T. F. *Nat. Chem.* **2017**, *9*, 453–456.
10. Wang, M.-Y.; Cao, Y.; Liu, X.; Wang, N.; He, L.-N.; Li, S.-H. *Green Chem.* **2017**, *19*, 1240–1244.
11. Meng, Q.; Schirmer, T. E.; Berger, A. L.; Donabauer, K.; König, B. *J. Am. Chem. Soc.* **2019**, *141*, 11393–11397.
12. Donabauer, K.; Maity, M.; Berger, A. L.; Huff, G. S.; Crespi, S.; König, B. *Chem. Sci.* **2019**, *10*, 5162–5166.
13. Osakada, K.; Sato, R.; Yamamoto, T. *Organometallics* **1994**, *13*, 4645–4647.
14. (a) Huang, K.; Sun, C.; Shi, Z. *Chem. Soc. Rev.* **2011**, *40*, 2435–2452; (b) Tortajada, A.; Juliá-Hernández, F.; Börjesson, M.; Moragas, T.; Martin, R. *Angew. Chem. Int. Ed.* **2018**, *57*, 15948–15982.

15. (a) Takahashi, K.; Cho, K.; Iwai, A.; Ito, T.; Iwasawa, N. *Chem. A Eur. J.* **2019**, *25*, 13504–13508; (b) Takahashi, K.; Hirataka, Y.; Ito, T.; Iwasawa, N. *Organometallics* **2020**, *39*, 1561–1572.
16. Zhang, L.; Cheng, J.; Ohishi, T.; Hou, Z. *Angew. Chem. Int. Ed.* **2010**, *49*, 8670–8673.
17. Song, L.; Fu, D.-M.; Chen, L.; Jiang, Y.-X.; Ye, J.-H.; Zhu, L.; Lan, Y.; Fu, Q.; Yu, D.-G. *Angew. Chem. Int. Ed.* **2020**, *59*, 21121–21128.
18. Matsumoto, T.; Uchijo, D.; Koike, T.; Namiki, R.; Chang, H. *Sci. Rep.* **2018**, *8*, 14623–14632.
19. Matsumoto, T.; Chang, H.; Wakizaka, M.; Ueno, S.; Kobayashi, A.; Nakayama, A.; Taketsugu, T.; Kato, M. *J. Am. Chem. Soc.* **2013**, *135*, 8646–8654.
20. Schmalzbauer, M.; Svejstrup, T. D.; Fricke, F.; Brandt, P.; Johansson, M. J.; Bergonzini, G.; König, B. *Chem* **2020**, *6*, 2658–2672.
21. Börjesson, M.; Moragas, T.; Gallego, D.; Martin, R. *ACS Catal.* **2016**, *6*, 6739–6749.
22. (a) Moragas, T.; Gaydou, M.; Martin, R. *Angew. Chem. Int. Ed.* **2016**, *55*, 5053–5057; (b) Gemmeren, M. V.; Bçrjesson, M.; Tortajada, A.; Sun, S.-Z.; Okura, K.; Martin, R. *Angew. Chem. Int. Ed.* **2017**, *56*, 6558–6562.
23. Osakada, K.; Sato, R.; Yamamoto, T. *Organometallics* **1994**, *13*, 4645–4647.
24. (a) Fujihara, T.; Nogi, K.; Xu, T.-H.; Tsuji, Y. *J. Am. Chem. Soc.* **2012**, *134*, 9106–9109; (b) Nogi, K.; Fujihara, T.; Terao, J.; Tsuji, Y. *J. Org. Chem.* **2015**, *80*, 11618–11623.
25. (a) Charboneau, D. J.; Brudvig, G. W.; Hazari, N.; Lant, H. M. C.; Saydjari, A. K. *ACS Catal.* **2019**, *9*, 3228–3241; (b) Yanagi, T.; Somerville, R. J.; Nogi, K.; Martin, R.; Yorimitsu, H. *ACS Catal.* **2020**, *10*, 2117–2123.
26. Shimomaki, K.; Murata, K.; Martin, R.; Iwasawa, N. *J. Am. Chem. Soc.* **2017**, *139*, 9467–9470.
27. Murata, K.; Numasawa, N.; Shimomaki, S.; Takaya, J.; Iwasawa, N. *Chem. Commun.* **2017**, *53*, 3098–3101.
28. Meng, H.-Y.; Wang, S.; König, B. *Angew. Chem. Int. Ed.* **2017**, *56*, 13426–13430.
29. (a) Shimomaki, K.; Nakajima, T.; Caner, J.; Toriumi, N.; Iwasawa, N. *Org. Lett.* **2019**, *21*, 4486–4489; (b) Bhunia, S. K.; Das, P.; Nandi, S.; Jana, R. *Org. Lett.* **2019**, *21*, 4632–4637.
30. Ishida, N.; Masuda, Y.; Liao, W.-Q.; Murakami, M. *Chem. Lett.* **2019**, *48*, 1316–1318.
31. Zhu, C.; Zhang, Y.-F.; Liu, Z.-Y.; Zhou, L.; Liu, H.-D.; Feng, C. *Chem. Sci.* **2019**, *10*, 6721–6726.
32. Sahoo, B.; Bellotti, P.; Julia-Hernandez, F.; Meng, Q.-Y.; Crespi, S.; Konig, B.; Martin, R. *Chem. A Eur. J.* **2019**, *25*, 9001–9005.
33. Jing, K.; Wei, M.-K.; Yan, S.-S.; Liao, L.-L.; Niu, Y.-N.; Luo, S.-P.; Yu, B.; Yu, D.-G. *Chin. J. Catal.* **2021**. https://doi.org/10.1016/S1872-2067(21)63859-7.
34. (a) Ouyang, K.; Hao, W.; Zhang, W.-X.; Xi, Z. *Chem. Rev.* **2015**, *115*, 12045–12090; (b) Wang, Q.; Su, Y.; Li, L.; Huang, H. *Chem. Soc. Rev.* **2016**, *45*, 1257–1272.
35. Liao, L.-L.; Cao, G.-M.; Ye, J.-H.; Sun, G.-Q.; Zhou, W.-J.; Gui, Y.-Y.; Yan, S.-S.; Shen, G.; Yu, D.-G. *J. Am. Chem. Soc.* **2018**, *140*, 17338–17342.
36. (a) Jin, Y.-S.; Toriumi, N.; Iwasawa, N. *ChemSusChem* **2022**, *15*, e202102095; (b) Ran, C.-K.; Niu, Y.-N.; Song, L.; Wei, M.-K.; Cao, Y.-F.; Luo, S.-P.; Yu, Y.-M.; Liao, L.-L.; Yu, D.-G. *ACS Catal.* **2022**, *12*, 18–24.
37. Fan, Z.-N.; Chen, S.-H.; Zou, S.; Xi, C.-J. *ACS Catal.* **2022**, *12*, 2781–2787.
38. Yan, S.-S.; Liu, S.-H.; Chen, L.; Bo, Z.-Y.; Jing, K.; Gao, T.-Y.; Yu, B.; Lan, Y.; Luo, S.-P.; Yu, D.-G. *Chem* **2021**, 7, 3099–3113.
39. (a) Koike, T.; Akita, M. *Chem* **2018**, *4*, 409–437; (b) Li, Y.; Liang, Y.; Dong, J.; Deng, Y.; Zhao, C.; Su, Z.; Guan, W.; Bi, X.; Liu, Q.; Fu, J. *J. Am. Chem. Soc.* **2019**, *141*, 18475–18485; (c) Kawamura, S.; Egami, H.; Sodeoka, M. *J. Am. Chem. Soc.* **2015**, *137*, 4865–4873.

40. Zhang, Z.; Ye, J.-H.; Ju, T.; Liao, L.-L.; Huang, H.; Gui, Y.-Y.; Zhou, W.-J.; Yu, D.-G. *ACS Catal.* **2020**, *10*, 10871–10885.
41. Lapidus, A. L.; Pirozhkov, S. D.; Koryakin, A. A. *Russ. Chem. Bull.* **1978**, *27*, 2513–2515.
42. (a) Yoon, T. P.; Ischay, M. A.; Du, J. *Nat. Chem.* **2010**, *2*, 527–532; (b) Narayanam, J. M. R.; Stephenson, C. R. J. *Chem. Soc. Rev.* **2011**, *40*, 102–113; (c) Osawa, M.; Nagai, H.; Akita, M. *Dalton Trans.* **2007**, *827−829*; (d) Kalyani, D.; McMurtrey, K. B.; Neufeldt, S. R.; Sanford, M. S. *J. Am. Chem. Soc.* **2011**, *133*, 18566–18569.
43. Murata, K.; Numasawa, N.; Shimomaki, K.; Takaya, J.; Iwasawa, N. *Front. Chem.* **2019**, *7*, 371.
44. Seo, H.; Liu, A.; Jamison, T. F. *J. Am. Chem. Soc.* **2017**, *139*, 13969–13972.
45. Meng, Q. Y.; Wang, S.; Huff, G. S.; König, B. *J. Am. Chem. Soc.* **2018**, *140*, 3198–3201.
46. Huang, H.; Ye, J.-H.; Zhu, L.; Ran, C.-K.; Miao, M.; Wang, W.; Chen, H.-J.; Zhou, W.-J.; Lan, Y.; Yu, B.; Yu, D.-G. *CCS Chem.* **2020**, *2*, 1746–1756.
47. Kang, G. W.; Romo, D. *ACS Catal.* **2021**, *11*, 1309–1315.
48. Yatham, V. R.; Shen, Y. Y.; Martin, R. *Angew. Chem. Int. Ed.* **2017**, *56*, 10915–10919.
49. Hou, J.; Ee, A.; Cao, H.; Ong, H. W.; Xu, J. H.; Wu, J. *Angew. Chem. Int. Ed.* **2018**, *57*, 17220–17224.
50. Fu, Q.; Bo, Z.-Y.; Ye, J.-H.; Ju, T.; Huang, H.; Liao, L.-L.; Yu, D.-G. *Nat. Commun.* **2019**, *10*, 3592.
51. Zhang, B.; Yi, Y.-P.; Wu, Z.-Q.; Chen, C.; Xi, C.-J. *Green Chem.* **2020**, *22*, 5961–5965.
52. Zhou, W.-J.; Wang, Z.-H.; Liao, L.-L.; Jiang, Y.-X.; Cao, K.-G.; Ju, T.; Li, Y.-W.; Cao, G.-M.; Yu, D.-G. *Nat. Commun.* **2020**, *11*, 3263.
53. Liao, L.-L.; Cao, G.-M.; Jiang, Y.-X.; Jin, X.-H.; Hu, X.-L.; Chruma, J. J.; Sun, G.-Q.; Gui, Y.-Y.; Yu, D.-G. *J. Am. Chem. Soc.* **2021**, *143*, 2812–2821.
54. Zhou, C.; Li, M.; Sun, J.-W.; Cheng, J.; Sun, S. *Org. Lett.* **2021**, *23* (8), 2895–2899.
55. Wang, H.; Gao, Y.-Z.; Zhou, C.-L.; Li, G. *J. Am. Chem. Soc.* **2020**, *142*, 8122–8129.
56. Ye, J.-H.; Miao, M.; Huang, H.; Yan, S.-S.; Yin, Z.-B.; Zhou, W.-J.; Yu, D.-G. *Angew. Chem. Int. Ed.* **2017**, *56*, 15416–15420.
57. Mello, R.; Arango-Daza, J. C.; Varea, T.; González-Núñez, M. E. *J. Org. Chem.* **2018**, *83*, 13381–13394.
58. Morgenstern, D. A.; Wittrig, R. E.; Fanwick, P. E.; Kubiak, C. P. *J. Am. Chem. Soc.* **1993**, *115*, 6470–6471.
59. Ju, T.; Zhou, Y.-Q.; Cao, K.-G.; Fu, Q.; Chen, L.; Liao, L.-L.; Yu, D.-G. *Nat. Catal.* **2021**, *4*, 304–311.
60. Hou, J.; Ee, A.; Feng, W.; Xu, J.-H.; Zhao, Y.; Wu, J. *J. Am. Chem. Soc.* **2018**, *140*, 5257–5263.
61. Fan, X.-Y.; Gong, X.; Ma, M.-Y.; Wang, R.; Walsh, P. J. *Nat. Commun.* **2018**, *9*, 4936.
62. Ju, T.; Fu, Q.; Ye, J.-H.; Zhang, Z.; Liao, L.-L.; Yan, S.-S.; Tian, X.-Y.; Luo, S.-P.; Li, J.; Yu, D.-G. *Angew. Chem. Int. Ed.* **2018**, *57*, 13897–13901.
63. Wang, S.; Cheng, B. Y.; Srsen, M.; König, B. *J. Am. Chem. Soc.* **2020**, *142*, 7524–7531.
64. Cao, G.-M.; Hu, X.-L.; Liao, L.-L.; Yan, S.-S.; Song, L.; Chruma, J. J.; Gong, L.; Yu, D.-G. *Nat. Commun.* **2021**, *12*, 3306.
65. Okumura, S.; Uozumi, Y. *Org. Lett.* **2021**, *23*, 7194–7198.
66. Inoue, S.; Takeda, N. *Bull. Chem. Soc. Jpn.* **1977**, *50*, 984–986.
67. Hirai, Y.; Aida, T.; Inoue, S. *J. Am. Chem. Soc.* **1989**, *111*, 3062–3063.
68. Komatsu, M.; Aida, T.; Inoue, S. *J. Am. Chem. Soc.* **1991**, *113*, 8492–8498.
69. Cocolios, P.; Guilard, R.; Bayeul, D.; Lecomte, C. *Inorg. Chem.* **1985**, *24*, 2058–2062.
70. Johnston, R. F.; Cooper, J. C. *Organometallics* **1987**, *6*, 2448–2449.

About the authors

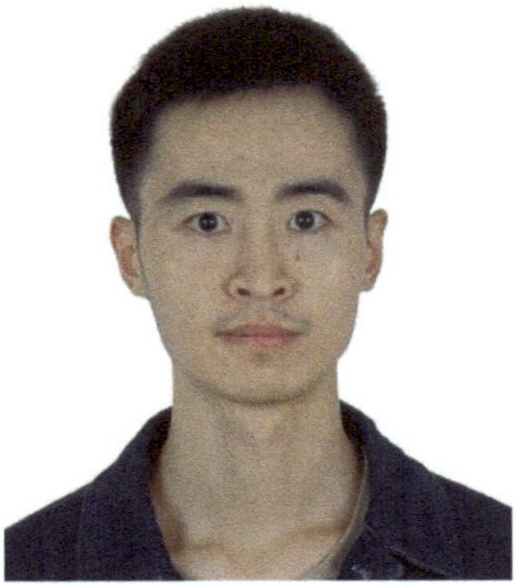

Hong Zhao was born in Anhui Province, China. In 2015, he received his B.S. from Jinan University at Canton. In 2017, he obtained his master degree from Universite de Rennes 1, France, majored in "Catalysis, Molecules and Green Chemistry". Since 2021, he purchases his PhD degree in Sichuan University under supervision of Prof. Dr. Da-Gang Yu. His research focuses on the photocatalytic polymerization with CO_2.

Wei Wang was born in Hubei Province, China. He received his B.S. from Northeast Forestry University in 2015. Since 2019, he has been pursuing doctoral degree in Sichuan University (with Prof. Dr. Da-Gang Yu). His research focuses on the photocatalytic conversion of CO_2.

Hai-Peng Zhang was born in 2000 at Chengdu, China. He joined Professor Yu's group in 2019 and received his BSc degree in chemistry from Sichuan University in 2022. He is currently a Ph. D. candidate in Shanghai Institute of Organic Chemistry, China Academy of Science.

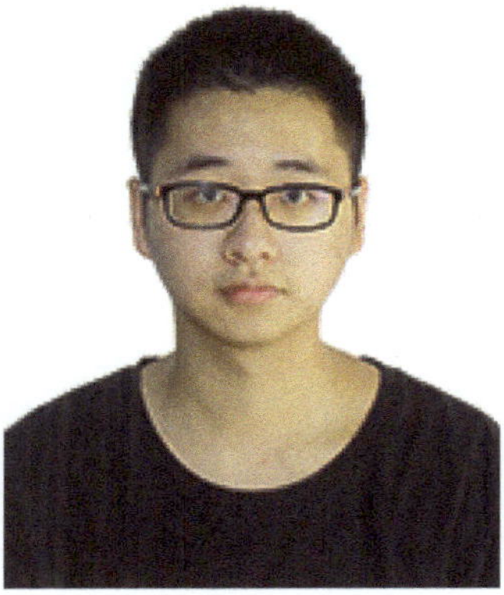

Yi He obtained his B.A. degree in 2022 at Sichuan University (P.R. China) under the supervision of Prof. Da-Gang Yu. Now he is currently a Ph.D. candidate in Tsinghua University. His research interests include NHC catalysis and C-H functionalization.

Dr. Si-Shun Yan was born in Chongqing, China. He received his B.S. in 2016 and Ph. D. in 2021 from Sichuan University under the supervision of Prof. Dr. Da-Gang Yu. During the Ph.D. study, his research interests mainly focus on reductive carboxylation reactions with CO_2. In 2022, he started to carry out postdoctoral research with Humboldt fellowship in the Leibniz Institute for Catalysis (LIKAT Rostock) with Prof. Dr. Matthias Beller. His current research is focused on developing new carbonylation reactions with CO.

Prof. Dr. Da-Gang Yu was born in Jiangxi Province, China. He received his B.S. from Sichuan University in 2007 and Ph.D. from Peking University in 2012 (with Prof. Dr. Zhang-Jie Shi). Then he carried out postdoctoral research with Humboldt fellowship in Muenster University (with Prof. Dr. Frank Glorius). Since 2015, he has been working independently in College of Chemistry, Sichuan University with support from "The Thousand Young Talents Plan" (2015) and National Natural Science Foundation of China–Outstanding Young

Scholars (2018). Since 2015, he has been committed to the CO2 chemistry by developing novel strategies for CO2 activation and new types of organic reaction with CO2. Recent accomplishments include Thieme Chemistry Journal Award (2017), Asian Core Program Lectureship Award (to Japan, 2017), Chinese Chemical Society Youth Award (2018), *Science China Chemistry* Emerging Investigator (2020), *Chemical Communications* Emerging Investigator (2020) and Chinese Catalytic Rookie Award (2021).

CHAPTER FIVE

Heterogeneous catalysts for the conversion of CO_2 into cyclic and polymeric carbonates

Francesca Milocco, Giulia Chiarioni, and Paolo P. Pescarmona*
Chemical Engineering group, Engineering and Technology institute Groningen (ENTEG), Faculty of Science and Engineering, University of Groningen, Groningen, The Netherlands
*Corresponding author: e-mail address: p.p.pescarmona@rug.nl

Contents

Abstract

In this chapter, a didactic and critical overview of the main classes of heterogeneous catalysts for the conversion of CO_2 into cyclic carbonates or polycarbonates is provided. The type of active sites, the catalytic mechanism and the state-of-the-art strategies to design active, selective and stable heterogeneous catalysts for these reactions are presented and discussed.

1. Introduction

The reaction of carbon dioxide with epoxides to generate cyclic or polymeric carbonates (Scheme 1) is an efficient path to convert a greenhouse gas as CO_2 into useful chemical products. The reaction has 100% atom efficiency, implying that theoretically no waste is generated, and has additional assets in the context of green chemistry as it utilizes a non-toxic, renewable and widely available feedstock as CO_2, and in general does not require a solvent. On the other hand, the most widely employed epoxides used for the reaction with CO_2, propylene oxide and ethylene oxide, are toxic, volatile and currently obtained from non-renewable petroleum feedstock. Current

Advances in Catalysis, Volume 70
ISSN 0360-0564
https://doi.org/10.1016/bs.acat.2022.07.001

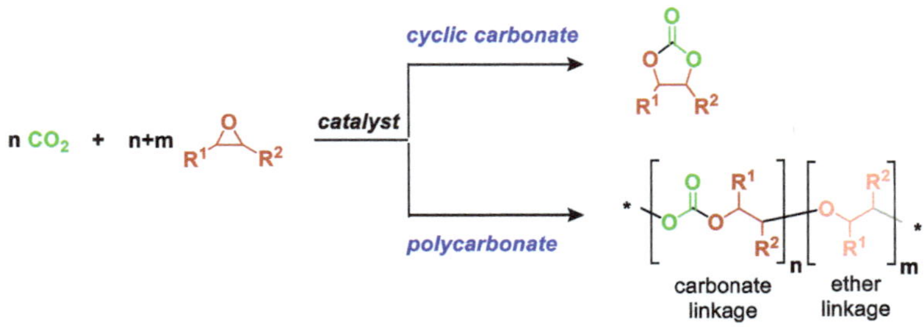

Scheme 1 The reaction of carbon dioxide with an epoxide producing either a cyclic carbonate or a polycarbonate.

research efforts are dedicated in identifying routes to produce these epoxides from bio-based resources and utilize other renewable, bio-based and less volatile epoxides as feedstock for the reaction with CO_2 *(1)*.

The reaction of CO_2 with epoxides is thermodynamically favorable as a consequence of the high free energy of epoxides, which counterbalances the high thermodynamic stability of CO_2. Among the two possible products, cyclic carbonates are thermodynamically favored but the formation of polycarbonates typically involves a lower activation energy. This means that the formation of cyclic carbonates is promoted by higher reaction temperatures, whereas carrying out the reaction at lower temperatures can increase the selectivity toward the polycarbonates by kinetic control *(2)*. Both classes of products have several potential applications, some of which are still under development while others already reached industrial implementation *(2)*. Cyclic carbonates find applications as green, polar aprotic solvents and plasticizers, but also as sustainable alternative to toxic reactants to prepare industrially-relevant products such as bisphenol-A-based polycarbonates and non-isocyanate poly(hydroxy)urethanes *(1–4)*. The polycarbonates prepared from CO_2 and epoxides have very different properties, and thus different applications, compared to the bisphenol-A-based polycarbonates used as engineering plastics. For example, poly(propylene carbonate) containing ether bonds has suitably low and tuneable glass transition temperature (T_g) to be used industrially as polyol building block for the production of polyurethanes *(3)*. Other potential applications of aliphatic polycarbonates are as packaging materials, polymeric coatings, polymeric electrolytes and biocompatible medical implants *(3,5–7)*.

In order to proceed at relevant rates, the reaction of carbon dioxide with epoxides requires the use of a catalyst. Additionally, the features of the catalytic system (together with the nature of the epoxide and the reaction temperature and pressure) can play a crucial role in determining the selectivity between the cyclic and the polymeric carbonate product. The most common and active catalysts for promoting the reaction of CO_2 with epoxides contain sites with Lewis acid character that coordinate the O atom of the epoxide (Scheme 2, species *1*), thus activating it toward a nucleophilic attack by a Lewis base, which results in the opening of the epoxide ring and formation of an alkoxide intermediate (Scheme 2, species *2*). This initial step is followed by the insertion of CO_2 in the alkoxide with formation of a carbonate intermediate (species *3*). This species can undergo intramolecular ring closure to yield the cyclic carbonate product (Scheme 2, *4a* and *5a*). Alternatively, the carbonate intermediate can act as a nucleophile and attack a new epoxide molecule, followed by another CO_2 insertion step, and so on until a polycarbonate is generated (Scheme 2, *4b*, *5b* and *6b*). The consecutive insertion of two epoxide molecules in the growing polymeric chain

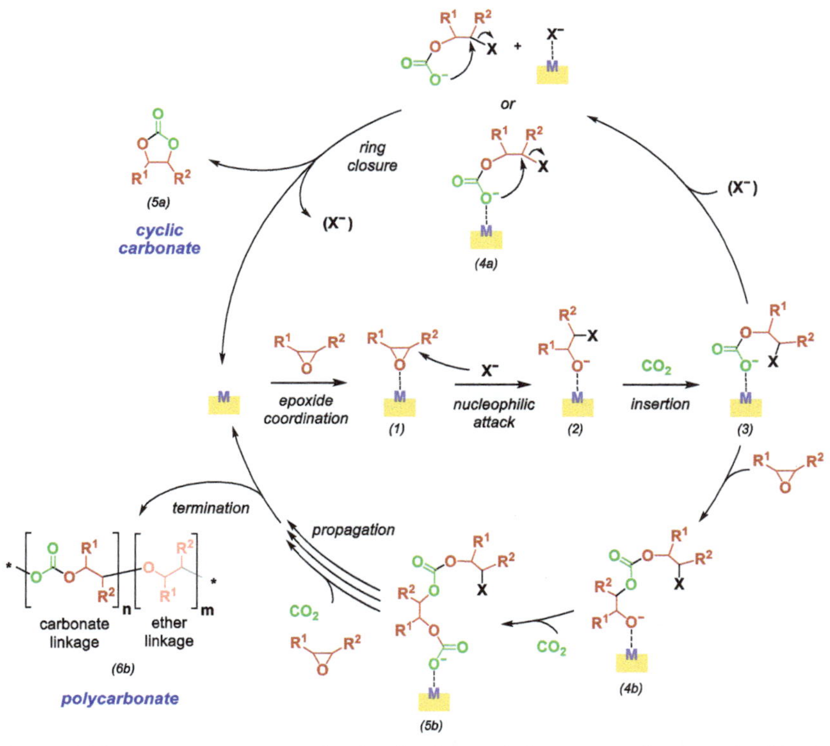

Scheme 2 The most common mechanism for the catalytic reaction of carbon dioxide with epoxides. X^- is a Lewis base acting as a nucleophile; M is a Lewis acid site (e.g., a metal ion or a hydrogen bond donor). In case the Lewis acid site is not a metal ion but a hydrogen bond donor, only the path leading to the cyclic carbonate product is followed.

can also occur, and this would lead to a polycarbonate containing ether linkages (Scheme 1). If the catalytic system does not contain a Lewis acid species, the same mechanism though without a species activating the epoxide can be followed, but higher temperatures are required (typically $\geq$100 °C) and the cyclic carbonate is the only product *(3)*. Though the catalytic mechanism presented in Scheme 2 is the one most commonly proposed for the reaction of CO_2 with epoxides, specific catalysts (e.g., zinc dicarboxylates or double metal cyanides, see Sections 2.2.1 and 2.2.2) have been proposed to follow related but yet different mechanisms *(8)*.

The catalytic sites with Lewis base character acting as nucleophiles are generally provided by halides (Cl^-, Br^-, I^-), by alkoxide (RO^-) or carboxylate ($RCOO^-$) groups, or by organic bases (e.g., 4-dimethylaminopyridine, DMAP; 1,5,7-triazabicyclodec-5-ene, TBD), while the Lewis acid sites, if present, can be provided by metal centers or by metal-free hydrogen bond donor species (e.g., -OH groups) *(9)*. When both types of active species are present in the catalytic system, they can be part of a bifunctional single-component catalyst or be in the form of a binary system consisting of two separate components. Complete selectivity toward the cyclic carbonate product is generally achieved with metal-free catalysts, whereas for metal-containing catalysts the selectivity between cyclic and polymeric carbonate typically depends on the nature of the epoxide, on the nature of the nucleophilic species and of the metal, on the nucleophile-to-metal ratio and on the reaction conditions (CO_2 pressure, temperature), all of which can be tuned to maximize the formation of the desired product *(2)*.

2. Heterogeneous catalysts

Although so far homogeneous catalysts for the reaction of CO_2 with epoxides have been more extensively studied and developed compared to their heterogeneous counterparts, the latter are a growing area of research. Particularly, with the increasing interest toward a broad industrialization of this route for the conversion of carbon dioxide, the advantages of heterogeneous catalysts in terms of ease of separation from the products and reutilization make them more attractive compared to homogeneous systems (despite the higher activity that the latter general display). This is especially relevant when the target product is a cyclic carbonate, which in many cases is a liquid product at room temperature (e.g., propylene carbonate, glycerol carbonate) or can be easily brought to the liquid phase by heating it above the melting point (typically $<$100 °C). In such cases, the solid nature of the heterogeneous catalyst enables a much easier separation from the reaction products compared to homogeneous catalysts. On the other hand, when the target product is a polycarbonate, the highly viscous or solid nature of these products implies that the separation is more cumbersome and typically requires dissolution and precipitation steps with the use of solvents also in the case of heterogeneous catalysts and often some residual catalyst remains trapped in the polymeric matrix *(5,10)*. Yet, the state-of-the-art heterogeneous catalysts for the copolymerization of CO_2 with epoxides still have relevant assets as they display good air and moisture stability (which is not the case for several metal complex homogeneous catalysts) and they can be prepared through relatively straightforward and inexpensive synthesis methods (see Sections 2.2.1 and 2.2.2).

The following sections will present and discuss the main classes of heterogeneous catalysts for the reaction of CO_2 with epoxides to produce either cyclic carbonates or polycarbonates.

2.1 Heterogeneous catalysts for the synthesis of cyclic carbonates

Most of the heterogeneous catalysts that have been studied for the synthesis of cyclic carbonate through the cycloaddition of carbon dioxide to epoxides have been proposed to follow the mechanism presented in Scheme 2. All these catalysts thus contain a Lewis base acting as nucleophile, which is the essential active species for catalyzing the reaction. The most advanced designs of heterogeneous catalysts for this reaction also include a Lewis acid species, either in the form of a metal or as a hydrogen bond donor, with the role of activating the epoxide toward the nucleophilic attack. The catalyst should be designed to guarantee proximity between these two types of active sites, in order to enable an efficient cooperation between them. Additionally, the active sites should be accessible, which is promoted by a large specific surface area in which pores, if present, are large enough to prevent diffusion limitations of reactants and/or products *(8)*. Finally, the catalyst should be stable under the reaction conditions and its separation from the reaction mixture should be viable, in order to enable its reusability.

The selectivity of these heterogeneous catalysts toward the formation of cyclic carbonate stems from one or more of the following features *(2,3,8)*:

- A relatively high reaction temperature ($\geq$100 °C), which promotes the formation of the thermodynamic product of the reaction of CO_2 with epoxides, i.e., the cyclic carbonate).

- The absence of a Lewis acid metal center that is able to coordinate the growing polycarbonate chain. This means that in all metal-free catalysts the carbonate intermediate tends to undergo ring closure with formation of the cyclic carbonate product (*3*→*5a* in Scheme 2).
- The nature of the nucleophile. A nucleophile that is a better leaving group (such as I^-) favors the ring closure (*3*→*5a* in Scheme 2), thus leading to higher selectivity toward the cyclic carbonate, whereas a worse leaving group (such as Cl^-) tends to remain bound to the intermediate, thus promoting the growth of the polycarbonate chain (*3*→*6b* in Scheme 2).
- For metal-based catalysts, a nucleophile-to-Lewis acid ratio > 1. An excess of Lewis base as nucleophile relative to the Lewis acid metal center can cause the displacement of the carbonate intermediate from the Lewis acid site, thus promoting the ring closure step and the consequent formation of the cyclic carbonate (*3*→*5a* in Scheme 2).

In many of the metal-free heterogeneous catalysts presented in this section, the nucleophilic species is a halide. It has been observed that the halide that gives the highest activity can vary between different catalysts. This can be understood considering that the activity depends on a balance between the nucleophilicity of the halide (relevant for step *1*→*2* in Scheme 2) and its leaving ability (relevant for step *4a*→*5a* in Scheme 2). In an aprotic medium as the one in which the reaction of CO_2 with epoxides is generally carried out (a solvent is typically not required and most epoxides are aprotic), the order of nucleophilicity is $Cl^- > Br^- > I^-$, whereas the leaving ability follows the opposite trend $I^- > Br^- > Cl^-$. Whether the leaving ability or the nucleophilicity weighs more in determining the activity depends also on whether a Lewis acid site that activates the epoxide toward the nucleophilic attack is present in the catalyst and on the strength of its interaction with epoxide and reaction intermediates. In practice, the highest catalytic activity in the cycloaddition of CO_2 to epoxides is generally observed with either iodide or bromide.

The different classes of materials that have been investigated as heterogeneous catalysts for the cycloaddition of carbon dioxide to epoxides to produce cyclic carbonates are presented below. For each type of catalyst design, the nature of the active sites and the advantages and limitations are discussed in the context of the desired features summarized above. It is worth mentioning that several catalytic systems consisting of a heterogenous material providing the Lewis acid active sites and a homogeneous organic halide (or another nucleophile source) have been reported to be active for the cycloaddition of CO_2 to epoxides. However, such systems are not fully heterogeneous and this complicates their separation from the reaction mixture and thus their large-scale applicability. Therefore, this class of catalysts is not presented in this chapter.

2.1.1 Heterogeneous catalysts based on immobilized active species

The first generation of heterogeneous catalysts for the reaction of CO_2 with epoxides to produce cyclic carbonates was designed and developed by immobilizing homogeneous catalytic species with Lewis base character on a solid support *(3,11)*. The types of active species that have been immobilized in this class of heterogenous catalysts range from organic halides (such as quaternary ammonium salts and phosphonium salts) *(12–14)*, to ionic liquids (typically imidazolium halides) *(15–19)* and non-ionic organic bases (such as 1,5,7-triazabicyclodec-5-ene, TBD) *(20,22,21)*. The most common supports that have been used in this application are polymers or silica-based materials, including ordered mesoporous silicas with high specific surface area (e.g., MCM-41 and SBA-15) *(17,22–24)*. The immobilization is typically realized through grafting to the support. In order to prevent the leaching of active species during the cycloaddition reaction, it is crucial that such grafting occurs through the formation of robust, covalent bonds *(8)*. An alternative approach to prepare polymers functionalized with catalytically active groups for the cycloaddition reaction is through the polymerization of monomers that contain the desired groups (or a precursor thereof). The two approaches can also be combined, as in the case of multilayered supported ionic liquids, which can be prepared by using bis-vinylimidazolium salts as precursors. The terminal vinyl groups of these compounds react with each other in the presence of a radical initiator to form a crosslinked oligomeric network of imidazolium units, but also react with the thiol-functionalized support (silica or polystyrene), leading to immobilization of the multilayered ionic liquid network by grafting (Fig. 1A) *(17,18)*. This led to catalysts with a high loading of imidazolium units per unit mass, which in turn allowed to reach improved productivity (as $g_{carbonate}\ g_{catalyst}^{-1}$). A similar strategy has been applied with less common supports as single-walled carbon nanotubes and carbon nanohorns *(25,26)*. Another approach to generate heterogeneous catalysts with high local concentration spots of imidazolium halide active sites consists in grafting polyhedral oligomeric silsesequioxanes with vinyl terminal groups on a silica support, followed by functionalization of the available vinyl groups with imidazolium halide groups *(27)*.

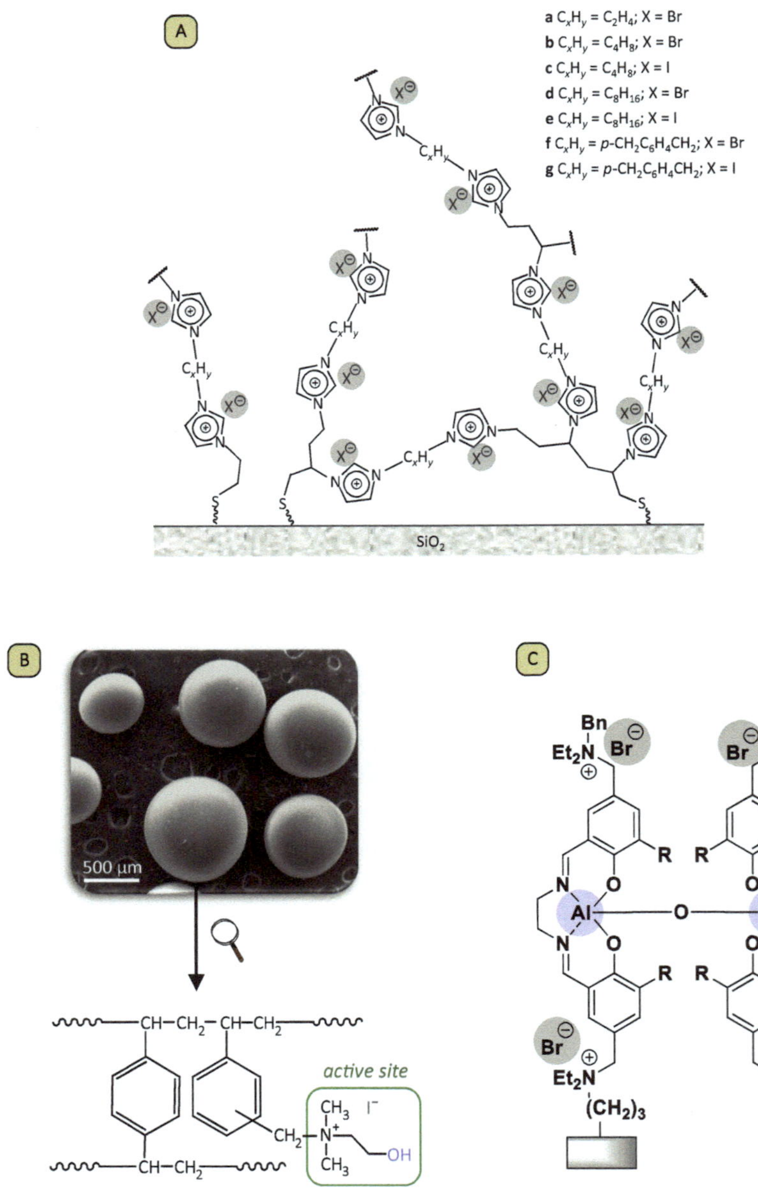

Fig. 1 Selected examples of heterogeneous catalysts based on immobilized active species: (A) A metal-free monofunctional catalyst; (B) A metal-free bifunctional catalyst; (C) A metal-containing bifunctional catalyst. *Images adapted from Agrigento, P.; Al-Amsyar, S. M.; Sorée, B.; Taherimehr, M.; Gruttadauria, M.; Aprile, C.; Pescarmona, P. P.,* Catal. Sci. Technol. ***2014**,* 4 *(6), 1598–1607; Alassmy, Y. A.; Asgar Pour, Z.; Pescarmona, P. P.,* ACS Sustainable Chem. Eng. ***2020**,* 8 *(21), 7993-8003; North, M.; Villuendas, P.; Young, C.,* Chem. Eur. J. ***2009**,* 15 *(43), 11454-11457.*

Within the class of heterogeneous catalysts based on immobilized Lewis bases, those in which the active species are organic halides or ionic liquids follow the mechanism in Scheme 2. On the other hand, if the active species is an immobilized non-ionic organic base, the proposed catalytic mechanism proceeds through the initial activation of CO_2 *via* a nucleophilic attack by the non-ionic base *(28,29)*. The generated CO_2-adduct acts as a nucleophile and attacks the epoxide, leading to the formation of a carbonate intermediate, which then undergoes ring closure to yield the cyclic carbonate product.

Since the heterogeneous catalysts discussed so far rely only on Lewis bases acting as nucleophiles as catalytic species without the promoting effect of Lewis acid sites, they generally require relatively high temperatures for the reaction of CO_2 with epoxides ($\geq$100 °C) in order to reach high products yields *(3)*.

The next generation of heterogeneous catalysts based on the immobilization of active species on a solid material was designed to boost the activity by combining a Lewis base acting as a nucleophile with a hydrogen bond donor species (such as a -OH or a -COOH group) operating as a Lewis acid to activate the epoxide toward the nucleophilic attack (Scheme 2) *(3,11)*. One of the first reports of this type of heterogenous catalysts consisted of silica-supported phosphonium salts, in which the surface silanol groups in the vicinity of the immobilized salt have been proposed to act as the hydrogen bond donors *(30)*. Another, more recent example of the exploitation of hydrogen bond donors present on the support is represented by bio-based compounds such as cellulose and chitosan, which can be functionalized with quaternary ammonium halides or imidazolium halides *(31–34)*. The hydroxyl groups that are present along the chain of these polysaccharides can act as hydrogen bond donors and enhance the activity of the halides in the cycloaddition of CO_2 to epoxides.

In order to increase the cooperation between the Lewis base and the Lewis acid sites, the proximity between the two species should be maximized. In this context, supported ionic liquids or quaternary ammonium halides have been developed in which the organic cationic part of the active moiety has been functionalized with a hydrogen bond donor species such as hydroxyl and carboxyl groups *(35–43)*. Also for these systems, the most common supports are polymers and silica. Similar types of bifunctional active sites can also be encountered in commercial polymers that are available in the format of macroscopic beads (typically 300–800 μm), as in the case of resins consisting of polystyrene cross-linked with divinylbenzene with dimethylethanolammonium chloride groups (Fig. 1B) *(44)*. The type of halide acting as nucleophile can be tuned through a straightforward ion-exchange step, allowing to optimize the catalytic activity. The optimum catalyst combined I^- as nucleophilic species and -OH groups as hydrogen-bond donors and displayed good activity at mild temperatures (45–80 °C) with a broad scope of epoxides. The bead format of several of these catalysts *(42,44)* is an asset for a potential large-scale application as it allows straightforward separation from the reaction mixtures.

The monofunctional (Lewis base only) or bifunctional (Lewis base and Lewis acid sites) catalysts described above are all metal-free materials. This is an advantage, as metal-free catalysts are generally non-toxic, low-cost and more sustainable compared to their metal-based counterparts *(9,45,46)*. Yet, a few metal-containing bifunctional heterogenous catalysts have been reported. This research was stimulated by the very high activity displayed by metal complexes acting as homogeneous catalysts for the reaction of CO_2 with epoxides *(3)*. Two main approaches have been investigated for these metal-based catalysts. The first consists in immobilizing ionic liquids or other active species on silicates containing Lewis acid metal centers in their framework (e.g., Sn, Ti, Zn) *(47–50)*. However, the presence of the metal centers led to only a minor enhancement of the catalytic activity compared to the counterparts using metal-free silica as support, possibly because of the relatively large distance between the metal centers and the nucleophile source. The second approach consists in the immobilization of metal complexes that were previously found to be promising homogeneous catalysts, such as the bifunctional Al-salen complexes with quaternary ammonium bromides covalently attached to the salen ligand, which can be supported on polystyrene or silica (Fig. 1C) *(51,52)*. However, the higher synthetic complexity and cost of these catalysts limited their applicability so far.

2.1.2 Metal organic frameworks

A class of materials that has been increasingly investigated as heterogeneous catalysts for the cycloaddition of CO_2 to epoxides is represented by metal-organic frameworks (MOFs) *(53–56)*. These materials consist of metal nodes interconnected through organic linkers in such a way that a porous, crystalline structure is generated. The metal nodes can be in the form of single metal ions, but more often are small clusters of metal ions interconnected through bridging oxygen atoms. The organic linkers are characterized by a rigid part, typically provided by an aromatic ring, and by two or more coordinating sites, for example in the form of carboxylic groups. The rigidity and the presence of multiple coordinating sites are essential features to yield a crystalline material. By tuning the nature of the metal species and of the linkers it has been possible to prepare a large variety of MOF structures, with specific pore sizes (typically in the micropore range, i.e., $d_{pore} < 2$ nm) and extremely high-specific surface area (>1000 m^2/g). The first reports of MOFs as catalysts for the cycloaddition of CO_2 to epoxides exploited the high CO_2 adsorption capacity stemming from the high surface area of these materials, combined with the coordinatively unsaturated sites with Lewis acid character that are often present in the metal nodes. However, in these catalytic systems the nucleophilic species that is essential to catalyze the reaction was generally provided by a homogeneous organic salt (e.g., a tetrabutylammonium halide, TBAX, with X=Cl, Br or I), which implied that these systems cannot be considered fully heterogeneous. This is a major drawback in terms of catalyst separation and reutilization *(57)*. Different approaches have been explored in order to overcome this limitation: (i) utilizing specific MOFs characterized by linkers with Lewis basic moieties (e.g., ZIF-8, ZIF-67, ZIF-68, ZIF-90 and Co-MOF-74) *(54,57–62)*; or (ii) tailored modification of the organic linkers to introduce functional groups containing a nucleophile (e.g., ammonium and phosphonium groups, pyridinium iodide, quaternized hydrazine) *(54,57,63–66)*. However, the latter approach leads to a decrease in the available pore volume and surface area within the MOF structure, which can cause diffusion limitations for reactants and/or products *(3)*. The organic linkers of MOFs can also be designed and synthesized so that they can act as ligands to form complexes with single metal sites *(67)*, thus providing an alternative strategy to introduce metal centers acting as Lewis acid sites into the structure of MOFs.

An original catalyst design that was reported recently involves the incorporation of a polymer with immobilized ionic liquid groups (see also Section 2.1.1.) into the pores of a MOF (MIL-101), yielding a heterogeneous catalyst in which the active sites provided by the ionic liquid moieties (Br^- as nucleophile) cooperate with those provided by the MOF (Cr centers with Lewis acid character), see Fig. 2A *(68,69)*. Notably, the obtained composite material displayed higher catalytic activity compared to that of a physical mixture of the polymeric ionic liquid and of the MOF. This enhanced activity was attributed to the increased accessibility of the active sites of the ionic liquid moieties that is brought about by the high specific surface area of the composite material (2462 m^2/g) *(68)*, though the proximity between the two types of active sites might also play a role.

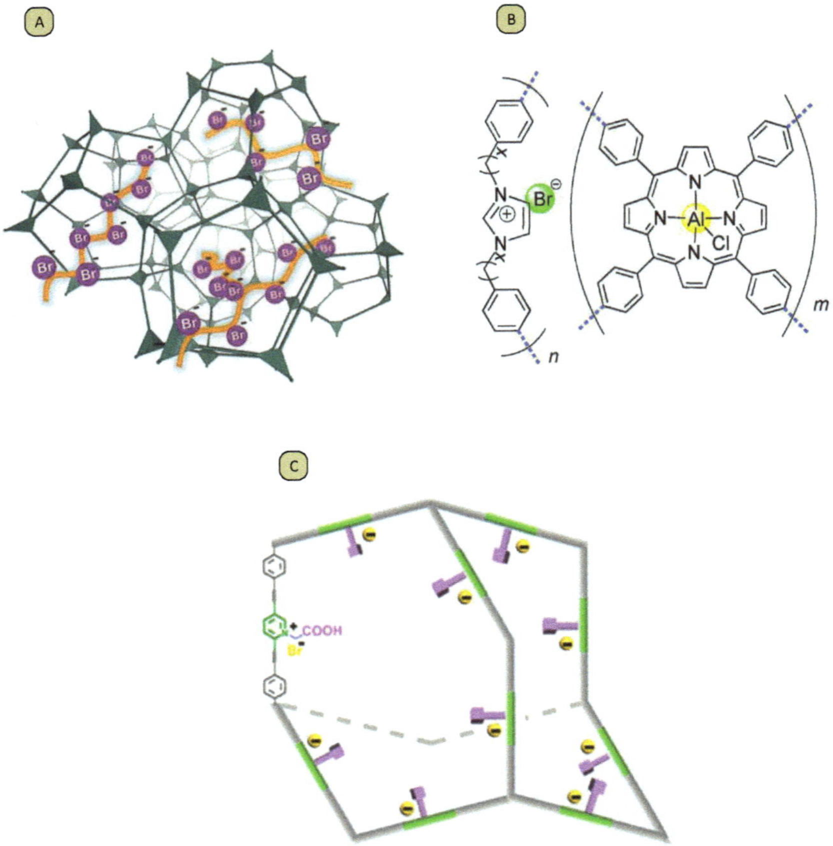

Fig. 2 Selected examples of porous heterogeneous catalysts based: (A) Composite of MOF MIL-101 and of polymer with ionic liquid groups; (B) Metal-POP based on immobilized Al-porphyrin and ionic liquid units; (C) Metal-free POP with active sites containing bromide anions as nucleophiles and carboxyl groups as hydrogen bond donors. *Images adapted from Ding, M.; Jiang, H. L.,* ACS Catal. ***2018****, 8 (4), 3194-3201; Chen, Y.; Luo, R.; Xu, Q.; Jiang, J.; Zhou, X.; Ji, H.,* ChemSusChem ***2017****, 10 (11), 2534-2541; Ma, D.; Liu, K.; Li, J.;Shi, Z.,* ACS Sustainable Chem. Eng. ***2018****, 6 (11), 15050-15055.*

Although promising results have been achieved with MOFs as heterogeneous catalysts for the cycloaddition of CO_2 to epoxides, many of these materials still suffer from one or more of the following limitations: (i) low stability under reaction conditions, leading to deactivation of the catalyst and thus limiting their reusability *(70)*; (ii) need for high reaction temperature or high catalyst loading in order to achieve good cyclic carbonate yields *(59-61,71)*; and (iii) diffusion limitations of reactants and/or products *(70)*.

Additionally, the synthesis of the most advanced, bifunctional MOFs described above typically requires complex, multiple steps, thereby making them too expensive for upscaling and industrial application.

2.1.3 Porous organic polymers

Metal-functionalized porous organic polymers (Me-POPs) are a class of materials that is related to MOFs (see Section 2.1.2) as they also consist of metal centers interconnected through organic moieties in such a way that a porous network is generated, but with the important difference that MOFs are crystalline materials with well-defined pore size whereas Me-POPs are typically amorphous and display disordered porosity *(3,54,56,72,73)*. Although the lack of ordered pores with uniform size might be considered as a drawback, Me-POPs have the advantage of a larger degree of synthetic flexibility. This important feature enables tuning the physicochemical properties that define their catalytic behavior, such as the type and amount of active sites, the specific surface area, the pore size range and volume *(74–76)*. Particularly, the possibility of tuning the porosity can prevent diffusion limitations of reactants and products *(74)*, which on the other hand is a potential issue with functionalized MOFs (see Section 2.1.2). As in the case of MOFs, a major challenge for the application of Me-POPs as heterogeneous catalysts has been the inclusion of both Lewis acid sites for activating the epoxide and Lewis bases acting as nucleophiles in proximity to each other within the material. Two main strategies have been explored to incorporate metal centers with Lewis acid properties in the structures of Me-POPs: (i) the synthesis of a porous organic polymer containing sites to which metal ions can be grafted afterwards; and (ii) the crosslinking of pre-synthesized metal complexes *(74–77)*. A notable example of the latter approach is the synthesis of bifunctional Me-POPs containing both immobilized metal complexes (e.g., Mg- or Al-porphyrins) *(75,76,78)* providing the Lewis acid sites and immobilized organic halides providing the nucleophilic species (Fig. 2B). Both the Mg-POP and the Al-POP catalyst displayed remarkably high activity under mild reaction conditions and with low catalyst loading.

As already highlighted in the previous sections, the presence of metal centers is not strictly necessary to achieve high catalytic activity. In this context, metal-free POPs containing halides as nucleophilic species and hydrogen bond donors to activate the epoxide can be prepared by carefully choosing and tuning the building blocks *(79)*. A notable example of this approach is the POP synthesized by a coupling reaction between two aromatic compounds as tetrakis(4-ethynylphenyl)methane and 2,5-dibromopyridine, followed by the functionalization of the obtained high surface area scaffold by reaction with bromoacetic acid. The obtained POP displayed a high loading of accessible sites (Br^- and -COOH) in close proximity to each other (Fig. 2C) *(79)*. This combination enabled this catalyst to achieve a good activity under mild reaction conditions (although still inferior compared to the best performing metal-containing POPs).

Covalent organic frameworks (COFs) are class of porous compounds that are related to POPs but differ from them for being crystalline. These materials can be considered as the metal-free counterpart of MOFs. The choice of the building blocks used to prepare COFs enables tailoring the size and shape of their pores as well as the nature of their active species. This is exemplified by covalent triazine frameworks, which can be prepared as halide-free materials from 1,4-dicyanobenzene *(80)* or by using a viologen-based building block that led to a material containing chlorides and characterized by larger pores *(81)*. Both materials are moderately active as heterogeneous catalysts for the cycloaddition of CO_2 to epoxides, with the halide-free system exploiting the basic sites of the triazine units as active sites and requiring higher temperature to achieve good product yields.

2.1.4 Carbon nitride and N-doped carbon materials

Graphitic carbon nitride (g-C_3N_4) and N-doped carbon are two classes of materials containing basic N sites that can act as catalytic sites for the cycloaddition of CO_2 with epoxides in a similar way as the non-ionic organic bases discussed in Section 2.1.1. The attractive feature of these catalysts is that they are metal- and halogen-free, with the latter feature having been proposed to be beneficial to prevent the risk of corrosion-related issues caused by acids that might form through the hydrolysis of organic halides *(82)*. Several heterogeneous catalysts belonging to this class have been reported *(83–85)*, including efforts to enhance the surface area of the active phase by preparing mesoporous g-C_3N_4 *(86)* or by supporting nanosized g-C_3N_4 on SBA-15 mesoporous silica *(87)*. However, the activity of these heterogeneous catalysts is only moderate and relatively high reaction temperature ($\geq$120 °C) is required to achieve good cyclic carbonate yields. Additionally, these catalysts were generally tested with a rather reactive substrate as epichlorohydrin *(1)* and displayed significantly lower activity when tested with other epoxides. It has been suggested that epichlorohydrin can react with the N-sites present in these materials *(88)*, generating a chloride-based catalyst that would operate similarly to the organic halide active sites discussed in Section 2.1.1, and which would thus not anymore classify as halogen-free catalyst.

2.2 Heterogeneous catalysts for the synthesis of polycarbonates

The copolymerization between CO_2 and epoxides to produce polycarbonates has been known since 1969 when Inoue and coworkers reported that a heterogeneous catalytic system deriving from a mixture of diethyl zinc and water was active in catalyzing the alternating copolymerization of carbon dioxide and propylene oxide yielding poly(propylene carbonate) *(89)*. Even if the majority of heterogeneous catalysts reported for the reaction of CO_2 with epoxides are selective toward cyclic carbonates (see Section 2.1), a few classes of industrially relevant metal-based heterogeneous systems capable of producing polycarbonates have been developed, namely, zinc dicarboxylates and double metal cyanides *(3,90)*.

The features of an ideal catalyst for the copolymerization of CO_2 with epoxides are *(91)*:

- high activity with a broad scope of epoxides, preferably at mild temperature and CO_2 pressure;
- high selectivity, yielding a copolymer with the minimum amount of cyclic carbonate by-product and the desired fraction of ether linkages (which can be different depending on target application, as ether linkages cause a decrease in the T_g of the polymer);
- control over molecular weight and microstructure;
- catalyst residues, if present, should not affect the polymer in terms of color and transparency;
- non-toxic.

Zinc dicarboxylates and double metal cyanides display several of these features, although the conditions in which they operate are not as mild as those that can be achieved with state-of-art homogeneous catalysts, the epoxide scope is more limited (particularly for zinc dicarboxylates), the presence of different types of active sites (*vide infra*) can lead to broad polydispersity and the formation of a fraction of ether linkages is hard to avoid *(5,10)*. On the other hand, zinc dicarboxylates and double metal cyanides are cheaper, easier to synthesize and more robust compared to most homogeneous catalysts for the copolymerization of CO_2 with epoxides, and they have low or no toxicity and do not cause coloring of the polymer *(92)*. These features make them a more industrially viable option than homogeneous systems.

2.2.1 Zinc dicarboxylates

Zinc dicarboxylates are industrially employed as heterogeneous catalysts for the synthesis of polycarbonates and the majority of commercial poly(propylene carbonate) is currently produced with the use of zinc glutarate *(93,94)*. Depending on the catalyst selectivity, i.e., the carbonate linkage percentage, up to 43% of the mass of poly(propylene carbonate) comes from CO_2 *(90)*.

The first study of zinc dicarboxylates as catalysts for the copolymerization of CO_2 and epoxides was reported by Inoue and coworkers in 1973 *(95)*, when they added the dialkylzinc-dicarboxylic acid systems to the series of previously reported organozinc compounds derived from mixing dialkylzinc with a compound with two active hydrogens such as water, a primary amine or a dihydric phenol *(89,96,97)*. However, in this report the zinc dicarboxylate was not isolated and only in 1981 Hattori and coworkers described the synthesis of zinc dicarboxylates from zinc hydroxide and aliphatic dicarboxylic acids and their application as heterogeneous catalysts for the copolymerization of CO_2 and epoxides *(98)*.

2.2.1.1 Synthesis

Among the zinc dicarboxylates, zinc glutarate is certainly the most studied and over the years several synthetic procedures have been investigated for its production, namely, using various zinc sources (e.g., ZnO, $Zn(OH)_2$, $Zn(NO_3)_2 \cdot 6H_2O$, $Zn(Et)_2 \cdot 6H_2O$, $Zn(OAc)_2$, $Zn(ClO_4)_2$) and carboxylate sources (e.g., glutaric acid, anhydride, methyl ester, glutaronitrile) *(91,92)*. Notably, even if these different synthetic routes yield products with the same structure according to X-ray powder diffraction analysis, their catalytic activity can vary significantly. The most widely employed route is the condensation reaction of zinc oxide and glutaric acid in toluene, which has been found to yield the zinc glutarate catalyst with the highest activity *(99–101)*.

2.2.1.2 Catalytic mechanism

In the most widely accepted mechanism, the heterogeneous zinc-dicarboxylates catalyze the copolymerization reaction through a bimetallic reaction pathway involving the interaction of two active sites on the surface of the zinc dicarboxylate (Scheme 3) *(102–104)*. This proposed mechanism originated from an analogy with the homogeneous β-diketiminate zinc complexes, which were the first catalysts to be proven to catalyze the reaction *via* a bimetallic mechanism *(105)*.

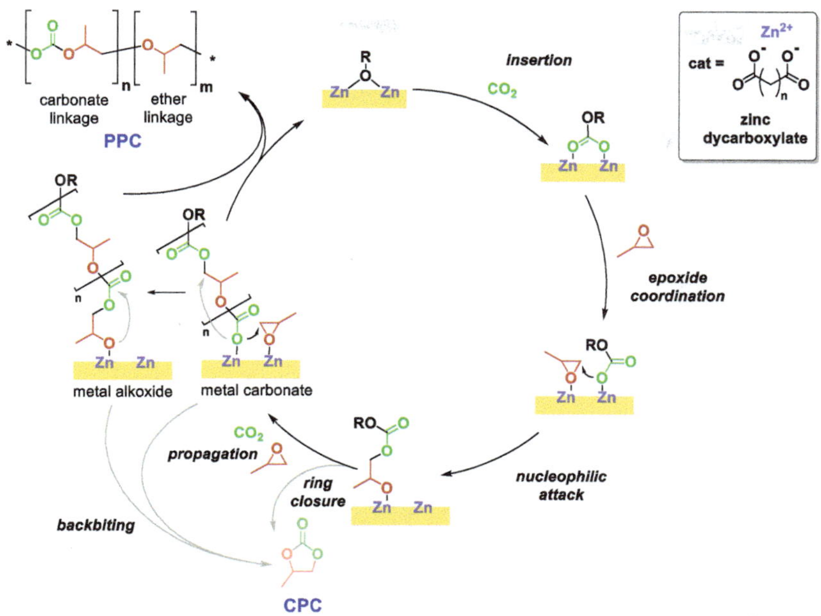

Scheme 3 Proposed bimetallic mechanism for the copolymerization of CO_2 and propylene oxide catalyzed by zinc dicarboxylates *(102–104)*. [OR = initiating group (carboxylate, alkoxide or hydroxide) or growing polymer chain; PPC = poly(propylene carbonate); CPC = cyclic propylene carbonate].

In such mechanism, the copolymerization proceeds *via* two iterative steps *(102–104)*:

(i) *CO_2 insertion* into a zinc-OR bond (where OR can be a carboxylate, an alkoxide or a hydroxide) forming a zinc-carbonate intermediate.

(ii) *Nucleophilic attack* of a precoordinated epoxide by the carbonate group, leading to the ring opening of the epoxide with the formation of a zinc-alkoxide species.

Besides, various side reactions can also take place *(90,92,102–104)*:

(iii) Formation of ether linkages due to nucleophilic attack of the metal alkoxide intermediate on an epoxide molecule, i.e., a *homopolymerization* step (note that the consecutive insertion of two CO_2 molecules is thermodynamically disfavored and has never been reported).

(iv) Formation of cyclic carbonates, which are the thermodynamic products of the reaction of CO_2 with epoxides, either if the ring closure occurs before the polymeric chain growth or *via backbiting* reactions from the metal alkoxide or metal carbonate intermediates.

(v) *Chain transfer* reactions due to the presence of protic reagents in the polymerization medium (e.g., water, alcohols or acids), which can lead to highly controlled polymerizations, albeit yielding polymers with lower molecular weights.

The mechanism is supported by the fact that carboxylate or alkoxide groups have been found at the end of the polymer chain and that carbon dioxide insertion into metal alkoxide bonds is well established *(102)*.

2.2.1.3 Structure, characterization and catalytic performance

Elucidation of the single crystal X-ray structure of zinc glutarate *(106,107)* revealed a 3D-network structure composed of layers of glutarate ligands and Zn atoms which are tetrahedrally coordinated to oxygen atoms (Zn—O bond length of 1.95–1.96 Å) *(99,106)* of four different dicarboxylate units (Fig. 3). The latter can be found in two types of folding: extended or bent (see Fig. 3C), which leads to the formation of two types of alternating metallacycles (A and B, see Fig. 3). This arrangement results in alternating Zn—Zn interatomic distances of 4.64 Å and 4.78 Å *(91,99)*. Although the bidentate carboxylate groups can bind to the Zn centers in various manners, in the crystal structure of zinc glutarate only the *syn-anti* binding mode was found *(107)*. Nevertheless, solid-state NMR and IR studies suggest the

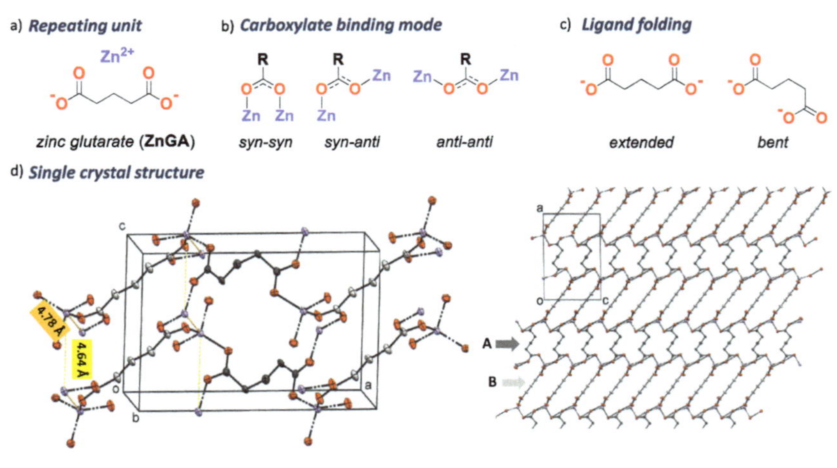

Fig. 3 Zing glutarate: (A) repeating unit; (B) possible binding modes of the carboxylate group; (C) folding of the dicarboxylate ligands; (D) single crystal structure showing 50% probability ellipsoids. CIF downloaded from ref. *(108)* (CCDC-No. 1267/413). The glutarate ligands are shown in light gray for the extended folding and in dark gray for the bended folding.

presence of the *syn-syn* binding mode (see Fig. 3B), even if the samples have a comparable powder X-ray diffractogram to the one calculated from the single crystal structure *(99,107,109)*.

The openings of two types of channels formed by the metallacycles are spatially limited due to the van der Waals radii of the constituent atoms, and this hampers the diffusion of carbon dioxide and propylene oxide (PO, the most common epoxide used with zinc glutarate). Hence, the catalytic activity is restricted to the outer surface of the zinc glutarate particles (it has been estimated that <5% of the total metal sites of zinc glutarate are at the surface and thus able to act as catalytic sites) *(106)*. Therefore, strategies to increase the activity of zinc glutarate focused on increasing the surface area or altering the surface morphology by one or more of the following methods *(91,92)*:

(i) optimization of the stirring method: magnetic, mechanic, ultrasonic *(110,111)*;
(ii) use of various supports *(112)* or amphiphilic templates (which allow obtaining catalysts with high surface area despite the large particle size) *(101)*;
(iii) post-modification: ball-milling *(28,36)* and ultrasonic fragmentation *(36)* (the latter leading to exceptionally high productivities of 3.54–4.01 $g_{PPC}/(g_{cat}\ h)$) *(111)*
(iv) surface modification *via* treatment with an acid (e.g., HCl, Lewis acid metal salt) *(113,114)*.

Nevertheless, and contrary to the typical trend observed in heterogeneous catalysis, the catalytic activity of zinc glutarate was found not to increase necessarily with increasing surface area *(99,100)*, suggesting that other factors such as the crystallinity *(99,107)*, the local structure (e.g., *syn-anti* vs other carboxylate binding modes) and the related distance between two neighboring zinc centers *(99)*, and the type of end-groups *(115,116)* play a more relevant role in determining the catalytic performance. It has been proposed that high crystallinity is a prerequisite for the monomers activation, whereas small particle size and large surface area help achieve high catalytic activity *(111)*. Moreover, high crystallinity has also been suggested to favor the formation of polymers with high molecular weight *(111)*. However, the relationship between the morphological and structural features and the catalytic activity of zinc glutarates is still a matter of debate, despite the extensive effort of several research groups and the deployment of a broad range of characterization techniques (infrared spectroscopy *(106,107,109)*, X-ray diffraction *(100,103,106–108)*, X-ray absorption *(99,117)*, X-ray photoelectron spectroscopy *(101,117)*, solid-state NMR spectroscopy *(99)*, electron microscopy *(103,107)*, surface area analysis *via* Brunauer-Emmet-Teller method *(100,101,103,113,118)*). Hence, the exact structure of the catalytically active species is still not well established due to the difficulties in analyzing the end-groups of zinc dicarboxylate heterogeneous catalysts and various initiating groups have been hypothesized: OH^-, $RCOO^-$, Et^-, $EtOSO^-$, Cl^- (see Scheme 4) *(92,115,116)*.

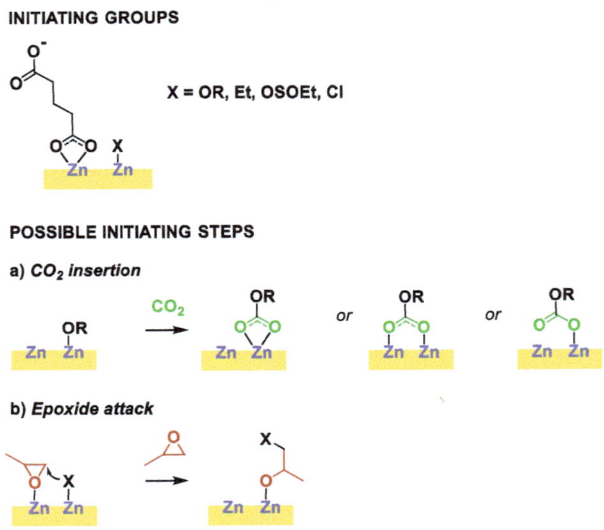

Scheme 4 Possible initiating steps *(92,115–117)* [OR=carboxylate, alkoxide or hydroxide].

The question whether the reaction is initiated by coordination of PO or CO_2 to the catalyst surface has not yet found a conclusive answer. On one hand, insertion of carbon dioxide into a Zn-OR bond has been reported and the reaction of ZnGA with CO_2 can lead to three possible coordination geometries: bidentate, bridging and unidentate. Near edge X-ray absorption fine structure (NEXAFS) studies suggested that CO_2 reversibly binds to the zinc glutarate surface and that the unidentate coordination is less likely to take place (Scheme 4A) *(117)*. On the other hand, it seems more likely that the reaction is initiated by the PO monomer due to the stronger basicity of the oxygen compared to the one of the very stable CO_2 molecule. This is supported by NEXAFS experiments, which revealed that also PO can bind reversibly to the catalyst surface leading to a stronger change in the NEFAXS spectra of ZnGA compared to CO_2 *(117)*. The coordination of the epoxide to the Lewis acid zinc center activates it toward the nucleophilic attack

(Scheme 4B). Other end groups than a carboxylate may also act as initiating groups and this might contribute to explaining the variation in catalytic activity of zinc glutarates prepared *via* different synthetic routes. Zinc hydroxyl groups may be formed as monocondensation product of ZnO and the dicarboxylic acid *(115),* when $Zn(OH)_2$ is used as zinc source or as a consequence of hydrolysis of the surface species, and they have been proposed to be catalytic active sites based on the fact that –OH terminated poly(propylene carbonate)s have been detected *via* MALDI-TOF mass spectrometry *(116)*. However, this could also originate from the presence of protic compounds in the reaction medium or by the tendency of carbonate end groups to convert into terminal hydroxyl group upon liberation of CO_2 *(2)*. Instead, zinc glutarate synthesized from $ZnEt_2$ may contain ethyl end groups, leading to low activity. The performance of these catalysts can be improved by post-activation with SO_2, which results in the formation of ethyl-sulfonate end groups *(92,115)*. Similarly, –Cl end groups may be formed *via* treatment with diluted hydrochloric acid solution *(113)*.

While several metals have been studied in homogenous catalytic systems (e.g., Mg, Al, Cr, Fe, Co, Zn) *(3,5),* only Zn-containing heterogeneous catalyst have been reported for the copolymerization of CO_2 with epoxides. Reports with other metals are limited only to surface modification of Zn-catalyst with Lewis metal salts *(114)*. The reason for the choice of Zn as preferred metal has not been established yet, but a plausible explanation is that Zn(II), having a d^{10} configuration, is redox inactive *(90)* and/or that zinc glutarate has a structure with an ideal metal-metal distance (*vide infra*) *(103)*. On the other hand, various dicarboxylate ligands have been investigated, e.g., succinic acid (SA, C_4), glutaric acid (GA, C_5) adipic acid (AA, C_6) or pimelic acid (PA, C_7) *(103,115,119,120)*, and the obtained zinc dicarboxylates are all active catalysts for the copolymerization of CO_2 and PO, with the only exception of zinc succinate *(92)*. Assuming a bimetallic catalytic pathway (*vide supra*), at least two Zn centers have to be in spatial proximity to ensure catalytic activity. Mechanistic studies suggested that the Zn—Zn distance of 4.6–4.8 Å is optimal *(105)* and, as can be seen in Fig. 3D, this structural feature is mainly due to the geometry with which the Zn centers are coordinated to the carboxylic groups rather than to the length of the aliphatic chain. Nevertheless, in the zinc succinate this distance is found only on one main *hkl* plane, while in the other three dicarboxylates it is present on each main *hkl* plane, thus explaining the poor activity of ZnSA. Among the three higher homologs, ZnGA is certainly the most studied. Nevertheless, ZnAA is economically more interesting because adipic acid is a large-scale industrial product (widely used for the production of nylon 6,6) *(121)* and, therefore, is significantly cheaper than glutaric acid.

The stability to humidity and the shelf-life of zinc dicarboxylates is an item that deserves further studies. Several works report the catalyst synthesis and the catalytic tests under anhydrous atmosphere, as well as the purification of the epoxide by drying and distillation prior to use in catalysis. The deactivation upon exposure to humidity has been reported by some authors *(122),* while other ones reported that a post synthetic treatment involving exposure of the zinc glutarate to a saturated water atmosphere followed by a drying step at 130 °C in a vacuum oven led to an increase in the catalytic activity *(103).* Furthermore, not much attention has been dedicated to the recycling of the catalyst, probably due to the difficulties encountered in separating it from the reaction mixture without deteriorating its structure. The few studies on the reuse of zinc dicarboxylate catalysts indicate a significant drop of activity upon recycling *(91,123),* highlighting the need of developing a suitable recycling procedure. The conditions typically used for the copolymerization reactions with zinc dicarboxylate catalysts are above 60 °C and above 20 bar as initial CO_2 pressure (before heating). The substrate scope for zinc dicarboxylates is mainly limited to propylene oxide and ethylene oxide (in the case of industrial studies). Nevertheless, zinc glutarate has also been demonstrated to be an active catalyst for the copolymerization of CO_2 with epoxides with electron-withdrawing groups, such as epichlorohydrin *(124).*

2.2.2 Double metal cyanides

Double metal cyanides (DMCs) are another important class of heterogeneous catalysts for the copolymerization of CO_2 with epoxides. DMCs are insoluble, sterically hindered inorganic complexes in which two different metals (M' and M") are connected by cyano bridges and are thus analogs of the Prussian blue pigment, $Fe_4^{III}[Fe^{II}(CN)_6]_3 \cdot xH_2O$, where x = 14–16 *(125).* Since 1966, DMCs have been widely used in industry, first as catalysts for the synthesis of polyether polyols and then as molecular sieves (pore size of 0.56 × 0.86 nm) *(126,127).* In 1985, Kruper et al. (Dow Chemical Company) reported for the first time the application of Zn-Fe(III) DMC as catalyst for the synthesis of poly(carbonate ether)s from carbon dioxide (CO_2) and several epoxides (ethylene oxide, propylene oxide, butene oxide and cyclohexene oxide) *(128).* Since then, various metal combinations were tested to optimize the activity and selectivity of DMC catalysts for the synthesis of polycarbonates. In these DMC structures, M' is a divalent metal (e.g., Zn^{2+}, Fe^{2+}, Co^{2+}, Ni^{2+}), while M" is a transition metal in an oxidation state between +2 and +4 (e.g., Ni^{2+}, Pd^{2+}, Pt^{2+}, Fe^{3+}, Co^{3+}, Cr^{3+}, Mn^{3+}, Mo^{4+}) *(129–137)*. Among these catalysts, Zn-Fe(III) and Zn-Co(III) are the most commonly employed ones, with Zn-Co(III) being the most promising one due to its higher activity and selectivity toward the synthesis of poly(carbonate ether)s *(129,138)*. Although with homogeneous systems it is possible to achieve perfectly alternating copolymerization of CO_2 and epoxides, poly(carbonate ether)s are the main products of the reaction when heterogeneous DMC catalysts are employed and the occurrence of ether linkages (achieved by subsequent epoxide insertion) cannot be completely avoided *(3,90)*.

2.2.2.1 Structure and synthesis

The most common DMC catalyst for the copolymerization of CO_2 with epoxides is Zn-Co(III), which can be indicated with the non-stoichiometric general formula: $Zn_3[Co(CN)_6]_2 \cdot xZnCl_2 \cdot yCA \cdot zH_2O$, where x, y and z can be both fractional and integer numbers and CA represents the complexing agent used in the synthesis (see below) *(125)*. The composition and 3D structure of DMCs are highly dependent on the synthetic method, which can lead to catalysts with significantly different activity and crystallinity *(137–160)*. In the cubic crystal structure of Zn-Co DMC (analog to that of Prussian blue), both Zn and Co are in octahedral coordination (Fig. 4).

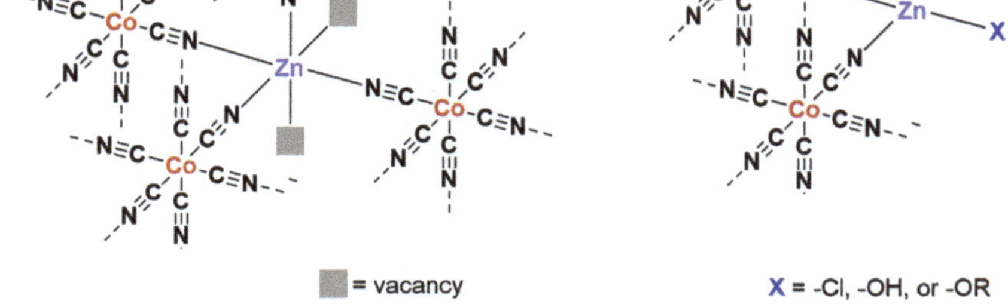

Fig. 4 Structure and active sites of Zn-Co DMC catalysts. *Adapted from* Zhang, Y. Y.; Zhang, X. H., Sustainable Polym. Biomass 2017, 279–313.

In this structure, one-third of all hexacyanocobaltates are absent as a result of the charge imbalance between Zn^{2+} and $[Co(CN)_6]^{3-}$, resulting in vacancies in the coordination sphere of the Zn^{2+} metal ions (on average tetra-coordinated) *(152)*. Moreover, due to the bulkiness of the $Co(CN)_6$ moiety, which is intact and inaccessible for the reacting substrates, the tetra-coordinated Zn^{2+} plays the role of the active site in DMCs (Fig. 4) *(159)*. The crystallinity may vary depending on the synthetic methods and in general, the less crystalline a DMC catalyst is, the more active and selective it is toward the synthesis of polycarbonates *(125,135,136)*.

Zn-Co(III) DMCs are most commonly synthesized by mixing an aqueous solution of the metal-halide salt (usually $ZnCl_2$) and an aqueous solution of a hexacyanometallate salt (usually potassium hexacyanocobaltate) under stirring, which leads to the precipitation of the desired heterogeneous DMC (Scheme 5). More recently, the synthesis of DMCs was achieved by mechanochemistry following ball milling or by a solvent-free grinding procedure *(137,138)*. These approaches, although less widespread, do not require the use of solvents and are, therefore, more environmentally friendly.

$$3ZnCl_2 + 2K_3Co(CN)_6 \longrightarrow Zn_3[Co(CN)_6]_2\downarrow + 6KCl$$

Scheme 5 Synthesis of Zn-Co DMC.

The nature of the zinc halide precursor (ZnX_2) and of the hexacyanometallate salt used to prepare the Zn-Co(III) DMCs has a crucial impact on the activity and selectivity of the catalyst. Different zinc halides have been used for the synthesis of DMCs, but in the case of ZnF_2 no reaction with $K_3Co(CN)_6$ was observed and thus no DMC was produced *(146)*. The type of halide (Cl^-, Br^- or I^-) used in the synthesis has been shown to affect the catalytic performance of the DMC in a different way for different epoxides *(136,146)*. In the copolymerization of cyclohexene oxide with CO_2, the nature of the halide was reported not to have a relevant impact on the activity of the DMC catalyst, while it had an influence on the selectivity (polymeric vs. cyclic carbonate) and on the CO_2 incorporation in the final poly(carbonate ether)s. In general, CO_2 fixation into the polymers decreased following the trend: $Cl^- \geq Br^- > I^-$ *(136,146)*. For what concerns the copolymerization of propylene oxide with CO_2, the nature of the halide had only minimal effect on the activity and selectivity of DMC *(136,146)*. However, it dramatically impacted the molecular weight of the final polymers, which followed the trend: $Cl^- \geq I^- > Br^-$ *(146)*. Regardless the type of halide, an excess of the zinc salt had, in general, beneficial effects on the final activity of the catalyst *(136,146)*.

In terms on catalytic performance, the choice of the metal cyanide is even more significant, as the electron cloud distribution surrounding the bridging cyanide ligands changes drastically when the transition metal in the cyanometallate precursor changes, influencing the activity of the Zn center.

Complexing agents (CAs) and co-complexing agents (co-CAs) might also be added during the synthesis of DMCs to tune their catalytic performance. CAs can significantly enhance the activity of these catalysts by decreasing their crystallinity and by improving $ZnCl_2$ retention in their 3D structure *(135)*. This makes the active sites more accessible and acidic, resulting in an overall improvement in catalytic performance *(125,136,138)*. Alcohols are commonly used as CAs *(125,135,136,140)*, although in more recent works also β-, γ-, and δ-dicarbonyl compounds (i.e., dicarboxylic acids such as oxalic, succinic, and glutaric acids; diketones such as 3,4-hexanedione, 2,4-pentanedione, and 2,5-hexanedione; and diesters such as diethyl oxalate, diethyl malonate, and diethyl succinate) have been investigated as CAs in the synthesis of DMCs *(141,142)*. Among the studied CAs, tert-butanol was found to be the most effective in boosting the catalytic activity of DMCs *(125,140)*. Small oligomers such as PEG or PTMEG can be also utilized in the synthesis as co-CAs *(143–145)*. The primary function of co-CAs is to promote the dispersion of nucleation centers in the synthesis mixture, thus limiting the growth of DMC particles and so increasing the surface area of the catalyst. However, an excess of co-CA might harm catalytic performance by blocking the active sites of the DMC catalyst *(125,136,144)*.

The immobilization of DMCs on solid supports has been investigated as a strategy to increase their activity and selectivity toward the synthesis of poly(carbonate ether)s. For instance, it was reported that supporting a nanosized Zn-Co DMC on silica (*via* sol-gel method) helped increase the reachability of its active sites and, therefore, the turnover frequency of the catalyst *(160)*. Instead, supporting a Zn-Co DMC on TiO_2 increased the selectivity of the catalyst for the production of poly(carbonate ether)s over cyclic carbonates by increasing the acidity of the active sites *(161)*.

2.2.2.2 Catalytic mechanism

Despite several studies and advances in disclosing the structure of the active sites of DMCs and in proposing plausible mechanisms, further investigation is still required to fully clarify the catalytic behavior of these materials *(135)*. The initial step in the copolymerization between CO_2 and epoxides differs depending on the nature of the active site (which can contain Zn-Co, Zn-OH or Zn-OR groups, see Fig. 4B) and on the presence or the absence of initiators (generally alcohols) used to expedite the reaction *(125,136)*. The mechanism that is most frequently reported involves a Zn-OH active site, which is able to interact both with the epoxide (nucleophilic attack after epoxide coordination, Scheme 6) and CO_2 (allowing its insertion, Scheme 6) *(160,162–164)*. The following steps have been proposed *(125,136,160,162–164)*:

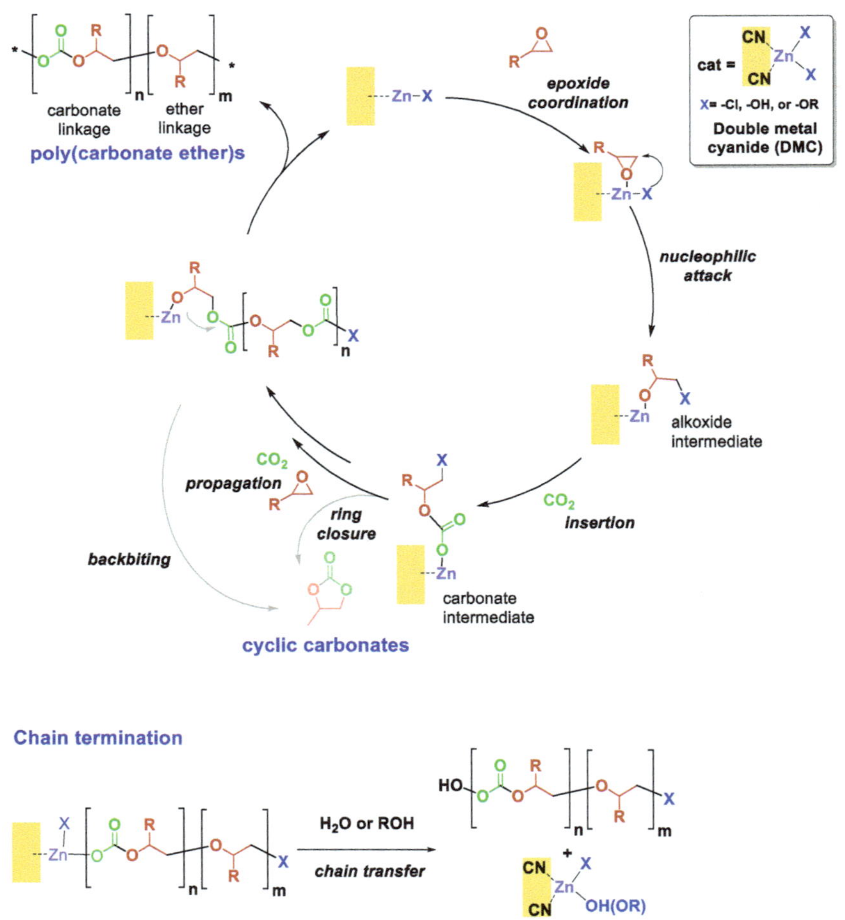

Scheme 6 Proposed mechanism for the copolymerization of CO_2 and propylene oxide catalyzed by Zn-Co DMC. [OR = alkoxide group originating from the complexing agent or from the initiator].

(i) *Epoxide coordination* to a zinc active center (Zn-X bond, where X=OH, Cl, or OR, with the latter originating from a CA or from an initiator).

(ii) *Nucleophilic attack* of the coordinated epoxide by the X^- substituent of zinc, leading to the formation of an alkoxide intermediate by ring opening of the epoxide.

(iii) *CO_2 insertion* to give the carbonate intermediate.

(iv) *Propagation* by iteration of steps (i)–(iii), leading to the growth of the polycarbonate chain.

As in the case of the zinc dicarboxylates catalysts (see Section 2.2.1), there are several competing reactions *(125,136,160,162–164)* that might take place leading to the formation of by-products:

(v) Subsequent *insertion of two (or more) epoxide molecules*, producing ether linkages in the polymer backbone.

(vi) *Ring closure* or *backbiting* of the carbonate intermediate leading to the formation of cyclic carbonates, which are the thermodynamic products of the reaction of CO_2 with epoxides *(165)*.

(vii) *Chain transfer* reactions arising from the presence of protic reagents (such as residual water and alcohols used as initiators, see bottom part of Scheme 6) in the polymerization mixture. In the presence of these chain transfer agents (CTAs), polymers with lower molecular weight but narrower molecular weight distribution (i.e., lower polydispersity) are achieved.

Steps (i) and (iii) can also be inverted since the active site of DMC is able to promote both the coordination of the epoxide and CO_2 insertion as the initial step of the reaction.

2.2.2.3 Catalytic performance

DMCs are generally employed for the copolymerization of CO_2 and epoxides in a temperature range between 65 C and 140 °C and under 5–50 bar of CO_2 *(135)*. DMC catalysts have a broader epoxide scope compared to zinc dicarboxylates. The most studied epoxides in the copolymerization with CO_2 in the presence of DMC catalysts are propylene oxide and cyclohexene oxide, but the Zn-Co DMC catalyst proved to be active also with long-chain aliphatic epoxides, bulky aliphatic epoxides, aromatic epoxides, epichlorohydrin and related glycidyl compounds, bio-based epoxides derived from fatty acids or oils and for terpolymerization reactions *(91,162,166–175)*. The glass transition temperatures (T_g) of the resulting CO_2/epoxide copolymers are strongly influenced by the nature of the epoxide, ranging from −44 °C to 109 °C *(125,162,174)*.

DMC catalysts have in general less control on the final structure of the polymeric backbone (low CO_2 fixation and tendency to form ether linkages) when compared to homogeneous systems, which are normally able to selectively yield polycarbonates with high to complete CO_2 incorporation in optimized conditions *(90, 91, 125, 136-181)*. On the other hand, DMCs are robust, cost-effective, air and moisture resistant catalysts, which can be synthesized with a one-step procedure that also allows particles size control *(90,125,136)*. The cobalt incorporated in the catalyst is stable and fully coordinated, meaning that no catalyst decomposition occurs during the reaction preventing the coloring of the final polycarbonates, which are instead highly transparent, and environmental pollution caused by heavy metal ions is avoided *(125)*. For these reasons, DMCs are relevant catalysts, especially for industrial, large-scale applications *(182–193)*.

3. Concluding remarks and perspectives

The investigation of heterogeneous catalysts for the reaction of CO_2 and epoxides to produce either cyclic carbonates or polycarbonates is a lively, growing field of research because this type of catalysts are more suited for large-scale, industrial applications compared to their homogeneous counterparts. In the context of sustainability, there is a rising interest for routes that enable the conversion of carbon dioxide into useful products. Cyclic carbonates and polycarbonates are excellent candidates in this sense, with a large number of potential or already industrially established applications. Therefore, it is crucial to design and develop active, selective and stable heterogeneous catalysts to enable the reaction of CO_2 with epoxides into these products. Recent research efforts allowed to develop heterogeneous catalysts with high activity with a broad scope of epoxides under mild reaction conditions, particularly when the cyclic carbonate is the target product. With the state-of-the-art catalysts, cyclic carbonates can be obtained with (nearly) full selectivity, with the diol formed by the reaction of the epoxide with (adventitious) water being sometimes observed as a minor side-product. On the other hand, the zinc-based heterogeneous catalysts presented in Section 2.2 display high selectivity toward the polycarbonate, but the formation of ether linkages within the polymer chain and of a small fraction of cyclic carbonate as side-product cannot be avoided. For these zinc-based heterogeneous catalysts, the separation from the polycarbonates can be achieved but the stability and reusability have not been well investigated. Conversely, the stability and reusability of most of the heterogeneous catalysts for the synthesis of cyclic carbonates (Section 2.1) have been studied, in general demonstrating a high degree of recyclability. Yet, a minor, gradual deactivation is often observed. It has been reported that the deactivation can be caused by loss of the halides acting as nucleophiles and that in such case regeneration can be achieved by a straightforward treatment with an aqueous solution of KI *(44)*. Most of the heterogeneous catalysts that have been studied for the reaction of carbon dioxide with epoxides were tested in batch reactors, with only few reports of the utilization of fixed-bed continuous-flow reactors for the synthesis of cyclic carbonates *(3,20,40,43)*. The latter reactor configuration might be preferable for the large-scale production of these compounds. In the perspective of the industrial production of cyclic carbonates and polycarbonates, it is also important to take into account the costs connected to the synthesis of the catalysts. This is an aspect that can be addressed in future research, particularly for the synthetically most complex among the heterogeneous catalysts for the conversion of CO_2 into cyclic carbonates.

References

1. Pescarmona, P. P. *Curr. Opin. Green Sustain. Chem.* **2021**, *29*, 100457.
2. Taherimehr, M.; Pescarmona, P. P. *J. Appl. Polym. Sci.* **2014**, *131* (21).
3. Kamphuis, A. J.; Picchioni, F.; Pescarmona, P. P. *Green Chem.* **2019**, *21* (3), 406–448.
4. Banitaba, S. N.; Semnani, D.; Heydari-Soureshjani, E.; Rezaei, B.; Ensafi, A. A. *Solid State Ion.* **2020**, *347*, 115252.
5. Coates, G. W.; Moore, D. R. *Angew. Chem. Int. Ed.* **2004**, *43* (48), 6618–6639.
6. Zhang, Y.-Y.; Wu, G.-P.; Darensbourg, D. J. *Trends Chem.* **2020**, *2* (8), 750–763.
7. Marbach, J.; Nörnberg, B.; Rahlf, A. F.; Luinstra, G. A. *Cat. Sci. Technol.* **2017**, *7* (13), 2897–2905.
8. Pescarmona, P. P.; Taherimehr, M. *Cat. Sci. Technol.* **2012**, *2* (11), 2169–2187.
9. Fiorani, G.; Guo, W.; Kleij, A. W. *Green Chem.* **2015**, *17* (3), 1375–1389.
10. Bahramian, B.; Ma, Y.; Rohanizadeh, R.; Chrzanowski, W.; Dehghani, F. *Green Chem.* **2016**, *18* (13), 3740–3748.
11. Guo, L.; Lamb, K. J.; North, M. *Green Chem.* **2021**, *23* (1), 77–118.
12. Wang, J. Q.; Kong, D. L.; Chen, J. Y.; Cai, F.; He, L. N. *J. Mol. Catal. A: Chem.* **2006**, *249* (1–2), 143–148.
13. Du, Y.; Cai, F.; Kong, D. L.; He, L. N. *Green Chem.* **2005**, *7* (7), 518–523.
14. Zhang, Q.; Zhang, S.; Li, S. *Macromolecules* **2012**, *45* (7), 2981–2988.
15. Han, L.; Park, S. W.; Park, D. W. *Energ. Environ. Sci.* **2009**, *2* (12), 1286–1292.
16. Xie, Y.; Zhang, Z.; Jiang, T.; He, J.; Han, B.; Wu, T.; Ding, K. *Angew. Chem. Int. Ed.* **2007**, *46* (38), 7255–7258.
17. Aprile, C.; Giacalone, F.; Agrigento, P.; Liotta, L. F.; Martens, J. A.; Pescarmona, P. P.; Gruttadauria, M. *ChemSusChem* **2011**, *4* (12), 1830–1837.
18. Agrigento, P.; Al-Amsyar, S. M.; Sorée, B.; Taherimehr, M.; Gruttadauria, M.; Aprile, C.; Pescarmona, P. P. *Cat. Sci. Technol.* **2014**, *4* (6), 1598–1607.
19. Ghazali-Esfahani, S.; Song, H.; Păunescu, E.; Bobbink, F. D.; Liu, H.; Fei, Z.; Dyson, P. J. *Green Chem.* **2013**, *15* (6), 1584–1589.

20. Zanda, N.; Sobolewska, A.; Alza, E.; Kleij, A. W.; Pericàs, M. A. *ACS Sustainable Chem. Eng.* **2021**, *9* (12), 4391–4397.
21. Jagtap, S. R.; Raje, V. P.; Samant, S. D.; Bhanage, B. M. *J. Mol. Catal. A: Chem.* **2007**, *266* (1–2), 69–74.
22. Barbarini, A.; Maggi, R.; Mazzacani, A.; Mori, G.; Sartori, G.; Sartorio, R. *Tetrahedron Lett.* **2003**, *44* (14), 2931–2934.
23. Udayakumar, S.; Lee, M. K.; Shim, H. L.; Park, S. W.; Park, D. W. *Cat. Com.* **2009**, *10* (5), 659–664.
24. Cheng, W.; Chen, X.; Sun, J.; Wang, J.; Zhang, S. *Catal. Today* **2013**, *200*, 117–124.
25. Buaki-Sogó, M.; Vivian, A.; Bivona, L. A.; García, H.; Gruttadauria, M.; Aprile, C. *Cat. Sci. Technol.* **2016**, *6*, 8418–8427.
26. Calabrese, C.; Liotta, L. F.; Carbonell, E.; Giacalone, F.; Gruttadauria, M. *ChemSusChem* **2017**, *10*, 1202–1209.
27. Calabrese, C.; Liotta, L. F.; Giacalone, F.; Gruttadauria, M. *ChemCatChem* **2019**, *11*, 560–567.
28. Zhai, G.; Liu, Y.; Lei, L.; Wang, J.; Wang, Z.; Zheng, Z.; Huang, B. *ACS Catal.* **2021**, *11* (4), 1988–1994.
29. Alves, M.; Grignard, B.; Mereau, R.; Jerome, C.; Tassaing, T.; Detrembleur, C. *Cat. Sci. Technol.* **2017**, 7, 2651–2684.
30. Takahashi, T.; Watahiki, T.; Kitazume, S.; Yasuda, H.; Sakakura, T. *Chem. Commun.* **2006**, *15*, 1664–1666.
31. Roshan, K. R.; Jose, T.; Kathalikkattil, A. C.; Kim, D. W.; Kim, B.; Park, D. W. *Appl. Catal., A* **2013**, *467*, 17–25.
32. Zhao, Y.; Tian, J. S.; Qi, X. H.; Han, Z. N.; Zhuang, Y. Y.; He, L. N. *J. Mol. Catal. A: Chem.* **2007**, *271* (1–2), 284–289.
33. Tharun, J.; Hwang, Y.; Roshan, R.; Ahn, S.; Kathalikkattil, A. C.; Park, D. W. *Cat. Sci. Technol.* **2012**, *2* (8), 1674–1680.
34. Sun, J.; Wang, J.; Cheng, W.; Zhang, J.; Li, X.; Zhang, S.; She, Y. *Green Chem.* **2012**, *14* (3), 654–660.
35. Sun, J.; Cheng, W.; Fan, W.; Wang, Y.; Meng, Z.; Zhang, S. *Catal. Today* **2009**, *148* (3–4), 361–367.
36. Han, L.; Choi, H. J.; Choi, S. J.; Liu, B.; Park, D. W. *Green Chem.* **2011**, *13* (4), 1023–1028.
37. Chen, X.; Sun, J.; Wang, J.; Cheng, W. *Tetrahedron Lett.* **2012**, *53* (22), 2684–2688.
38. Watile, R. A.; Deshmukh, K. M.; Dhake, K. P.; Bhanage, B. M. *Cat. Sci. Technol.* **2012**, *2* (5), 1051–1055.
39. Zhang, W.; Wang, Q.; Wu, H.; Wu, P.; He, M. *Green Chem.* **2014**, *16* (11), 4767–4774.
40. Whiteoak, C. J.; Henseler, A. H.; Ayats, C.; Kleij, A. W.; Pericàs, M. A. *Green Chem.* **2014**, *16* (3), 1552–1559.
41. Sainz Martinez, A.; Hauzenberger, C.; Sahoo, A. R.; Csendes, Z.; Hoffmann, H.; Bica, K. *ACS Sustainable Chem. Eng.* **2018**, *6* (10), 13131–13139 (C. Kohrt, Dr. T. Werner, ChemSusChem 2015, 8, 2031 – 2034).
42. Kohrt, C.; Werner, T. *ChemSusChem* **2015**, *8* (12), 2031–2034.
43. Jose, T.; Cañellas, S.; Pericàs, M. A.; Kleij, A. W. *Green Chem.* **2017**, *19* (22), 5488–5493.
44. Alassmy, Y. A.; Asgar Pour, Z.; Pescarmona, P. P. *ACS Sustainable Chem. Eng.* **2020**, *8* (21), 7993–8003.
45. Cokoja, M.; Wilhelm, M. E.; Anthofer, M. H.; Herrmann, W. A.; Kühn, F. E. *ChemSusChem* **2015**, *8* (15), 2436–2454.
46. Saptal, V. B.; Bhanage, B. M. *Curr. Opin. Green Sustain. Chem.* **2017**, *3*, 1–10.

47. Marciniak, A. A.; Lamb, K. J.; Ozorio, L. P.; Mota, C. J. A.; North, M. *Curr. Opin. Green Sustain. Chem.* **2020**, *26*, 100365.
48. Liu, M.; Liu, B.; Liang, L.; Wang, F.; Shi, L.; Sun, J. *J. Mol. Catal. A: Chem.* **2016**, *418*, 78–85.
49. Comès, A.; Collard, X.; Fusaro, L.; Atzori, L.; Cutrufello, M. G.; Aprile, C. *RSC Adv.* **2018**, *8* (45), 25342–25350.
50. Comès, A.; Fiorilli, S.; Aprile, C. *J. CO_2 Util.* **2020**, *37*, 213–221.
51. Meléndez, J.; North, M.; Villuendas, P. *Chem. Commun.* **2009**, *18*, 2577–2579.
52. North, M.; Villuendas, P.; Young, C. *Chem.: Eur. J.* **2009**, *15* (43), 11454–11457.
53. He, H.; Perman, J. A.; Zhu, G.; Ma, S. *Small* **2016**, *12* (46), 6309–6324.
54. Liang, J.; Huang, Y. B.; Cao, R. *Coord. Chem. Rev.* **2019**, *378*, 32–65.
55. Pal, T. K.; De, D.; Bharadwaj, P. K. *Coord. Chem. Rev.* **2020**, *408*, 213173.
56. Song, X.; Wang, J.; Yang, L.; Pan, H.; Zheng, B. *Inorg. Chem. Commun.* **2020**, *121*, 108197.
57. Singh Dhankhar, S.; Ugale, B.; Nagaraja, C. M. *Chem.: Asian J.* **2020**, *15* (16), 2403–2427.
58. Miralda, C. M.; Macias, E. E.; Zhu, M.; Ratnasamy, P.; Carreon, M. A. *ACS Catal.* **2012**, *2* (1), 180–183.
59. Yang, L.; Yu, L.; Diao, G.; Sun, M.; Cheng, G.; Chen, S. *J. Mol. Catal. A: Chem.* **2014**, *392*, 278–283.
60. Mousavi, B.; Chaemchuen, S.; Moosavi, B.; Luo, Z.; Gholampour, N.; Verpoort, F. *New J. Chem.* **2016**, *40* (6), 5170–5176.
61. Tharun, J.; Mathai, G.; Kathalikkattil, A. C.; Roshan, R.; Won, Y. S.; Cho, S. J.; Park, D. W. *ChemPlusChem* **2015**, *80* (4), 715–721.
62. Cho, H. Y.; Yang, D. A.; Kim, J.; Jeong, S. Y.; Ahn, W. S. *Symposium Catal.* **2012**, *185*, 35–40.
63. Tharun, J.; Bhin, K. M.; Roshan, R.; Kim, D. W.; Kathalikkattil, A. C.; Babu, R.; Park, D. W. *Green Chem.* **2016**, *18* (8), 2479–2487.
64. Jose, T.; Hwang, Y.; Kim, D.-W.; Kim, M.-I.; Park, D.-W. Catal. Today. **2015**, *245*, 61–67.
65. Zhou, X.; Zhang, Y.; Yang, X.; Zhao, L.; Wang, G. *J. Mol. Catal. A: Chem.* **2012**, *361*, 12–16.
66. Ma, D.; Li, B.; Liu, K.; Zhang, X.; Zou, W.; Yang, Y.; Feng, S. *J. Mater. Chem. A* **2015**, *3* (46), 23136–23142.
67. Liu, T. T.; Liang, J.; Xu, R.; Huang, Y. B.; Cao, R. *Chem. Commun.* **2019**, *55* (28), 4063–4066.
68. Ding, M.; Jiang, H. L. *ACS Catal.* **2018**, *8* (4), 3194–3201.
69. Aguila, B.; Sun, Q.; Wang, X.; O'Rourke, E.; Al-Enizi, A. M.; Nafady, A.; Ma, S. *Am. Ethnol.* **2018**, *130* (32), 10264–10268.
70. Taherimehr, M.; Van de Voorde, B.; Wee, L. H.; Martens, J. A.; De Vos, D. E.; Pescarmona, P. P. *ChemSusChem* **2017**, *10* (6), 1283–1291.
71. Guillerm, V.; Weseliński, Ł. J.; Belmabkhout, Y.; Cairns, A. J.; D'Elia, V.; Wojtas, Ł.; Eddaoudi, M. *Nat. Chem.* **2014**, *6* (8), 673–680.
72. Huang, K.; Zhang, J. Y.; Liu, F.; Dai, S. *ACS Catal.* **2018**, *8* (10), 9079–9102.
73. Luo, R.; Chen, M.; Liu, X.; Xu, W.; Li, J.; Liu, B.; Fang, Y. *J. Mater. Chem. A* **2020**, *8* (36), 18408–18424.
74. Li, H.; Li, C.; Chen, J.; Liu, L.; Yang, Q. Chem.– Asian J. **2017**, *12*, 1095–1103.
75. Wang, W.; Wang, Y.; Li, C.; Yan, L.; Jiang, M.; Ding, Y. *ACS Sustainable Chem. Eng.* **2017**, *5* (6), 4523–4528.
76. Chen, Y.; Luo, R.; Xu, Q.; Jiang, J.; Zhou, X.; Ji, H. *ChemSusChem* **2017**, *10* (11), 2534–2541.

77. Li, C.; Wang, W.; Yan, L.; Wang, Y.; Jiang, M.; Ding, Y. *J. Mater. Chem. A* **2016**, *4* (41), 16017–16027.
78. Luo, R.; Chen, M.; Zhou, F.; Zhan, J.; Deng, Q.; Yu, Y.; Fang, Y. *J. Mater. Chem. A* **2021**, *9*, 25731–25749.
79. Ma, D.; Liu, K.; Li, J.; Shi, Z. *ACS Sustainable Chem. Eng.* **2018**, *6* (11), 15050–15055.
80. Roeser, J.; Kailasam, K.; Thomas, A. *ChemSusChem* **2012**, *5* (9), 1793–1799.
81. Buyukcakir, O.; Je, S. H.; Talapaneni, S. N.; Kim, D.; Coskun, A. *ACS Appl. Mater. Interfaces* **2017**, *9* (8), 7209–7216.
82. Shaikh, R. R.; Pornpraprom, S.; D'Elia, V. *ACS Catal.* **2018**, *8* (1), 419–450.
83. Su, Q.; Sun, J.; Wang, J.; Yang, Z.; Cheng, W.; Zhang, S. *Cat. Sci. Technol.* **2014**, *4* (6), 1556–1562.
84. Huang, Z.; Li, F.; Chen, B.; Yuan, G. *Cat. Sci. Technol.* **2016**, *6* (9), 2942–2948.
85. Samikannu, A.; Konwar, L. J.; Mäki-Arvela, P.; Mikkola, J. P. *Appl Catal B* **2019**, *241*, 41–51.
86. Xu, J.; Wu, F.; Jiang, Q.; Li, Y. X. *Cat. Sci. Technol.* **2015**, *5* (1), 447–454.
87. Huang, Z.; Li, F.; Chen, B.; Lu, T.; Yuan, Y.; Yuan, G. *Appl Catal B* **2013**, *136*, 269–277.
88. Song, X.; Wu, Y.; Pan, D.; Zhang, J.; Xu, S.; Gao, L.; Xiao, G. *J. CO_2 Util.* **2018**, *28*, 326–334.
89. Inoue, S.; Koinuma, H.; Tsuruta, T. *Die Makromolekulare Chemie* **1969**, *130* (1), 210–220.
90. Trott, G.; Saini, P. K.; Williams, C. K. *Philos. Trans. A Math. Phys. Eng. Sci.* **2016**, *2061* (374).
91. Luinstra, G. *Polym. Rev.* **2008**, *48* (1), 192–219.
92. Klaus, S.; Lehenmeier, M. W.; Anderson, C. E.; Rieger, B. *Coord. Chem. Rev.* **2011**, *255* (13–14), 1460–1479.
93. Lee, B. Y.; Cyriac, A. *Nat. Chem.* **2011**, *3* (7), 505–507.
94. Empower Materials: Home, www.empowermaterials.com, accessed 21 February 2021.
95. Kobayashi, M.; Inoue, S.; Tsuruta, T. *J. Polym. Sci., Polym. Chem. Ed.* **1973**, *11* (9), 2383–2385.
96. Kobayashi, M.; Inoue, S.; Tsuruta, T. *Macromolecules* **1971**, *4* (5), 658–659.
97. Kobayashi, M.; Tang, Y.-L.; Tsuruta, T.; Inoue, S. *Die Makromolekulare Chemie* **1973**, *169* (1), 69–81.
98. Soga, K.; Imai, E.; Hattori, I. *Polym. J.* **1981**, *13* (4), 407–410.
99. Kim, J. *J. Catal.* **2003**, *218* (1), 209–219.
100. Ree, M.; Bae, J. Y.; Jung, J. H.; Shin, T. J. *J. Polym. Sci., Part A: Polym. Chem.* **1999**, *37* (12), 1863–1876.
101. Kim, J.-S.; Kim, H.; Yoon, J.; Heo, K.; Ree, M. *J. Polym. Sci., Part A: Polym. Chem.* **2005**, *43* (18), 4079–4088.
102. Darensbourg, D. *Coord. Chem. Rev.* **1996**, *153*, 155–174.
103. Klaus, S.; Lehenmeier, M. W.; Herdtweck, E.; Deglmann, P.; Ott, A. K.; Rieger, B. *J. Am. Chem. Soc.* **2011**, *133* (33), 13151–13161.
104. Kernbichl, S.; Rieger, B. Aliphatic polycarbonates derived from epoxides and CO_2. In *Engineering Solutions for CO_2 Conversion*; 2021; pp. 385–406.
105. Moore, D. R.; Cheng, M.; Lobkovsky, E. B.; Coates, G. W. *J. Am. Chem. Soc.* **2003**, *125* (39), 11911–11924.
106. Kim, J.-S.; Kim, H.; Ree, M. *Chem. Mater.* **2004**, *16* (16), 2981–2983.
107. Ang, R.-R.; Sin, L. T.; Bee, S.-T.; Tee, T.-T.; Kadhum, A. A. H.; Rahmat, A. R.; Wasmi, B. A. *Chem. Eng. J.* **2017**, *327*, 120–127.
108. Zheng, Y. Q.; Lin, J. L.; Zhang, H. L. *Z. Krist-New Cryst. St.* **2000**, *215* (4), 535–536.
109. Ree, M.; Bae, J. Y.; Jung, J. H.; Shin, T. J.; Hwang, Y. T.; Chang, T. *Polym. Eng. Sci.* **2000**, *40* (7), 1542–1552.

110. Wang, S. J.; Du, L. C.; Zhao, X. S.; Meng, Y. Z.; Tjong, S. C. *J. Appl. Polym. Sci.* **2002**, *85* (11), 2327–2334.
111. Meng, Y. Z.; Du, L. C.; Tiong, S. C.; Zhu, Q.; Hay, A. S. *J. Polym. Sci., Part A: Polym. Chem.* **2002**, *40* (21), 3579–3591.
112. Zhu, Q.; Meng, Y. Z.; Tjong, S. C.; Zhao, X. S.; Chen, Y. L. *Polym. Int.* **2002**, *51* (10), 1079–1085.
113. Padmanaban, S.; Kim, M.; Yoon, S. *J. Ind. Eng. Chem.* **2019**, *71*, 336–344.
114. Padmanaban, S.; Yoon, S. *Catalysts* **2019**, *9* (11).
115. Eberhardt, R.; Allmendinger, M.; Zintl, M.; Troll, C.; Luinstra, G. A.; Rieger, B. *Macromol. Chem. Phys.* **2004**, *205* (1), 42–47.
116. Chisholm, M. H.; Navarro-Llobet, D.; Zhou, Z.; Poly(propylene carbonate). *Macromolecules* **2002**, *35* (17), 6494–6504.
117. Kim, J. *J. Catal.* **2003**, *218* (2), 386–395.
118. Zhong, X.; Dehghani, F. *Appl Catal B* **2010**, *98* (3–4), 101–111.
119. Zhu, Q.; Meng, Y. Z.; Tjong, S. C.; Zhang, Y. M.; Wan, W. *Polym. Int.* **2003**, *52* (5), 799–804.
120. Wang, J. T.; Shu, D.; Xiao, M.; Meng, Y. Z. *J. Appl. Polym. Sci.* **2006**, *99* (1), 200–206.
121. de Jong, E.; Higson, A.; Walsh, P.; Wellisch, M. *Bio-Based Chemicals Value Added Products from Biorefineries*; IEA Bioenergy, 2012. Task42 Biorefinery.
122. Nörnberg, B.; Spottog, C.; Rahlf, A.; Korashvili, R.; Berlin, C.; Luinstra, G. A. *Macromol. Symp.* **2013**, *333* (1), 190–196.
123. Caroll, W. E.; Motika, S. A. *Regeneration of Metallo-Organic Catalyst for the Carbon Dioxideepoxide Copolymerization*; 1990.
124. Sudakar, P.; Sivanesan, D.; Yoon, S. *Macromol. Rapid Commun.* **2016**, *37* (9), 788–793.
125. Zhang, Y. Y.; Zhang, X. H. *Sustainable Polym. Biomass* **2017**, 279–313.
126. Herold, R.J., US Patent, 3,278,459, 1966.
127. Shokouhimehr, M.; Soehnlen, E. S.; Khitrin, A.; Basu, S.; Huang, S. D. *Inorg. Chem. Commun.* **2010**, *13* (1), 58–61.
128. Kruper, W.J.;Swart, D.J., US Patent 450,704, 1985.
129. Alferov, K.; Wang, S.; Li, T.; Xiao, M.; Guan, S.; Meng, Y. *Catalysts* **2019**, *9* (8), 632.
130. Qiang, L.; Zhifang, G.; Lisha, P.; Xue, X. *Cat. Com.* **2015**, *64*, 114–118.
131. Robertson, N. J.; Qin, Z.; Dallinger, G. C.; Lobkovsky, E. B.; Lee, S.; Coates, G. W. *Dalton Trans.* **2006**, *45*, 5390–5395.
132. Coates, G.W.; Lee, S., Qin, Z.; Robertson, N.J., US Patent 0,051,554 A1, 2008.
133. Zhang, X. H.; Chen, S.; Wu, X. M.; Sun, X. K.; Liu, F.; Qi, G. R. *Chin. Chem. Lett.* **2007**, *18* (7), 887–890.
134. Srivastava, R.; Srinivas, D.; Ratnasamy, P. *J. Catal.* **2006**, *241* (1), 34–44.
135. Grefe, L.; Mejía, E. *Tetrahedron* **2021**, *98*, 132433.
136. Sebastian, J.; Srinivas, D. *Sustainable Polym. Biomass* **2017**, 315–345.
137. Zhang, W.; Lin, Q.; Cheng, Y.; Lu, L.; Lin, B.; Pan, L.; Xu, N. *J. Appl. Polym. Sci.* **2012**, *123* (2), 977–985.
138. Guo, Z.; Lin, Q. *J. Mol. Catal. A: Chem.* **2014**, *390*, 63–68.
139. Zhang, X. H.; Hua, Z. J.; Chen, S.; Liu, F.; Sun, X. K.; Qi, G. R. *Appl. Catal., A* **2007**, *325* (1), 91–98.
140. Chen, S.; Hua, Z.; Fang, Z.; Qi, G. *Polymer* **2004**, *45* (19), 6519–6524.
141. Lim, J.; Yun, S. H.; Kim, M. R.; Kim, I. *J. Nanosci. Nanotechnol.* **2017**, *17* (10), 7507–7514.
142. Tran, C. H.; Pham, L. T. T.; Jang, H. B.; Kim, S. A.; Kim, I. *Catal. Today* **2021**, *375*, 429–440.
143. Lee, I. K.; Ha, J. Y.; Cao, C.; Park, D. W.; Ha, C. S.; Kim, I. *Catal. Today* **2009**, *148* (3–4), 389–397.

144. Yoon, J. H.; Lee, I. K.; Choi, H. Y.; Choi, E. J.; Yoon, J. H.; Shim, S. E.; Kim, I. *Green Chem.* **2011**, *13* (3), 631–639.
145. Sebastian, J.; Darbha, S. *RSC Adv.* **2015**, *5* (24), 18196–18203.
146. Kim, I.; Yi, M. J.; Lee, K. J.; Park, D. W.; Kim, B. U.; Ha, C. S. *Catal. Today* **2006**, *111* (3–4), 292–296.
147. Molina-Maldonado, P.; Ruíz-Guerrero, R.; Hernández-Fuentes, C. *Catalysts* **2019**, *9* (11), 905.
148. Seo, Y. H.; Hyun, Y. B.; Lee, H. J.; Baek, J. W.; Lee, H. C.; Lee, J. H.; Lee, B. Y. *J. CO_2 Util.* **2021**, *53*, 101755.
149. Kuyper, J.; Lednor, P.W.; Pogany, G.A., US Patent 4,826,953, 1989.
150. Darensbourg, D. J.; Adams, M. J.; Yarbrough, J. C.; Phelps, A. L. *Inorg. Chem.* **2003**, *42* (24), 7809–7818.
151. Rodríguez-Hernández, J.; Reguera, E.; Lima, E.; Balmaseda, J.; Martínez-García, R.; Yee-Madeira, H. *J. Phys. Chem. Solid* **2007**, *68* (9), 1630–1642.
152. Parvulescu, V. I.; Kemnitz, E. *New Materials for Catalytic Applications*; Elsevier, 2016.
153. Zhang, X. H.; Wei, R. J.; Sun, X. K.; Zhang, J. F.; Du, B. Y.; Fan, Z. Q.; Qi, G. R. *Polymer* **2011**, *52* (24), 5494–5502.
154. Marquez, C.; Simonov, A.; Wharmby, M. T.; Van Goethem, C.; Vankelecom, I.; Bueken, B.; De Baerdemaeker, T. *Chem. Sci.* **2019**, *10* (18), 4868–4875.
155. An, N.; Li, Q.; Yin, N.; Kang, M.; Wang, J. *Appl. Organomet. Chem.* **2018**, *32* (11), e4509.
156. Sebastian, J.; Srinivas, D. *Appl. Catal., A* **2018**, *482*, 300–308.
157. Kim, I.; Anas, K.; Lee, S.; Ha, C. S.; Park, D. W. *Catal. Today* **2008**, *131* (1–4), 541–547.
158. Pinilla-de Dios, M.; Andrés-Iglesias, C.; Fernández, A.; Salmi, T.; Galdámez, J. R.; García-Serna, J. *Eur. Polym. J.* **2017**, *88*, 280–291.
159. Wojdeł, J. C.; Bromley, S. T.; Illas, F.; Jansen, J. C. *J. Mol. Model.* **2007**, *13* (6), 751–756.
160. Marquez, C.; Rivera-Torrente, M.; Paalanen, P. P.; Weckhuysen, B. M.; Cirujano, F. G.; De Vos, D.; De Baerdemaeker, T. *J. Catal.* **2017**, *354*, 92–99.
161. Subhani, M. A.; Gürtler, C.; Leitner, W.; Müller, T. E. *Eur. J. Inorg. Chem.* **2016**, 1944–1949.
162. Zhang, X. H.; Wei, R. J.; Zhang, Y. Y.; Du, B. Y.; Fan, Z. Q. *Macromolecules* **2015**, *48* (3), 536–544.
163. Li, Y.; Zhang, Y. Y.; Liu, B.; Zhang, X. H. *Chin. J. Polym. Sci.* **2018**, *36* (2), 139–148.
164. Stahl, S. F.; Luinstra, G. A. *Catalysts* **2020**, *10* (9), 1066.
165. Dharman, M. M.; Yu, J. I.; Ahn, J. Y.; Park, D. W. *Green Chem.* **2009**, *11* (11), 1754–1757.
166. Gao, Y.; Qin, Y.; Zhao, X.; Wang, F.; Wang, X. *J. Polym. Res.* **2012**, *19* (5), 1–9.
167. Sun, X. K.; Zhang, X. H.; Chen, S.; Du, B. Y.; Wang, Q.; Fan, Z. Q.; Qi, G. R. *Polymer* **2010**, *51* (24), 5719–5725.
168. Chen, S.; Qi, G. R.; Hua, Z. J.; Yan, H. Q. *J. Polym. Sci., Part A: Polym. Chem.* **2004**, *42* (20), 5284–5291.
169. Kim, I.; Yi, M. J.; Byun, S. H.; Park, D. W.; Kim, B. U.; Ha, C. S. *Macromol. Symp.* **2005**, *224* (1), 181–192.
170. Sun, X. K.; Zhang, X. H.; Liu, F.; Chen, S.; Du, B. Y.; Wang, Q.; Qi, G. R. *J. Polym. Sci., Part A: Polym. Chem.* **2008**, *46* (9), 3128–3139.
171. Zhang, Y. Y.; Wei, R. J.; Zhang, X. H.; Du, B. Y.; Fan, Z. Q. *J. Polym. Sci., Part A: Polym. Chem.* **2015**, *53* (6), 737–744.
172. Wei, R. J.; Zhang, X. H.; Zhang, Y. Y.; Du, B. Y.; Fan, Z. Q.; Qi, G. R. *RSC Adv.* **2014**, *4* (7), 3188–3194.

173. Wei, R. J.; Zhang, X. H.; Du, B. Y.; Sun, X. K.; Fan, Z. Q.; Qi, G. R. *Macromolecules* **2013**, *46* (9), 3693–3697.
174. Zhang, Y. Y.; Zhang, X. H.; Wei, R. J.; Du, B. Y.; Fan, Z. Q.; Qi, G. R. *RSC Adv.* **2014**, *4* (68), 36183–36188.
175. Shaarani, F. W.; Bou, J. J. *Sci. Total Environ.* **2017**, *598*, 931–936.
176. Sun, X. K.; Zhang, X. H.; Wei, R. J.; Du, B. Y.; Wang, Q.; Fan, Z. Q.; Qi, G. R. *J. Polym. Sci., Part A: Polym. Chem.* **2012**, *50* (14), 2924–2934.
177. Poland, S. J.; Darensbourg, D. J. *Green Chem.* **2017**, *19* (21), 4990–5011.
178. Liang, X.; Tan, F.; Zhu, Y. *Front. Chem.* **2021**, *9*, 150.
179. Huang, J.; Worch, J. C.; Dove, A. P.; Coulembier, O. *ChemSusChem* **2020**, *13* (3), 469–487.
180. Wang, Y.; Darensbourg, D. J. *Coord. Chem. Rev.* **2018**, *372*, 85–100.
181. Ang, R. R.; Sin, L. T.; Bee, S. T.; Tee, T. T.; Kadhum, A. A. H.; Rahmat, A. R.; Wasmi, B. A. *J. Clean. Prod.* **2015**, *102*, 1–17.
182. Xu, Y.; Lin, L.; Xiao, M.; Wang, S.; Smith, A. T.; Sun, L.; Meng, Y. *Prog. Polym. Sci.* **2018**, *80*, 163–182.
183. Dienes, Y.; Leitner, W.; Müller, M. G.; Offermans, W. K.; Reier, T.; Reinholdt, A.; Müller, T. E. *Green Chem.* **2012**, *14* (4), 1168–1177.
184. Liu, N.; Gu, C.; Wang, Q.; Zhu, L.; Yan, H.; Lin, Q. *RSC Adv.* **2021**, *11* (15), 8782–8792.
185. Wang, J.; Zhang, H.; Miao, Y.; Qiao, L.; Wang, X.; Wang, F. *Green Chem.* **2016**, *18* (2), 524–530.
186. Blasco, M. P. C.; Limiñana, M.Á. P.; Silvestre, C. R.; Calpena, E. O.; Aís, F. A. *Polymer* **2022**, *14* (2), 284.
187. Haider, K. W.; McDaniel, K. G.; Hayes, J. E.; Shen, J., U.S. Patent No. 7,977,501, 2011.
188. Langanke, J.; Wolf, A.; Hofmann, J.; Böhm, K.; Subhani, M. A.; Müller, T. E.; Gürtler, C. *Green Chem.* **2014**, *16* (4), 1865–1870.
189. Langanke, J.; Wolf, A. *Org. Process Res. Dev.* **2015**, *19* (7), 735–739.
190. Gu, G.; Dong, J.; Duan, Z.; Liu, B. *Polymer* **2021**, *13* (16), 2765.
191. Gunatillake, P. A.; Meijs, G. F.; Rizzardo, E.; Chatelier, R. C.; McCarthy, S. J.; Brandwood, A.; Schindhelm, K. *J. Appl. Polym. Sci.* **1992**, *46* (2), 319–328.
192. Gürtler, C.; Hofmann, J.; Müller, T.E.; Wolf, A.; Grasser, S.; Köhler, B., 2011 BayerMaterialScience AG, WO2011117332A1,**2011**.
193. Lewis, S.; Uthe, P.; Koshut, W., 2019 U.S. Patent Application No. 15/793,340.

About the authors

Francesca Milocco was born in Udine (Italy) and studied at the University of Trieste, where she obtained her MSc in Chemistry in 2016. She then moved to the Netherlands where she obtained her PhD in Chemistry in 2021 at the University of Groningen, where she then worked as a post-doc at the Engineering and Technology Institute Groningen. Since June 2022 she works as R&D Project Manager at Forbo Flooring System in Assendelft.

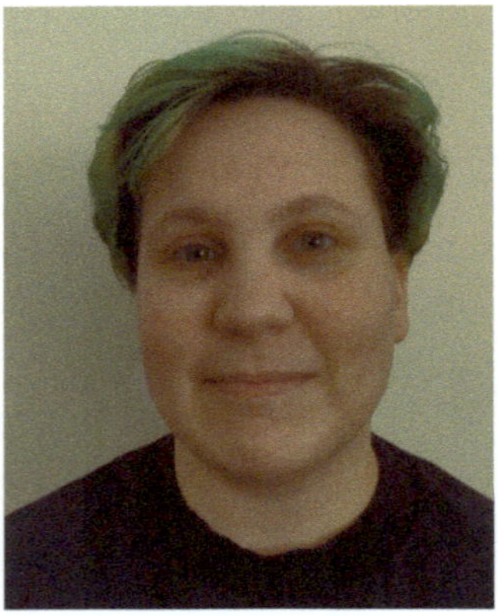

Giulia Chiarioni was born in Ferrara (Italy). She obtained her BSc in Chemistry from the University of Ferrara and her MSc in Chemistry from the University of Groningen in 2020. In May 2020, she started her PhD in the group of Professor Pescarmona and in collaboration with Covestro. Her research focuses on the synthesis of polycarbonates from (bio-based) epoxides and CO_2.

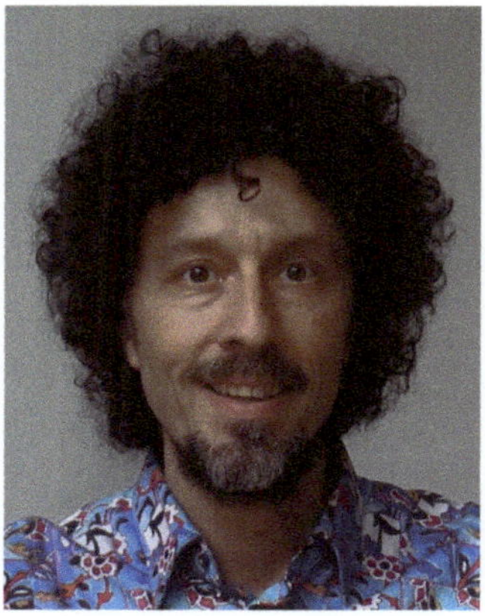

Paolo P. Pescarmona was born in Torino (Italy). He obtained his MSc in Chemistry from the University of Torino and his PhD from the TU Delft, in the Netherlands. Then, he worked as post-doc and as assistant professor at the KU Leuven (Belgium), before moving to the University of Groningen in 2014 where he is currently full professor in Catalysis & Sustainability at the Engineering and Technology Institute Groningen. His research interests focus on the rational design and development of catalytic materials for applications of academic, industrial and societal relevance, with special attention to green chemistry and sustainability.

CHAPTER SIX

Catalytic synthesis of bio-sourced organic carbonates and sustainable hybrid materials from CO_2

Felipe de la Cruz-Martínez[a], José A. Castro-Osma[b], and Agustín Lara-Sánchez[a,*]

[a]Universidad de Castilla-La Mancha, Departamento de Química Inorgánica, Orgánica y Bioquímica-Centro de Innovación en Química Avanzada (ORFEO – CINQA), Facultad de Ciencias y Tecnologías Químicas, Ciudad Real, Spain

[b]Universidad de Castilla-La Mancha, Departamento de Química Inorgánica, Orgánica y Bioquímica-Centro de Innovación en Química Avanzada (ORFEO – CINQA), Facultad de Farmacia, Albacete, Spain

*Corresponding author: e-mail address: agustin.lara@uclm.es

Contents

Abstract

The sustainable synthesis of organic molecules and polymeric materials from carbon dioxide is one of the biggest challenges the scientific community and the chemicals industry are facing. In this context, the preparation of cyclic organic carbonates and polycarbonates from epoxides and carbon dioxide are among the most studied processes for carbon dioxide utilization. Recent developments have showed a shift from the use of fossil fuel resources toward the use of bio-derived starting materials for

Advances in Catalysis, Volume 70
ISSN 0360-0564
https://doi.org/10.1016/bs.acat.2022.07.003

the production of bio-derived organic or polymeric carbonates. In this chapter, we cover recent advances on the use of bio-based feedstocks and carbon dioxide as a sustainable C_1 building block for the synthesis of bio-sourced organic carbonates and sustainable hybrid materials.

1. Introduction

The rapid growth of the world population has resulted in a substantial increase of energy demand, which is currently based on fossil fuels (Fig. 1) *(1)*. The excessive use of fossil fuels is the responsible for the large increase of greenhouse gases (GHG) in the atmosphere, which plays an important role in global warming, especially carbon dioxide (CO_2) (Fig. 2) *(2)*. In fact, CO_2 emissions account for more than 70% of total GHG emissions. Therefore, there is an urgent need to minimize CO_2 emissions in order to reduce the impact of anthropogenic emissions on the temperature of the planet since it is predicted that the temperature of Earth will increase 1.5 °C by 2052 if carbon dioxide emissions do not decrease *(3)*.

The most promising strategies to reduce CO_2 emissions is to apply carbon capture technologies such as carbon capture and storage (CCS) which captures and stores CO_2 using a range of sequestration ways, and carbon capture and utilization (CCU) which is based on the use of waste CO_2 to produce a range of chemical products without the need of CO_2 storage *(4,5)*. CCU has shown to be a feasible route to reduce CO_2 emissions by

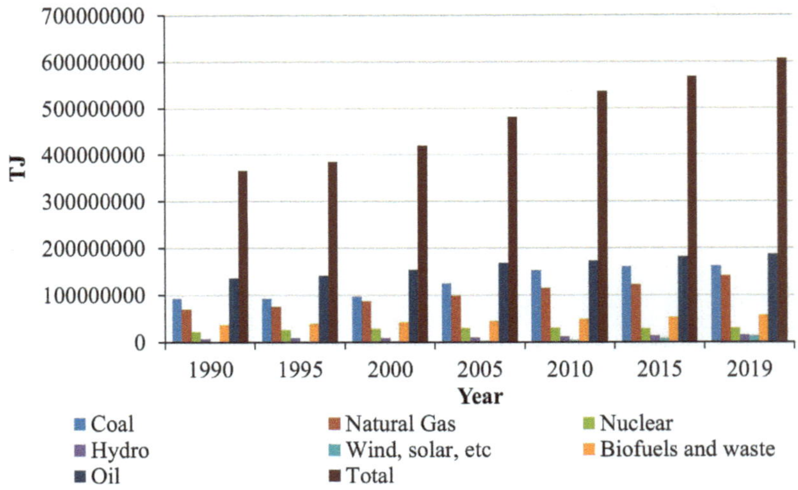

Fig. 1 Total energy supply by source *(1)*.

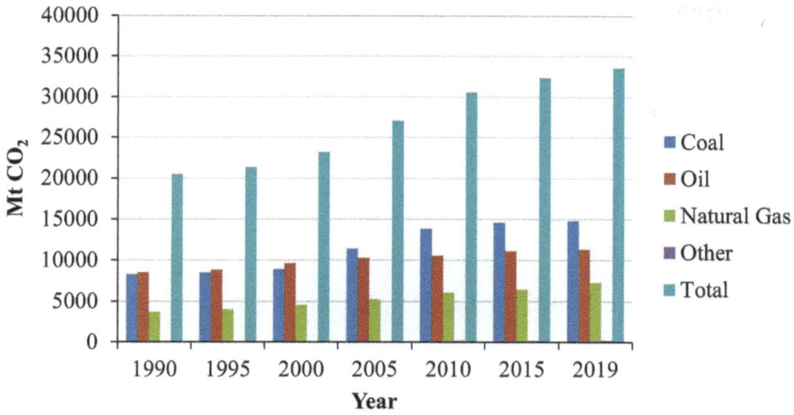

Fig. 2 CO_2 emissions by energy source *(2)*.

producing a broad range of organic compounds, polymers and fuels *(6)*. However, the use of CO_2 to synthesize chemical products must increase at least one order of magnitude to increase in importance since current CO_2 emissions outrun CO_2 utilization *(6,7)*. CCU technologies must be applied together with the use of renewable energy sources such as water, wind, sun, biomass and Earth to reach net zero CO_2 emissions by 2050.

1.1 A political view of CO_2

The way of life that today's society demands constitutes an environmental risk to the planet. The accumulation of waste products or residues in soils, seas, oceans, and even in the atmosphere itself, is causing the deterioration of our environment. Currently, there is an almost generalized scientific consensus around the idea that our mode of energy production and consumption is generating a global climate alteration, which will cause, in turn, serious impacts on both the planet and socioeconomic systems. The conference on climate change (COP26 2021) held in Glasgow (United Kingdom) in November 2021 has clearly marked the importance of fighting climate change and global warming *(8)*. Almost 200 countries agreed the Glasgow Climate Pact, which will accelerate action on climate this decade to accomplish the climate targets set by the Paris Climate Agreement *(8)*. The COP26 agreement set the actions to reduce the impacts of climate change for the next decade by: (1) reducing carbon dioxide and methane emissions to try to keep temperature rises within 1.5 °C; (2) reducing the use of coal, which is responsible for around 40% of total carbon dioxide emissions; (3) increasing funding for developing countries to make the switch from fossil fuel-based energy to a clean and sustainable one *(8)*. All alliance members are committed to the same goal: achieving carbon neutrality by 2050.

The rising consumption of fossil fuels is the responsible for the large increase of carbon dioxide concentration in the atmosphere. Even though CO_2 is essential for life since it keeps the Earth temperature warm enough to preserve life, the concentration of CO_2 has reached a record level of 412.5 ppm in 2020 *(9)*, which is around 130 ppm higher than when the industrial revolution started (Fig. 3).

The European Union has set a European Green Deal to tackle climate and environmental-related challenges by transforming the European Union's economy to a modern, resource-efficient and competitive one *(10)*. The EU plans to achieve climate neutrality by 2050 by setting out a broad range of ambitious policy initiatives which include actions in sectors such as industry, transport and mobility, energy and finance, ensuring the transition is cost-effective, socially balanced and fair. Reaching a carbon neutral EU will not only help to tackle global warming by reducing GHG emissions, but also will benefit people and the environment. In this sense, scientists are challenged to design sustainable production methodologies on the one hand and generate new environmentally friendly materials to contribute to a more sustainable world *(11–14)*. One of the most promising processes is the use of renewable sources such as CO_2 and products derived from biomass residues as starting platform chemicals for the design of new potentially biodegradable materials and more sustainable compounds of industrial interest by using catalysis *(15–18)*.

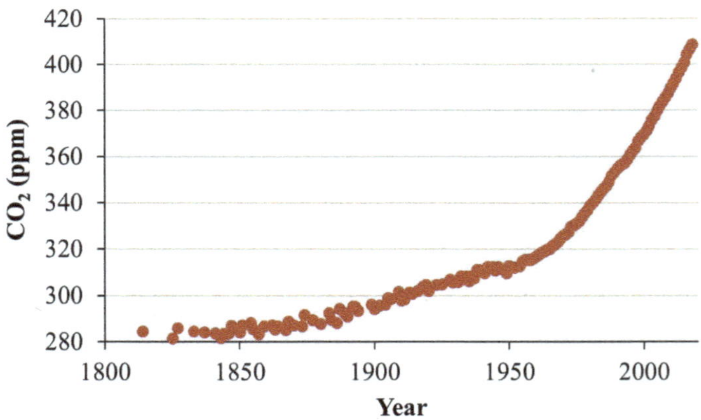

Fig. 3 Atmospheric CO_2 emissions *(9)*.

1.2 Industrial CO_2 utilization

Currently, there are few relevant chemical processes that use CO_2 as chemical feedstock on an industrial scale, however, their importance and social impact cannot be underestimated. Thus, the synthesis of urea from ammonia and CO_2 (Scheme 1A) with an annual production of more than 100 million tons, is the industrial process that uses the greatest amount of carbon dioxide as starting material *(6)*. On the other hand, the synthesis of salicylic acid (Scheme 1B) for the preparation of aspirin®, with an annual production of approximately 20,000 tons, has become one of the most relevant processes that use CO_2 as a reagent in the pharmaceutical industry *(6)*. Nevertheless, current industrial processes that use carbon dioxide as starting material utilize <1% of total CO_2 emissions for the synthesis of these two products mainly *(5,19–21)*. The main limitation to implement CCU technologies is the high stability of CO_2 *(7,22,23)*. Carbon dioxide is a triatomic linear molecule in which the carbon atom is in its highest oxidation state (+4) and it is bonded to two oxygen atoms through two double bonds with a C=O bond length of 1.16 Å. Even though both C=O bonds are polar, CO_2 is a non-polar molecule since the dipole moments of the C=O bonds are directed to opposite directions *(7,24)*. The reactivity of CO_2 is influenced by the HOMOs and LUMOs, which are localized on the oxygen atoms and the carbon atom respectively. Therefore, the oxygen atoms exhibit a Lewis base character, and the carbon atom displays a Lewis acid character. Nevertheless, the electrophilic character of carbon in the CO_2 molecule dominates its reactivity instead of the nucleophilicity of the oxygen atoms *(24)*.

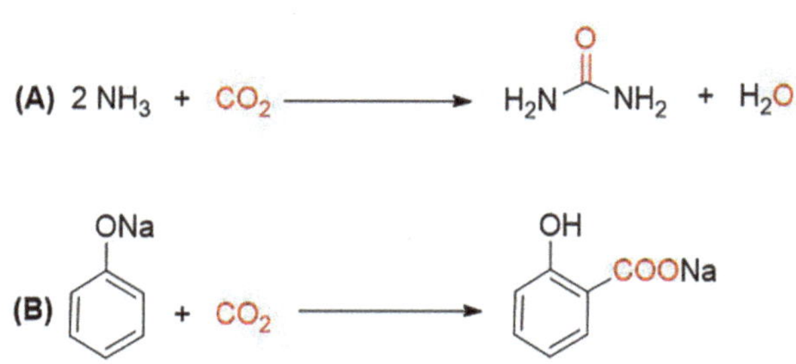

Scheme 1 Synthesis of (A) urea and (B) salicylic acid sodium salt.

It is expected that CCU technologies that use CO_2 as a sustainable C_1 synthon for the production of valuable products continue its progression and make a considerable impact on CO_2 recycling and reutilization *(25)*. Those processes not only should use CO_2 but also renewable energy resources and energy efficiency should be considered for industrial processes development. Therefore, there is an urgent need for the development of large-scale CO_2 utilization industrial processes to produce fuels, fine chemicals, organic and inorganic carbonates and advanced materials to significantly reduce the CO_2 content in the atmosphere *(26–29)*.

1.2.1 Fuels

The synthesis of CO_2-derived fuels is one of the most promising strategies for large-scale CO_2 utilization since fuels can be used for transportation, power generation or fuel cells. Methanol, formic acid and methane are among the most important fuels that can be synthesized from carbon dioxide *(30–33)*.

1.2.1.1 Methanol

Methanol is one of the fuels that is produced on a larger scale and is currently synthesized from syngas which is derived from fossil fuels. Methanol is not only used as a fuel but also as a chemical intermediate for the preparation of a wide variety of chemical products. However, the catalytic hydrogenation of CO_2 (Scheme 2) is an alternative and more sustainable route for the preparation of methanol *(34)*.

$$CO_2 + 3\,H_2 \xrightleftharpoons{\text{catalyst}} CH_3OH + H_2O$$

Scheme 2 Hydrogenation of CO_2 for the synthesis of methanol.

1.2.1.2 Formic acid

There is great interest in the development of formic acid synthesis processes (Scheme 3). Formic acid is widely used in the textile and food industries, and to a lesser extent, in the medical sector *(6)*. One of the most promising applications of formic acid is that it can act as a hydrogen carrier molecule for fuel cells since its decomposition generates H_2 and CO_2, being a promising alternative to the use of fossil resources *(35)*.

$$CO_2\,(g) + H_2\,(g) \xrightarrow{\text{catalyst}} HCOOH\,(l)$$

Scheme 3 Hydrogenation of CO_2 for the synthesis of formic acid.

1.2.1.3 Methane

Methane is the main component of natural gas, which is extracted from natural reservoirs. It is currently used as a fuel for heating within residential and as a starting material for the chemical industry, mainly for the production syngas. CO_2 can also be used for the preparation of the so-called methane-rich gas by catalytic hydrogenation (Scheme 4A), which can be converted *via* dry reforming of methane into a renewable alternative of syngas (Scheme 4B) *(36)*.

$$\textbf{(A)}\ CO_2\,(g) + 4H_2\,(g) \xrightleftharpoons{\text{catalyst}} CH_4 + 2H_2O$$

$$\textbf{(B)}\ CH_4(g) + CO_2\,(g) \xrightarrow{\text{catalyst}} 2\,CO\,(g) + 2\,H_2\,(g)$$

Scheme 4 (A) Hydrogenation of CO_2 for the synthesis of methane. (B) Synthesis of syngas from methane.

1.2.2 Fine chemicals

The chemical incorporation of CO_2 into fine chemicals has attracted much attention from academia and industry and many metal-catalyzed processes have been reported for the efficient conversion of CO_2 into fine chemicals as a more sustainable approach (Scheme 5) *(37,38)*. Thus, a great variety of catalytic processes for the chemical reaction of CO_2 with alkenes, alkynes, amines or alcohols, among others have allowed the synthesis of a wide range of functional molecules. For instance, the direct carboxylation of methane with CO_2, two important GHG, could afford acetic acid which is an important chemical intermediate.

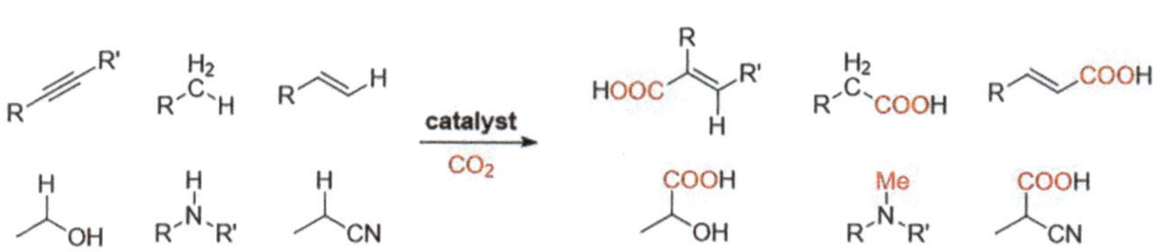

Scheme 5 Synthesis of fine chemicals from CO_2.

1.2.3 Carbonates

Both inorganic and organic carbonates can be prepared from carbon dioxide, and it is one of the CCU technologies that has received significant attention *(39)*. Inorganic carbonates can be prepared by reaction of metal oxides with CO_2 to form the corresponding metal carbonate, which are thermodynamically stable and thus could be used to store anthropogenic CO_2 emissions while reducing industrial solid waste *(39)*.

On the other hand, the fixation of CO_2 into both linear and cyclic carbonates is among the most promising routes for CCU *(40)*. Dimethyl carbonate (DMC) can be obtained by direct reaction of CO_2 and methanol, but this process has several limitations since the equilibrium is shifted toward the formation of CO_2 and methanol. A more sustainable alternative for the synthesis of DMC is the transesterification process of ethylene carbonate (EC), which is prepared by reaction of ethylene oxide and CO_2, with methanol *(41,42)*. DMC is an important organic product used in the manufacture of polycarbonates, as an alkylating agent and as a solvent in the chemical industry and as an electrolyte in lithium-ion batteries. The synthesis of cyclic carbonates from CO_2 has also received great interest from an industrial point of view (Scheme 6) *(43)*. It has been a commercial process for over 50 years and is currently an area of substantial research activity due to the low catalytic activity displayed by tetraalkylammonium salts which are used as catalysts and require high temperatures and pressures *(42)*.

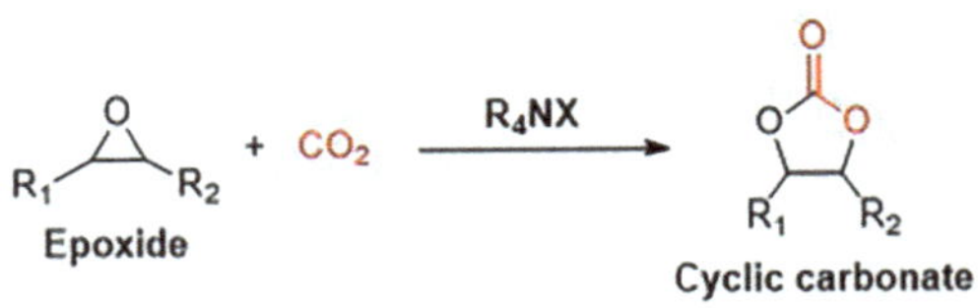

Scheme 6 Synthesis of organic carbonates from CO_2.

1.2.4 Polymers

The chemical industry and the scientific community have devoted much effort to designing new CO_2-based polymers *(44–46)*. It is worth noting several synthetic routes have been developed for polycarbonate synthesis containing aromatic or aliphatic substituents, using CO_2 as a renewable feedstock (Scheme 7) *(44)*. Among them, the copolymerization of CO_2 and epoxides to obtain aliphatic polycarbonates (Scheme 7E) is the most promising CCU technology and it is among the most studied polymerization processes in the last 20 years *(44–48)*. The industrial production of polyethylene and polypropylene carbonate from CO_2 and ethylene and propylene oxide as well as the production of polyether-polyols for the preparation of non-isocyanate polyurethanes (NIPUs), have already been developed on a pilot-plant scale *(44)*. These polymers have the potential to replace commercial polymeric materials whose market is increasing in the last decades and that are currently prepared from fossil fuel resources. Therefore, these CO_2-based polymers can be used as suitable materials to store carbon dioxide for long periods and are sustainable alternatives for the plastics industry.

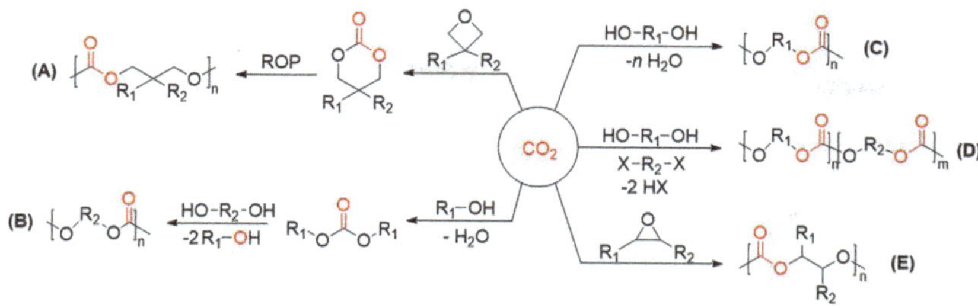

Scheme 7 Synthesis of polycarbonates from CO_2.

1.3 Conversion of CO_2 and bio-derived feedstocks into high value-added products

The use of renewable raw materials as a source of carbon from different types of resources, in some cases from waste products or emissions, requires the development and improvement of catalytic processes with high atom economy. Thus, the development and use of efficient catalysts is required to activate and transform the different types of raw materials into high added value chemical products *(18)*. Therefore, finding bio-based raw materials, which are not food relevant has become an urgent need for the preparation of industrially relevant chemical compounds and for the design of more sustainable polymeric materials *(49–52)*. Different routes to convert biomass into chemicals have been reported, but one methodology that has attracted considerable attention is the conversion into the so-called platform chemicals *(53–56)*. These are rather simple building blocks that can be prepared from renewable resources in a single step, either by fermentation or by thermochemical conversion in good yields, but still retain sufficient functionality for further conversion into compounds of industrial interest *(57)*. For example, the U.S. Department of Energy has reported that up to 12 types of compounds can be obtained from sugars *(25,58)*.

The use of waste-derived feedstocks and GHG carbon dioxide as starting materials for the highly selective preparation of chemical compounds requires the development of highly efficient catalytic processes *(25,58)*. Among the most important chemical transformation of CO_2, it is worth highlighting its reaction with epoxides to afford either cyclic- or polycarbonates, which uses CO_2 as a sustainable C_1 building block (Scheme 8). A broad range of epoxides has been used as starting materials for the preparation of both cyclic- and polycarbonates. However, the use of epoxides derived from renewable resources is mandatory in order to improve the sustainability of the process *(47,59,60)*.

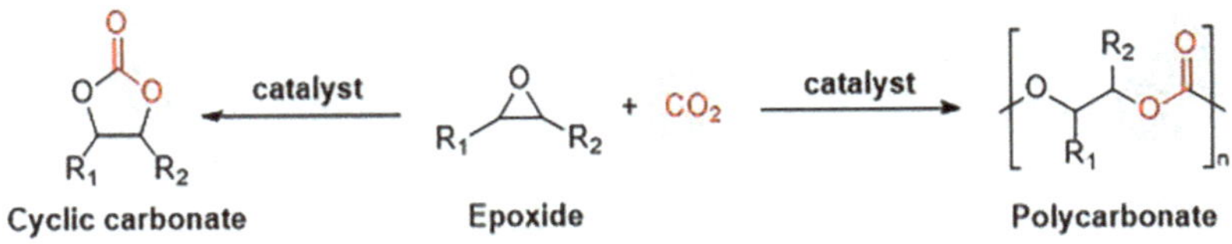

Scheme 8 Synthesis of cyclic- or polycarbonates.

In this contribution, we cover the most important developments on the use of homogeneous catalysts for the chemical transformation of carbon dioxide as a sustainable C_1 building block and bio-derived starting materials into bio-sourced cyclic organic carbonates and sustainable hybrid materials. The choice of the catalytic systems is key in order to increase their catalytic activity and selectivity as well as to control the physical and thermal properties of the polymeric materials obtained.

2. Bio-sourced cyclic carbonates

2.1 Introduction

Catalytic preparation of heterocyclic compounds still remains a challenge in the field of organic synthesis. In this context, the synthesis of cyclic carbonates from CO_2 is receiving growing attention because they are high added value chemicals, and their preparation and applications are framed within the principles of sustainable chemistry. These compounds are currently of particular interest in academic and industrial research due cyclic carbonates can be used as solvents in chemical processes and in batteries *(61–64)*, industrial lubricants *(65)* and useful intermediates for the preparation of a large variety of products *(66–70)*. Furthermore, it should be noted that some of these applications are possible due to the high stability of these compounds, that ensure CO_2 fixation for a long time as opposed to other well-known products such as urea that releases CO_2 more easily when it is used, for instance, in agriculture *(71)*. However, despite these advantages, there are important challenges that still need to be addressed in terms of catalyst synthesis, renewability, and scope of applications of cyclic carbonates. In this section, the state-of-the-art of cyclic carbonate synthesis is summarized with special focus on the preparation and applications of bio-sourced compounds.

2.2 Synthesis of cyclic carbonates

Cyclic carbonates can be obtained from different building blocks, including phosgene *(72)*, dimethyl carbonate *(73)*, urea *(74)*, carbon monoxide *(75)*, halohydrins *(76)*, alkenes *(77)* or propargyl alcohols *(78)*. However, the cycloaddition of CO_2 to constrained heterocycles such as epoxides, represents the most attractive route in terms of sustainability and atom-economy (Scheme 9). Indeed, this reaction is thermodynamically favored due to the release of the ring-strain energy contained in the epoxide substrate *(79)*. Thus, cyclic carbonates can be produced even under room temperature and one bar CO_2 pressure when highly active catalysts are employed *(43,80,81)*.

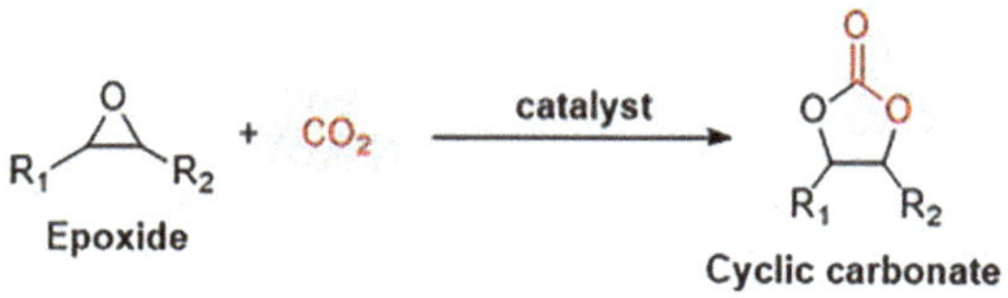

Scheme 9 Synthesis of cyclic carbonates from epoxides and CO_2.

The cycloaddition reaction of epoxides and CO_2 is also the most common method to afford cyclic carbonates at industrial level. Thus, in the Shell Omega process, ethylene carbonate (EC) is obtained by the reaction of ethylene oxide (EO) and CO_2 and subsequently hydrolyzed to produce ethylene glycol (EG) (Scheme 10) *(66,67)*. Also, in the Asahi process for the synthesis of bisphenol-A polycarbonate, ethylene carbonate is obtained from ethylene oxide and CO_2, and later reacted with methanol to afford dimethyl carbonate and ethylene glycol (Scheme 10) *(42)*. Currently, this process is implemented around the world with a total annual production of approximately 900,000 tons of polycarbonate, equivalent to approximately 16% of the global polycarbonate production capacity *(82)*. Nevertheless, it should be noted that ethylene oxide is a highly toxic gas that prevents a safe production of cyclic carbonate. Moreover, in terms of sustainability, this compound is considered a non-renewable feedstock since is produced from petroleum-based ethylene. Therefore, new sustainable routes based on the use of bio-based resources need to be considered in order to be efficiently implemented on an industrial scale.

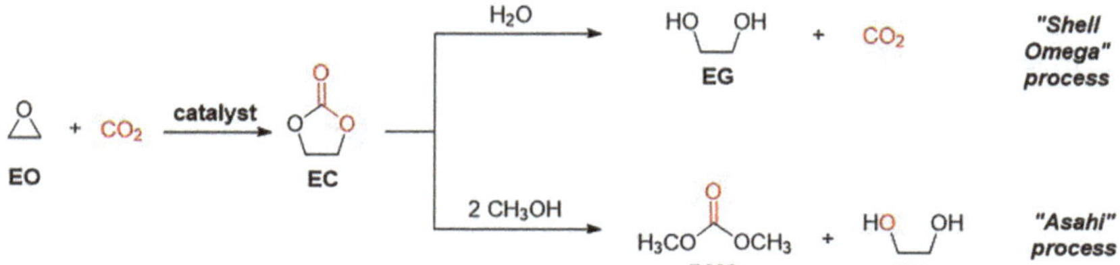

Scheme 10 Industrial approaches involving the cycloaddition reaction of ethylene oxide and CO_2.

2.3 Catalytic synthesis of bio-sourced cyclic carbonates

In recent years, the preparation of cyclic carbonates from renewable resources has received increasing attention both at academic and industrial level. As previously described, the most prominent way to synthesize these compounds is the catalytic coupling reaction of epoxides and CO_2, although depending on the nature of the starting material, other synthetic routes can be considered. It is worth highlighting that cyclic carbonates are only considered fully bio-sourced when feedstocks used for their preparation are derived from microorganisms, plants or animals. Thus, substrates such as terpenes, vegetable oils and carbohydrates, are among the most important renewable building blocks for the synthesis of these heterocyclic compounds.

2.3.1 Terpene-based cyclic carbonates

Terpenes are a class of organic products derived from isoprene (2-methyl-1,3-butadiene) which are produced predominately by plants *(83)*. The presence of double bonds which can be easily oxidized, allows the use of these compounds as starting material for the synthesis of structurally versatile cyclic carbonates. Since the isolation of the natural based-terpene 4-muurolen-7,15-diol-7,15-carbonate in the early 1990s *(84)*, several research groups have focused their efforts on the search of terpene platforms for cyclic carbonate formation. Undoubtedly, among them, limonene is likely the most studied terpene for this transformation. (*R*)-Limonene, the most abundant enantiomer isolated from citrus fruits *(85)*, can be selectively oxidized to afford either 1,2-limonene oxide or limonene dioxide. The successive cycloaddition of CO_2 to the epoxide groups in these compounds leads to the formation of the corresponding limonene-based cyclic carbonates (Scheme 11).

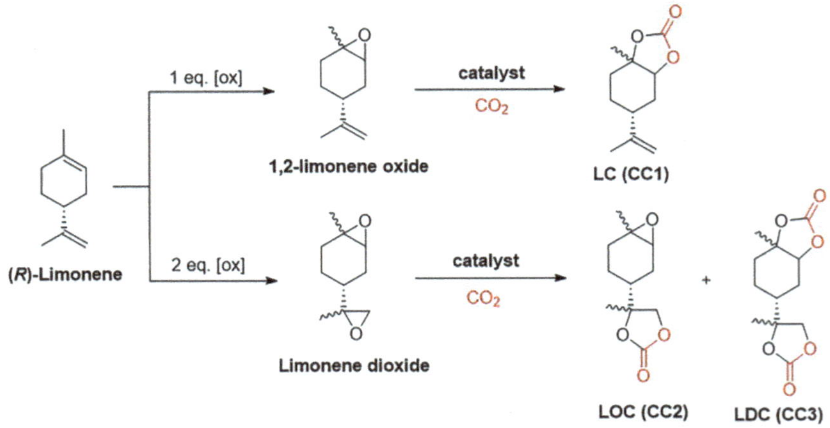

Scheme 11 Synthesis of limonene-based cyclic carbonates.

A wide range of catalytic systems have been developed for the coupling reaction of 1,2-limonene oxide and CO_2. Fig. 4 illustrates some of the most active catalysts used for this transformation *(86–90)*. In general, high temperatures (70–100 °C) and CO_2 pressures (10–50 bar), long reaction times (16–66 h) and the use of an appropriate nucleophile as cocatalyst are needed in order to obtain moderate to high conversions (57–80%). Furthermore, in some cases, the use of a solvent such as 2-butanone or acetonitrile is needed to perform the transformation *(86,88)*. Inexpensive and non-air/moisture sensitive compounds such as tetrabutylammonium salts (TBAX) have also been tested as catalysts for this transformation, but harsher reaction conditions were required *(91)*. On the other hand, it should be noted that the process is highly selective toward the formation of *trans* 1,2-limonene carbonate in most cases (>92%), suggesting a different reactivity of the starting epoxide isomers.

The coupling reaction of limonene dioxide and CO_2 has also been explored by using some of the catalysts showed in Fig. 4 *(86–89)*. In this case, a mixture of the mono- and dicarbonate compounds is obtained in all cases (Scheme 11), with moderate to high isolated yields (27–78%) for the dicarbonate product *(86–89)*. Undoubtedly, the major challenge at this point is finding the optimal catalytic system and reaction conditions in order to increase those isolated yields since this fully renewable compound with

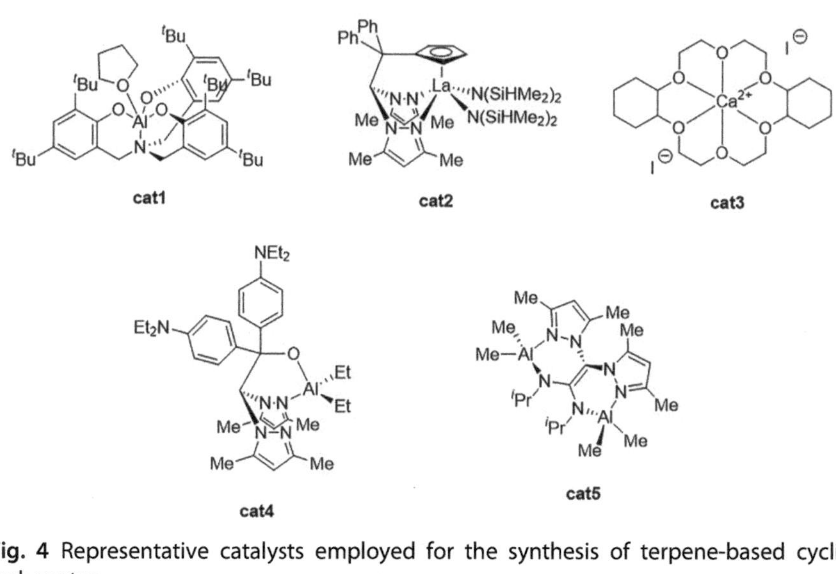

Fig. 4 Representative catalysts employed for the synthesis of terpene-based cyclic carbonates.

two cyclic carbonate groups can find application as building block for the synthesis of non-isocyanate polyurethanes *(92,93)*.

In addition to the aforementioned limonene-based products, catalysts collected in Fig. 4 have also been employed for the coupling reaction of CO_2 and other terpenes-based epoxides. For instance, **cat1** was used as catalyst in combination with bis(triphenylphosphine)iminium chloride (PPNCl) for the synthesis cyclic carbonates derived from carvone (**CC4**), menthene (**CC11**), citronellyl acetate (**CC16**), geranyl acetate (**CC17**), linalyl acetate (**CC18**) and neryl acetate (**CC19**) (Fig. 5) in moderate yields *(87)*. Moreover, bicyclic products **CC2,4,6** were obtained with high stereoselectivity and crystal structures for **CC2**, *trans*-**CC3** and *cis*-**CC4** were elucidated by X-ray diffraction analysis for the first time. In a later work, menthene-based cyclic carbonate **CC11** could also be isolated in high yield using the catalytic system comprised by **cat3** (Fig. 4) and triphenylphosphine as cocatalyst *(88)*. This system was also used for the synthesis of citronellyl and geranyl derivatives (**CC16,17**), but low yields were achieved even under more drastic conditions *(88)*.

More recently, new terpene-based cyclic carbonates have been synthesized using the combination of aluminum complex **cat4** (Fig. 4) and tetrabutylammonium chloride (TBAC) *(89)*. For instance, new carvone (**CC5,6**) and carvyl-based derivatives (**CC7,8**) (Fig. 5) were obtained under relatively mild reaction conditions. Among them, heterocyclic compounds **CC5** and **CC7** could be isolated as enantiomerically pure compounds after crystallization. Furthermore, it was observed that *endo* carvone-based cyclic carbonates decomposed under ambient conditions to the corresponding diol with elimination of CO_2. Finally, a higher catalyst loading was required for the production of new cyclic carbonates **CC9,10,12–15** in moderate yields (Fig. 5) *(89)*.

2.3.2 Vegetable oils-based cyclic carbonates

Vegetable oils are compounds of interest due to their availability in large quantities as renewable feedstock from food and crops waste *(94,95)*. The high number of double bonds accessible to be oxidized into their structure makes them attractive starting materials for the production of fully renewable polymers. Thus, for instance, epoxidized vegetable oils can be appropriately carbonated using CO_2 and subsequently employed as building blocks for the synthesis of NIPUs *(96–98)*. Soybean, sunflower and linseed oils, but also olive, castor, cottonseed and vernonia oils to a lesser extent, have been extensively explored for the synthesis of oil-based cyclic

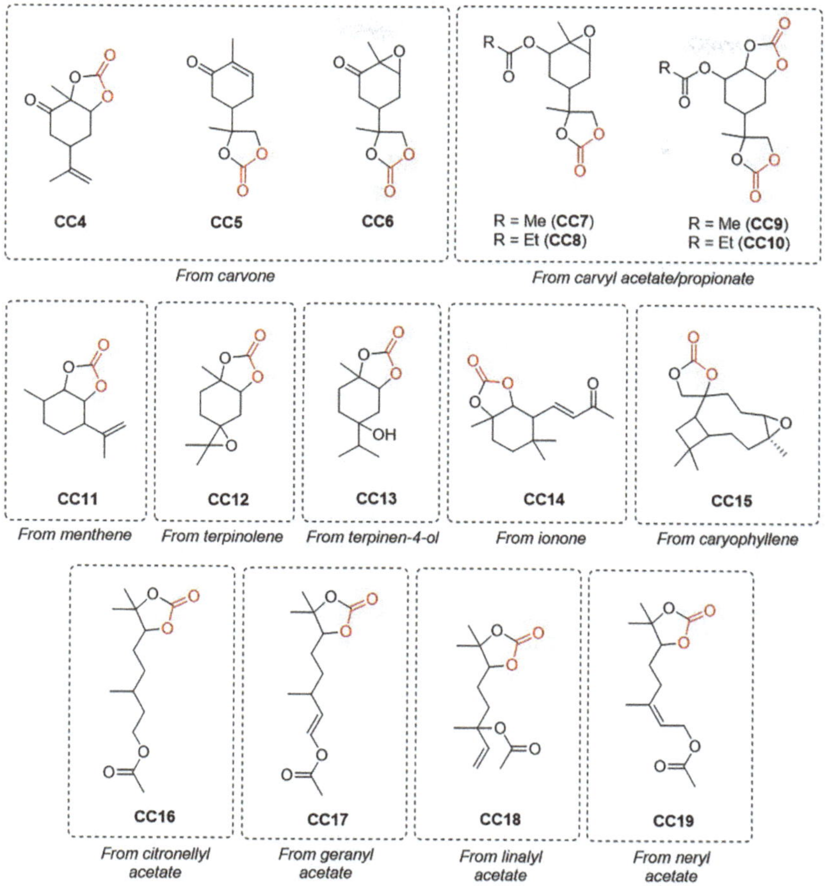

Fig. 5 Representative terpene-based cyclic carbonates.

carbonates *(99)*. Considering the number of double bonds per triglyceride unit, vegetable oils can be classified based on their composition in three different groups: oleic (18:1), linoleic (18:2) and linolenic (18:3). Thus, soybean oil can be considered as oleic due to this fatty acid is predominant in its structure (~90%).

After suitable oxidation of these oils, which is commonly carried out with hydrogen peroxide, carbonation with CO_2 can be performed (gas or supercritical) using an appropriate catalyst. This transformation has been widely studied using tetrabutylammonium bromide (TBAB) as catalyst, achieving in most cases good to high yields under different reaction conditions *(100–105)*. Notably, conversions above 77% could be achieved for the carbonatation of soybean and linseed oils even when the process was carried out at one bar of CO_2 pressure *(103–105)*. Moreover, it has been observed that the addition of appropriate amounts of water increased the reaction rate, and a further acceleration could be accomplished by combining the addition of water to the use of a microwave reactor *(103,104)*. In addition to TBAB, some organocatalysts have also been studied for the efficient conversion of epoxidized vegetable oils into the corresponding carbonated ones (Fig. 6) *(106–110)*. Thus, **cat6** and **cat7**/TBAC systems showed high catalytic activity for the synthesis of different carbonated vegetable oils under moderate temperature and/or CO_2 pressure, achieving high selectivity in most cases when ascorbic acid-based catalytic system **cat7** was used *(106,107)*. Carbonated soybean oil could be obtained in quantitative yields by using low amounts of the combination of organocatalyst **cat8** and TBAB under supercritical CO_2 conditions *(108)*. In addition, organic phosphorus salts have also proved to be selective catalysts for this transformation. Thus, catalytic system comprised by **cat9** and iron trichloride as additive led to the synthesis of carbonated linseed and sunflower oils in high yields *(109)*. More recently,

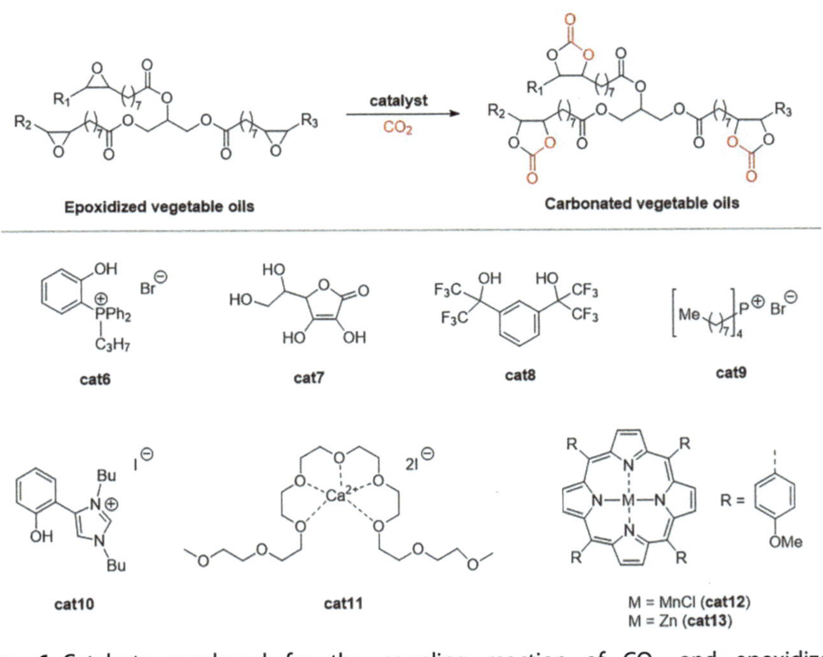

Fig. 6 Catalysts employed for the coupling reaction of CO_2 and epoxidized vegetables oils.

a wide range of epoxidized vegetable oils from different sources, such as oleaginous seeds, nuts and fruits, have been successfully carbonated under relatively mild reaction conditions by employing the bifunctional imidazole-based organocatalyst **cat10** *(110)*.

Some catalytic systems based on metals have also been investigated for this reaction. For instance, tin chloride compound $SnCl_4{\cdot}5H_2O$ in combination with TBAB allowed the quantitative production of carbonated soybean oil under moderate CO_2 pressure, however, high temperature was required *(111)*. Also, the simple alkaline earth salt $CaCl_2$ in combination with TBAB displayed a catalytic activity similar to the previously mentioned catalytic systems for the synthesis of carbonated soybean oil although at a significantly higher reaction temperature *(112)*. The carbonation of sunflower oil by using calcium iodide in combination with a (poly)ether complexing agent (**cat11**, Fig. 6) *(113)*, and the synthesis of carbonated sunflower, soybean and linseed oils using the previously mentioned catalytic system **cat3**/Ph_3P (Fig. 4) were performed under milder reaction conditions *(88)*. Finally, metalloporphyrin complexes **cat12–13** (Fig. 6) have also been tested for this transformation in the presence of TBAB. However, while manganese-based system **cat12** displayed high catalytic activity for the carbonation of epoxidized sunflower oil, low conversions were obtained when zinc analogue **cat13** was used under identical reaction conditions *(114)*.

2.3.3 Carbohydrate-based cyclic carbonates

Carbohydrates are another important and abundant source of natural compounds for the production of bio-derived cyclic carbonates. Due to the multiple alcohol groups bearing on its structure, a wide range of architecturally diverse cyclic carbonates can be prepared from them. It is noteworthy that some of these heterocyclic compounds, especially 6-membered cyclic carbonates, have acquired a particular interest in recent years due to they can be employed as monomers for the production of sugar-based polycarbonates *via* Ring-Opening Polymerization (ROP) processes *(115,116)*.

Traditionally, the synthesis of these heterocyclic compounds has involved the use of phosgene or phosgene derivatives, which are toxic reagents synthesized in an intensive process. Alternative methodologies to this process have consisted on the use of non-sustainable routes, also achieving low to moderate yields in most cases *(117–119)*. Therefore, new more sustainable approaches have been developed in recent years.

Thus, 1,3-butanediol and 2,5-pentanediol could be properly reacted with CO_2 to form the corresponding six-membered cyclic carbonates by using 1,8-diazabiciclo[5.4.0]undeca-7-eno (DBU) as catalyst (Scheme 12) *(120)*. In this process, an ionic carbonate intermediate is formed in the first step, and the subsequent reaction with tosyl-chloride (TsCl) in the presence of triethylamine (NEt_3) led to the formation of the final product. Furthermore, it should be noted that the stereochemistry was maintained during the reaction, suggesting that the carbonate intermediate group is tosylated instead of the residual free alcohol *(120)*.

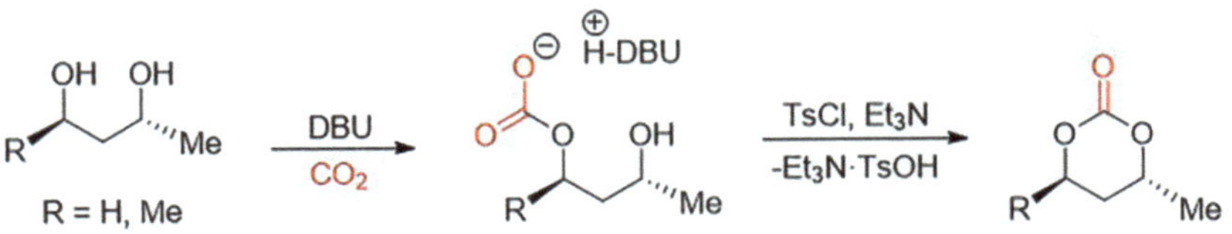

Scheme 12 Synthesis of 6-membered cyclic carbonates from 1,3-diols.

This methodology was applied for the synthesis of D-xylose-derived cyclic carbonate (Fig. 7, **CC20**), but only 14% isolated yield was obtained. The process was later improved when NEt_3 or 2,2,6,6-tetramethylpiperidine (TMP) were used as catalyst, achieving higher yields for the synthesis of the aforementioned D-xylose cyclic carbonate, but also for the production of three new D-mannose derivatives (Fig. 7, **CC20–23**) *(121)*. Pyranose based-cyclic carbonates have also been reacted with CO_2 using a similar route to afford the corresponding heterocyclic compound. For instance, a D-ribose-derived cyclic carbonate (Fig. 7, **CC24**) could be obtained in good yield by using DBU as catalyst. In this case, a previous step based on the selective pre-activation of the secondary alcohol at the 3-position of the 2-deoxy-D-ribose was required, resulting in the inversion of the stereochemistry observed in the process *(122)*. Later on, a new methodology was designed in order to yield cyclic carbonates with retention of the stereochemistry. Thus, by the selective protection and bromination of the 6-position of the corresponding carbohydrate as starting material, followed by the reaction with CO_2 in the presence of DBU as catalyst, new six-membered cyclic carbonates from sugars, such as D-glucose, D-mannose and D-galactose, could be produced (Fig. 7, **CC25–27**) *(123)*. A similar approach was applied for the regioselective synthesis of 5-membered cyclic carbonates from D-mannose and D-galactose (Fig. 7, **CC28,29**), through a simple one-step reaction involving dibromomethane *(124)*.

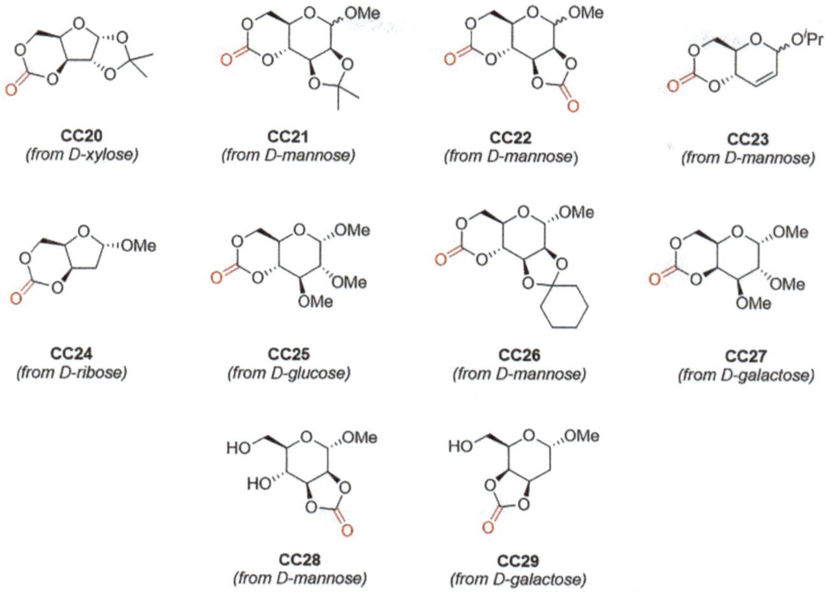

Fig. 7 Carbohydrate-based cyclic carbonates.

3. Fully bio-renewable polymers from CO_2

3.1 Introduction

Currently, there is an environmental problem caused by highly durable or toxic polymers, most of them are derived from fossil raw materials. The contribution of these materials to the development of the Society must be recognized, although their chemical properties are causing environmental problems, damage to plants, animals and even affects human health. In January 2018, the European Union released its vision for a more sustainable polymer industry to be achieved by 2030 *(125)*. In fact, the word of the year 2018 was "microplastic," reflecting the importance that this problem has acquired. Chemistry must take on the challenge of designing new biodegradable and bio-renewable polymeric materials that can replace fossil-based polymers in most of their applications *(126)*.

Polymeric materials are very important for the welfare state of the Society, since they are necessary for basic uses such as packaging, insulators, adhesives among others and for advanced applications such as automotive industry, medicine, medical devices, or technology known as engineering polymers. Therefore, polymers and consequently their monomers play an important role in our everyday life *(57)*. Hence, new strategies to yield innovative monomers from renewable bio-based feedstocks and platform chemicals are necessary to replace most of the fossil-fuel derived monomers currently used. In the last decades, the development of completely recyclable polymers has become an emerging frontier in polymer chemistry; however, the efficient synthesis of monomers or/and polymers from renewable feedstocks under environmentally friendly conditions with desirable physical properties and mechanical strength required for practical uses still represents a challenge.

As previously discussed, CO_2 is a renewable carbon feedstock to replace fossil resources in polymeric materials *(44,127–130)*. The advantages of using CO_2 as a carbon source are its abundance, low toxicity and low price, but the transformations should be oriented to a large-scale synthesis of bulk and fine chemicals *(131)*. The combination of CO_2 and platform molecules from bio-renewable resources to yield polymeric materials would be a sustainable route for the design of more friendly polymers which could reduce greenhouse gases emissions and increase resource efficiency for polymer production *(132)*.

This section aims to collect the most relevant CO_2-containing biopolymers and highlight the most important conceptual routes that have been developed for the chemical fixation of CO_2 together with other bio-renewable starting materials into polycarbonates, one of the most important examples of hybrid materials derived from CO_2 which have shown great interest to industry and the technological development of society. Two strategies for incorporating CO_2 into sustainable polymers have been reported: (a) the direct utilization of CO_2 as a co-monomer for the copolymerization process, and (b) the (co)polymerization of CO_2-based platform molecules as building blocks (Scheme 13).

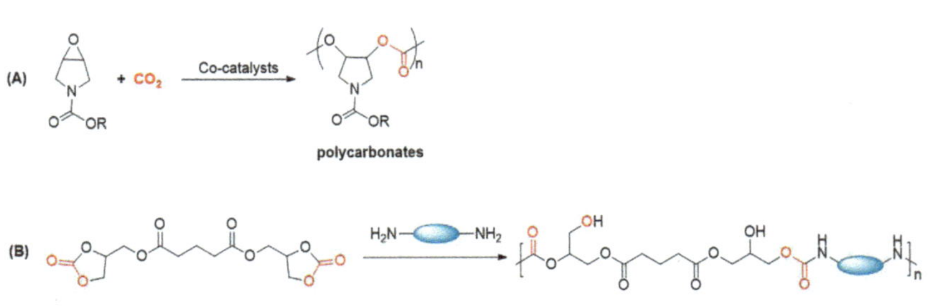

Scheme 13 Strategies for incorporating CO_2 into polymers; (A) direct utilization of CO_2 *(133)*; (B) (co)polymerization of CO_2-based building blocks *(89)*.

Even though various renewable CO_2-based polymers have shown to be potential alternatives to current non-bio-derived polymeric materials such as polyurethanes, polyesters and terpolymers, special attention will be devoted to discussing polycarbonates (Fig. 8).

3.2 Synthesis of polycarbonates

Polycarbonates (PCs) are a type of polymer with carbonate group (-O-CO_2R-) repeat units within their main backbone and for which CO_2 can be used as building block for their synthesis *(44)*. During the last few years, polycarbonates have arisen a great interest within the scientific and industrial community due to their wide range of applications and properties. Polycarbonate materials can be divided into aliphatic and aromatic polycarbonates. Currently, aromatic PCs are the most interesting for industry due to their excellent mechanical and thermal properties such as high impact resistance, stiffness, toughness, good thermal stability, transparency and flame retardancy. These materials are used as engineering plastics in automotive, electrical and electronic devices and in construction. The most important aromatic PC is prepared in industry by melt polycondensation of bisphenol-A and phosgene (Scheme 14A) *(134–137)*. In 2002, the Japanese company Asahi Kasei Corporation launched the first industrial phosgene-free route for polycarbonate production *(42)*. As previously discussed, this synthetic route uses CO_2 as a C_1 synthon to react with ethylene oxide resulting in the formation of ethylene carbonate, which is then converted into dimethyl carbonate by reaction with MeOH. Then, the acyclic carbonate is treated with phenol to afford diphenyl carbonate, which is finally used as a building block for the synthesis of a CO_2-sourced aromatic PC by melt transesterification reaction with bisphenol-A (Scheme 14B) *(42,138,139)*.

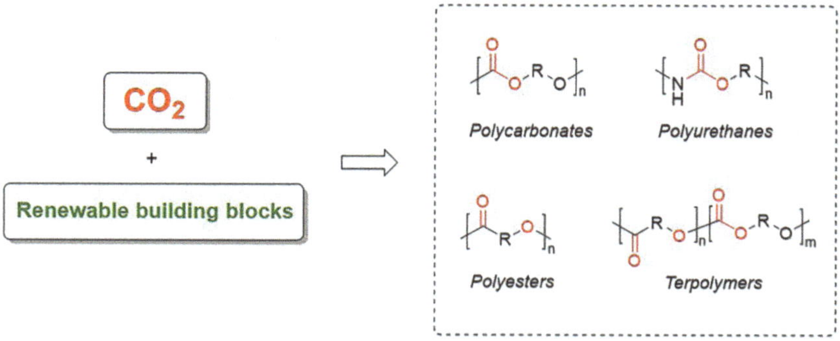

Fig. 8 Some fully renewable CO_2-based polymeric materials.

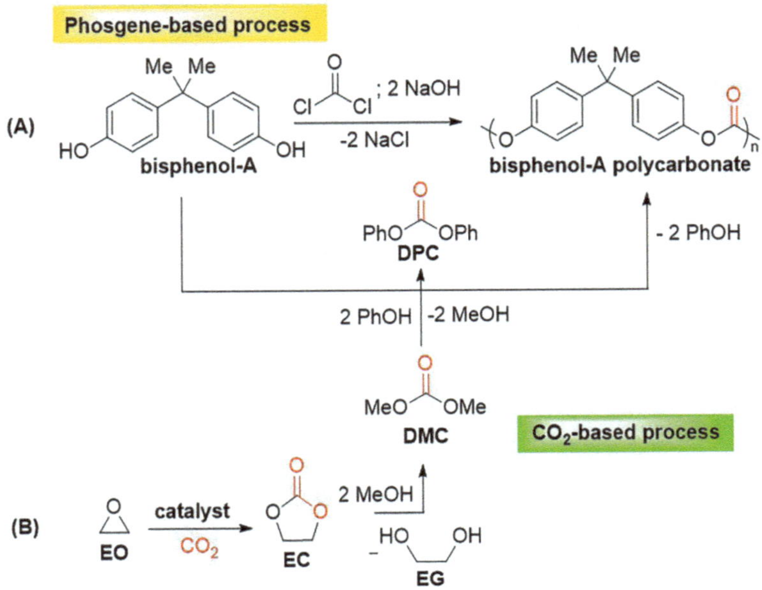

Scheme 14 Synthesis of aromatic PCs by a phosgene- or a CO_2-based process.

On the other hand, aliphatic PCs exhibit poor mechanical properties which restrict their scope of application to packaging, as a binder in ceramics or as polyols for the formulation of polyurethane foams *(140–144)*. Nevertheless, these polymers have very interesting properties such as low toxicity and good biodegradability and biocompatibility which make them useful candidates in the biomedical sector *(145–149)*. In recent years, these properties have drawn the attention of the scientific community and industry to produce new potentially biodegradable and bio-renewable materials with mechanical properties similar to those of aromatic PCs *(48)*. It is worth highlighting that some of these aliphatic polycarbonates have been designed as fully bio-derived polymers *(47)*. Aliphatic PCs are largely produced in industry by melt polycondensation of diols and alkyl carbonates (Scheme 14B).

There are different alternative routes for the industrial PCs synthesis that utilize CO_2-sourced building blocks or CO_2 as monomers (Scheme 7) *(44)*. Among all of them, the ring-opening copolymerization (ROCOP) of CO_2 with oxiranes is the most used for the preparation of polycarbonates fully derived from renewable sources. To achieve this objective two routes are possible: (a) the use of a CO_2-sourced building blocks obtained from a metabolite, or (b) the use of a precursor directly obtained from renewable feedstocks (Fig. 9). For instance, ethanol is a metabolite which is directly

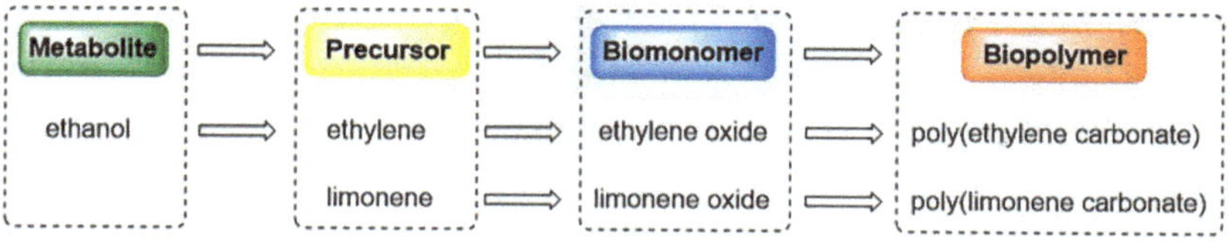

Fig. 9 Biopolymer production from metabolites and precursors.

produced in the metabolism. Thus, while ethanol can be directly obtained from renewable resources, it must go through additional reaction steps to yield an epoxide monomer. However, it should be noted that each of these extra steps decreases the greenness of the renewable monomer. Alternatively, the ideal process would be using an olefin precursor, such as limonene, which can be extracted from renewable resources and converted into the epoxide in a direct one-step process.

3.2.1 Synthesis of bio-poly(carbonate)s from CO_2-sourced building blocks

Fully bio-based PCs have been synthesized by catalytic ROP of cyclic carbonates. The ROP of 5-membered ring CCs (5CCs), such as EC or propylene carbonate (PrC) is thermodynamically disfavored because of a high positive polymerization enthalpy *(44)*. However, the ROP of 5CCs is possible if accompanied by some degree of decarboxylation, which induces the formation of ether units at the chain end, increasing the entropy of the system and affording a poly(ether-*co*-carbonate) material *(150,151)* (Scheme 15).

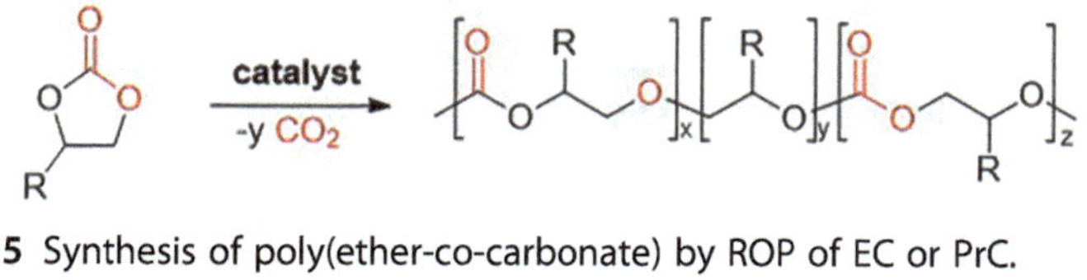

Scheme 15 Synthesis of poly(ether-co-carbonate) by ROP of EC or PrC.

Since EC or cyclohexene carbonate (CHC) could be obtained from renewable precursors *(57)*, their ROP could generate fully renewable PCs. Low molecular weight copolymers with a carbonate content of 10–50% have been obtained with bases, Lewis acids and transesterification catalysts for the ROP of EC and some degree of decarboxylation was required *(150,151)*. On the other hand, the ROP of *trans*-CHC could be carried out without decarboxylation. For example, isotactically pure poly(cyclohexene carbonate) has been prepared with M_n of up to

21,000 g mol^{-1} by ROP of enantiopure *trans*-(*R*,*R*)-CHC at 60 °C using a bulky Zn(diphenolate) complex as a catalyst and benzyl alcohol as an initiator *(152)*. High molecular mass copolymers (up to 82,000 g mol^{-1}) have been prepared by a ROCOP strategy, which facilitates EC polymerization when is copolymerized with different cyclic esters (Scheme 16) *(153)*. It is worth highlighting that the synthesis of a series of copolymers with M_n of up to 90,000 g mol^{-1} and an EC incorporation of up to 38% without loss of CO_2 could be carried out by controlled ROCOP of EC with various lactones such as *β*-butyrolactone (*β*-BL), *δ*-valerolactone (*δ*-VL), *ε*-caprolactone (*ε*-CL) or L-lactide (L-LA) using zinc complexes or 1,5,7-triazabicyclo[4.4.0] dec-5-ene (TBD) and a protic source as an initiator (Scheme 16) *(154)*.

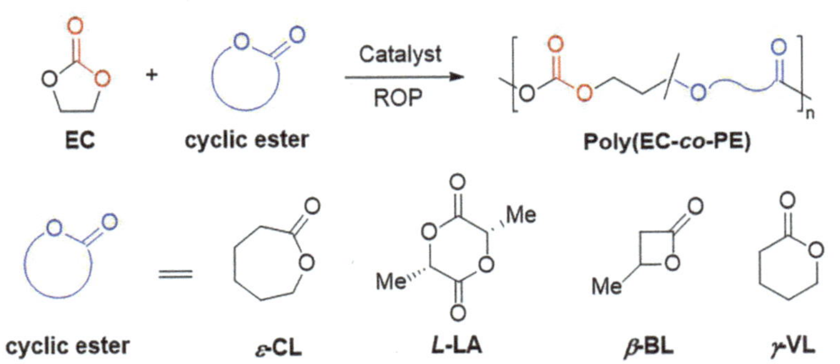

Scheme 16 ROCOP of EC with cyclic esters.

Anionic ROP of 4,6-O-benzylidene-2,3-O-carbonyl-*R*-D-glucopyranoside, a cyclic carbonate derived from a bio-renewable precursor, yielded low molecular weight PCs under mild reaction conditions using DBU as a catalyst at 30 °C for 12 h (Scheme 17) *(155)*.

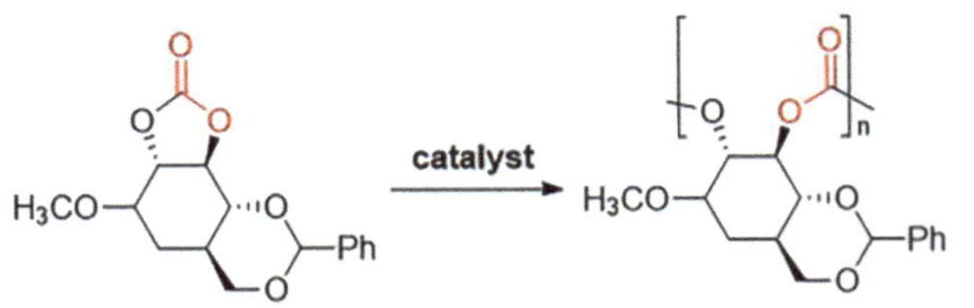

Scheme 17 Anionic ring-opening polymerization of a sugar-based cyclic carbonate.

ROP of 6-membered ring cyclic carbonates (6CCs) as monomers is thermodynamically more favored with respect to 5CCs ones because the ring-strain of 6CCs such as trimethylene carbonate (TMC) or functionalized analogs enables their ROP under relatively mild reaction conditions (Scheme 18) *(156)*. Different mechanisms including cationic, anionic, coordination–insertion, enzymatic or organocatalytic approaches, have been reported for the ROP of 6CCs to yield a wide range of aliphatic PCs with high molecular weights and controlled microstructures *(145,157,158)*. However, the synthesis of 6CCs can be tedious and only a handful of effective methods have been reported so far *(159)*. As previously mentioned for EC and CHC, TMC could be obtained from renewable metabolites, so TMC-based PCs could be considered as bio-renewable *(57)*, although these monomers are currently obtained from mineral sources.

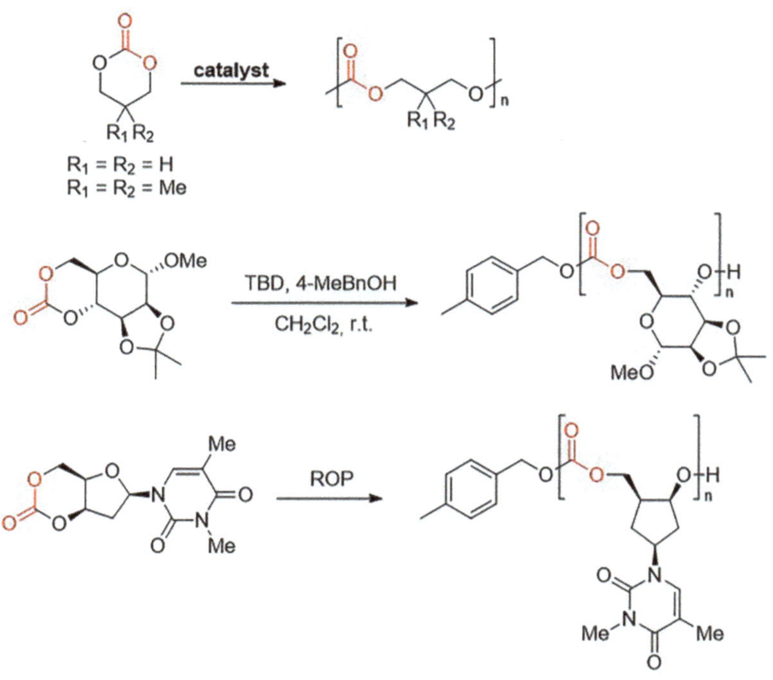

Scheme 18 Aliphatic PCs by ROP of 6CCs.

Recently, several PCs with different structures have been prepared at room temperature from different CO_2-sourced sugar-based monomers by using organocatalysts. These aliphatic polycarbonates exhibited high-temperature resistance and demonstrated possibility for post-polymerization functionalization, displaying potential application as high-performance commodity and biomedical materials (Scheme 18) *(115,122,123)*. Different low molecular weight PCs could be isolated from D-xylose-based monomers by ROP of a six-membered cyclic carbonate or by one-step ring-opening copolymerization process of a four-membered cyclic ether with carbon dioxide leading to linear polycarbonates (Scheme 19) *(160)*.

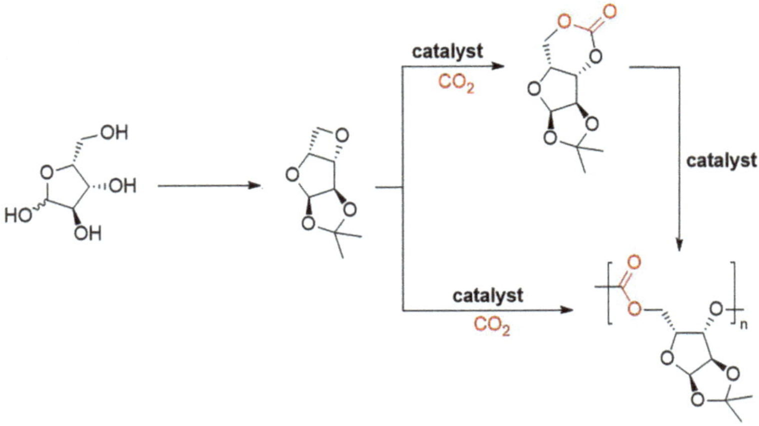

Scheme 19 Synthetic approach to produce linear polycarbonate derived from naturally sourced D-xylose.

3.2.2 Synthesis of bio-poly(carbonate)s using CO_2 as a monomer

Chemical transformations which incorporate CO_2 as an environmentally benign carbon feedstock and waste derived feedstocks in non-reductive and highly selective pathways are poised to become extremely important in the coming years. Different ways for PCs synthesis from CO_2 have been identified over the years, including the ROCOP of CO_2 and epoxides, the step-growth polymerization between diols and CO_2, and the stepwise polymerization between CO_2, dihalides and diols (Scheme 7). Among these transformations, the copolymerization of CO_2 and epoxides is the most intensively studied process since it fulfills the green chemistry principles (Scheme 20) *(44,47,48)*.

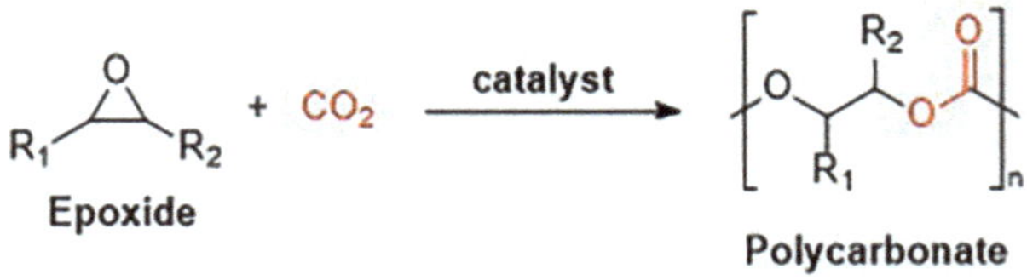

Scheme 20 Polycarbonates by ROCOP of epoxides and CO_2.

Although there are numerous epoxides that can be effectively copolymerized with CO_2, it is necessary the use of epoxidized monomers derived from renewable resources to improve the sustainability of this process *(47)*. Many copolymerization processes of CO_2 with epoxides that can be obtained from a metabolite (e.g., alkylene oxides from bio-ethanol) have been reported *(57)*; however, there is a limited number of ROCOP processes that use bio-sourced epoxides as starting materials. Among them, there are few examples of PC synthesis for which the epoxide precursors are directly obtained from renewable raw materials (Fig. 10) *(47,131,161)*.

Since Inoue et al. reported in 1969 the synthesis of poly(propylene carbonate) from propylene oxide and CO_2 using a heterogeneous zinc catalyst *(162)*, the ROCOP of epoxides and CO_2 has been extensively developed, both in academia and industry. Chemical companies like Empower *(144)* and Novomer *(163)* have industrially produced PCs from ethylene or propylene oxide and CO_2 to produce low molecular weight polycarbonate-polyols as replacements for polyether-polyols in the synthesis of polyurethanes.

A wide scope of highly active and selective metal-based catalysts with the general formula "L_nMX" (L_n is a multidentate ligand; X is commonly an anionic ligand which act as initiating group; M is typically a di- or trivalent metal center) has been developed for the ROCOP of epoxides with CO_2 *(48)*. Recently, organocatalysts have provided an alternative for the metal-free synthesis of PCs *(164–166)*. The ROCOP reaction has interesting features: (i) high atom-economy; (ii) linear increase of the polymer molecular weight with respect to epoxide conversion; (iii) the initiation step is faster than the propagation step leading to low polydispersity PCs; (iv) design of PCs with tunable properties depending on the epoxide used. The mechanism for these ROCOP processes has been well-established and most of them follow a coordination-insertion mechanism (Scheme 21). First, a Lewis acidic center activates the epoxide through coordination, after which the activated monomer undergoes ring-opening *via* nucleophilic attack of the initiating species (X). The resulting metal-alkoxide is then converted into a metal carbonate by a formal CO_2 insertion (Scheme 21, initiation). In a second step, chain propagation occurs by repetitive and sequential insertion of epoxides and CO_2 leading to fully alternating PCs. Depending on the metal center, ligand and/or epoxide structure, the chain propagation is accompanied by potential reactions which may include: (i) two or more consecutive chain enchainments that involve epoxide

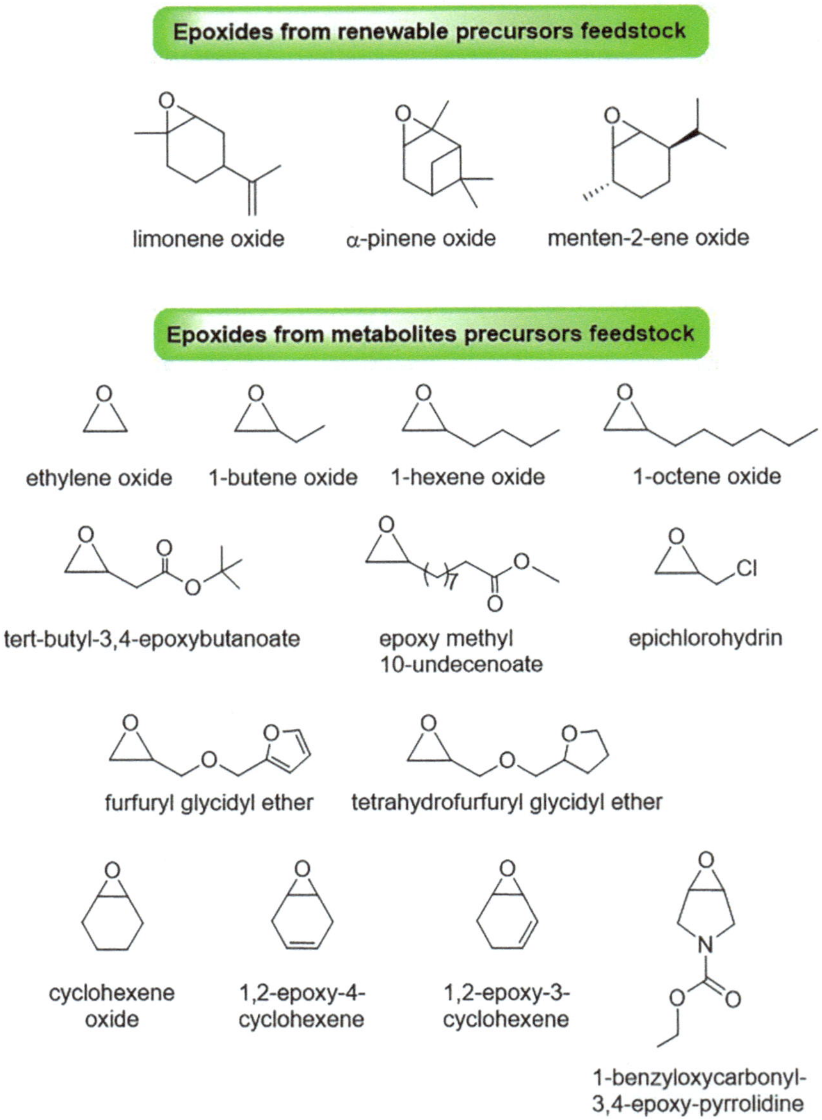

Fig. 10 Epoxides that have been reacted with CO_2 to yield polycarbonates from metabolites and from renewables precursors.

monomers resulting thus in the formation of ether linkages; (ii) intramolecular back-biting reactions that involve alkoxide- or carbonate-terminated chain-ends producing thermodynamically more stable five-membered cyclic carbonates as by-products; (iii) the presence of protic impurities (such as ROH or H_2O) may induce chain transfer reactions that affect the molecular weights and the polydispersity values of the produced PCs *(48)*.

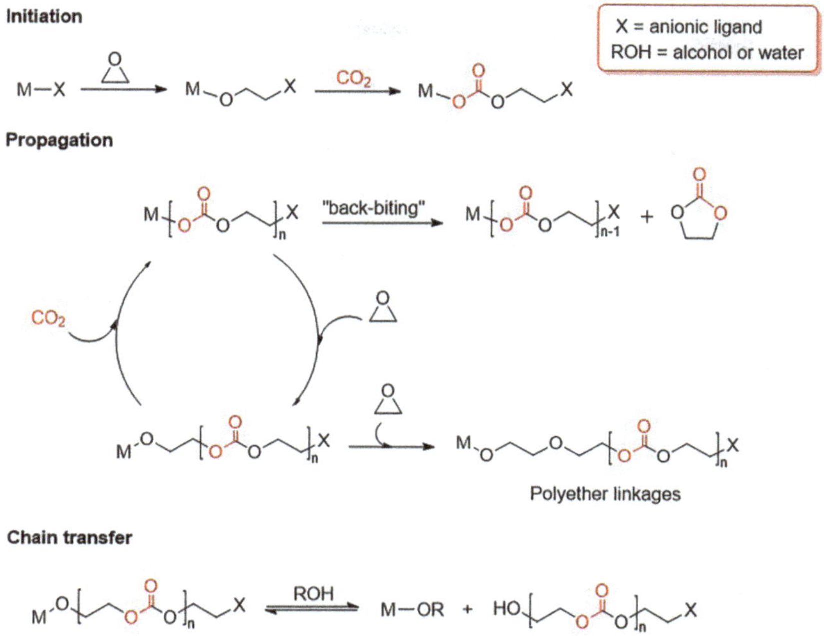

Scheme 21 Coordination-insertion mechanism for the synthesis of polycarbonates from epoxides and CO_2.

In the case of terminal epoxides, the ring-opening reaction can occur at both C—O bonds. Thus, when the attack is at the methylene C—O bond, the adjacent methine carbon retains the stereochemistry after the copolymerization with CO_2. On the other hand, either retention or inversion of the original stereochemistry are possible for the ring-opening reaction when occurs at the methine C—O bond (Scheme 22). The PC properties are also influenced by the regioselectivity of the epoxide ring-opening and the stereochemistry of the carbonate sequences. Thus, depending on the regioselectivity in terminal epoxide ring-opening reactions, head-to-head (H–H) and tail-to-tail (T–T) carbonate linkages can be produced by sequential ring-opening at the methine C—O bond followed by ring-opening at the methylene C—O bond, or head-to-tail (H–T) linkages, obtained by successive ring-opening reactions at the same carbon atom of the epoxide monomer. The stereochemistry of the PCs is also affected by the regiochemistry of the epoxide ring-opening. Thus, three different stereochemistries can be possible affording either isotactic, syndiotactic or atactic polymer domains (Scheme 22) *(44,48,167)*.

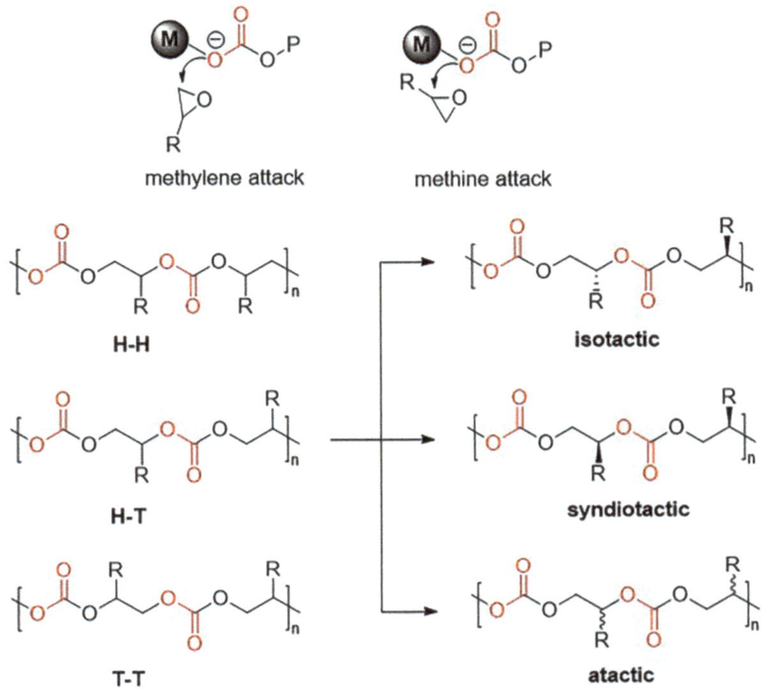

Scheme 22 Regio- and stereoselectivity in the copolymerization processes of epoxides and CO_2.

From an industrial point of view, the most important PCs obtained by this synthetic route are those obtained from EO and PO. Although these monomers can be obtained from different biomass feedstocks, they are currently derived from petroleum sources when are used in industry. On the other hand, cyclohexene oxide is the most used monomer in ROCOP processes to afford PC materials, and it has become a reference substrate to check the activity of the catalysts developed for this process. For a more comprehensive overview, readers are encouraged to consult the excellent reviews dedicated to the ROCOP processes of these substrates *(44,48,167)*. In this section, only copolymerization processes that involve epoxides derived from fully renewable precursors as substrates and CO_2 to produce fully bio-based polycarbonates will be discussed. To date, only three epoxides derived from terpenes have been used as bio-renewable substrate for the synthesis of PCs (Fig. 11). These terpenes are bio- and non-food-based resources *(168,169)*, a very important fact to be considered as platform molecules to replace fossil carbon sources.

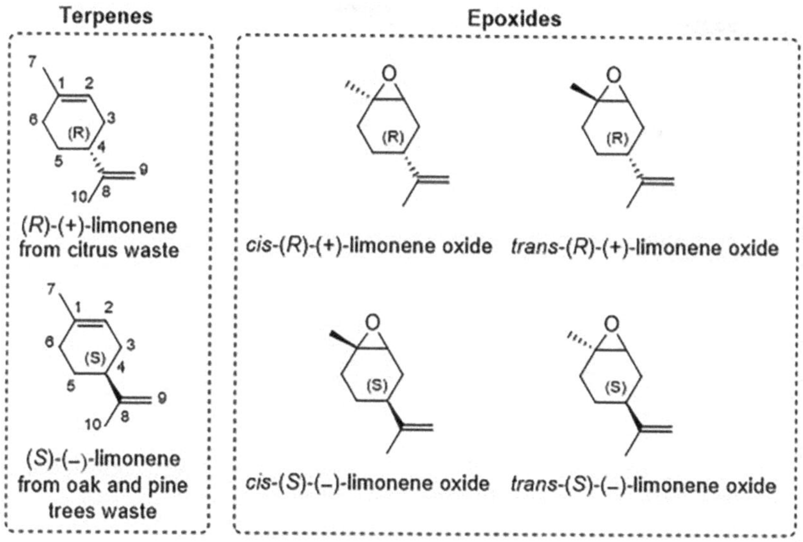

Fig. 11 Isomers of limonene and limonene oxide.

Limonene oxide (LO) has been by far the most used bio-derived epoxide for the synthesis of PCs by ROCOP with CO_2. Limonene, a terpene precursor, is the major component of orange oil (>90%), it is an abundant and readily available side product of the orange industry, with an annual production of roughly 500 kt per year (Fig. 11) *(170)*. Limonene has been recognized as a safe substance by the US FDA and furthermore it contains a chiral center and two functionalizable olefin bonds. Therefore, it is a very important renewable platform for the circular economy and the biorefinery concept *(170)*. After epoxidation, there are four possible isomers of limonene oxide, which are shown in Fig. 11. (*R*)-Limonene oxide, which is derived from (+)-limonene, is significantly the most used since it is the most economical isomer due to the large availability of (+)-limonene. Both (*R*)-LO and (*S*)-LO are sold as a mixture of *cis/trans* isomers in a ratio close to 40/60, respectively *(47)*. These geometric isomers can be enriched by various methods. For instance, *trans*-(*R*)-LO can be enriched by a two-step process wherein (*R*)-limonene is first reacted with N-bromosuccinimide to produce the *trans*-bromohydrin derivative (Scheme 23). Then, the *trans*-bromohydrin is ring-closed using aqueous NaOH to yield 85% of *trans*-(*R*)-LO isomer *(171)*. The use of enantiomerically enriched starting materials is a very interesting approach to produce polymeric materials with high regio- or stereoregularity.

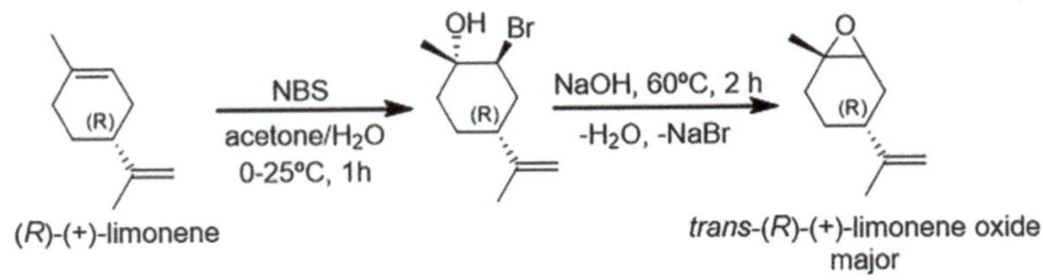

Scheme 23 Synthesis of enriched *trans*-(*R*)-LO.

The first successful ROCOP process of LO and CO_2 was reported in 2004 by using a β-diiminate acetate zinc complex as catalyst (**cat14**, Scheme 24) *(172)*. After this outstanding result, other functionalized β-diiminate zinc catalysts have been developed over the last decade (**cat14-cat17**, Scheme 24) *(171,173–175)*. These zinc catalysts were active under mild reaction conditions (25 °C, 10 bar) yielding PLC with moderate to high molecular weights and polydispersity (PDI) values close to one. Although the catalytic activity for these zinc catalysts was moderate, neither production of limonene carbonate byproduct nor polyether linkages were observed. The catalytic productivity was highly dependent on the ligands used, thus the incorporation of electron-withdrawing groups in the imine moiety and the use of a silylamide, $N(SiMe_3)_2$, as initiator group instead of an acetate group, resulted in a significant enhancement of the catalytic activity displayed by these complexes. Thus, the zinc complex containing two electron-withdrawing CF_3 groups in the imine and $N(SiMe_3)_3$ has shown to be the most active catalyst for PLC production to date, reaching a maximum TOF of 310 h^{-1} at 60 °C (**cat17**, Scheme 24) *(176)*. All Zn-based catalysts have shown preferential reactivity toward *trans*-(*R*)-LO isomer.

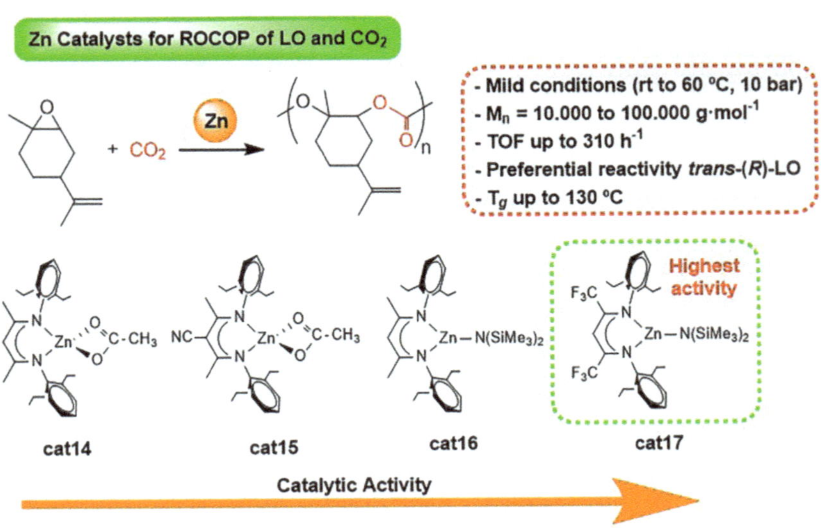

Scheme 24 Zn catalysts used in the ROCOP of limonene oxide and CO_2.

The combination of aminotriphenolate aluminum complexes and PPNX (X = Cl, Br) as a cocatalyst has also displayed high catalytic activity for the synthesis of PLC (Scheme 25) *(177)*. In general, these Al catalysts showed lower catalytic activity than the *β*-diiminate zinc complexes but they were active for PLC synthesis under relatively mild reaction conditions (25–45 °C and 0.5–10 bar) giving rise to moderate molecular weight PLCs with low polydispersity values. Even though zinc catalysts preferentially react with *trans*-(*R*)-LO, the catalyst system comprised by an aminotriphenolate aluminum complex and PPNCl can react with either isomer, but the reaction is faster and more selective toward *cis*-(*R*)-LO *(177)*.

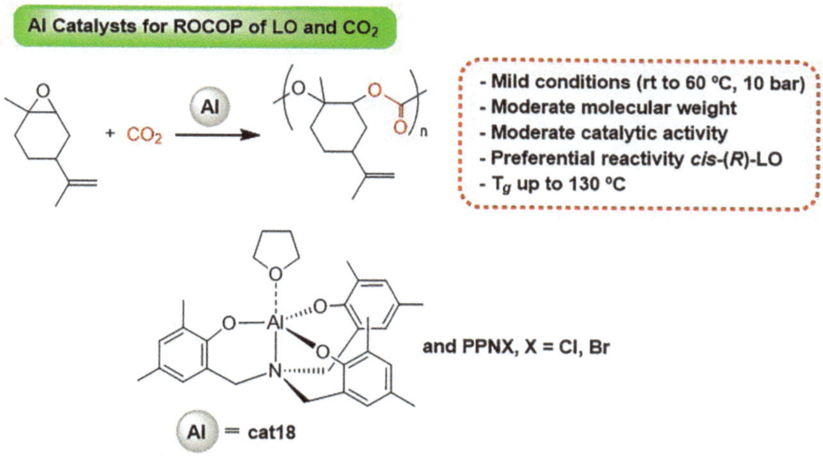

Scheme 25 Al catalysts used in the ROCOP of limonene oxide and CO_2.

The thermal properties for PLC are highly dependent on the molecular weight of the isolated polymer. Thus, glass transition temperatures can vary from as low as 71 °C for low molecular weight polymers *(178)* to up to 130 °C for high molecular weight copolymers *(171)*. On the other hand, thermal decomposition of the PLCs obtained have ranged from 180 °C *(177)* to up to 249 °C *(172)*.

The mechanical properties of poly(limonene carbonate) have recently been investigated *(171)*. PLC has a density of 1.09 g cm^{-3}, similar to polyolefins. Mechanical performances of PLC are believed to be between those of polystyrene and BPA-polycarbonate, 15% elongation at break, displaying *E*-modulus of 0.95 GPa and a pencil hardness of B. Furthermore, amorphous PLC has extremely good optical properties, with light transmission of 95% from 400 to 1000 nm (240 μm thick film) *(171)*. In addition to these optical properties, poly(limonene carbonate) shows good permeability to gases such as CO_2 and O_2 *(179)*. Compared to materials used as window glass substitutes, it is approximately 10 times more permeable than BPA-polycarbonate and 1000 times more permeable than poly(methyl methacrylate). These properties as "breathable glass" make it a totally renewable material that can be used in residential, commercial, and public buildings or greenhouses where this type of gas permeability is desired *(179)*.

Furthermore, the remaining double bond in the limonene moiety can be functionalized in different ways to produce a broader range of PLC derivatives with a wide range of thermal and mechanical properties which increases the applicability of these fully bio-renewable polymers (Scheme 26) *(47,180)*.

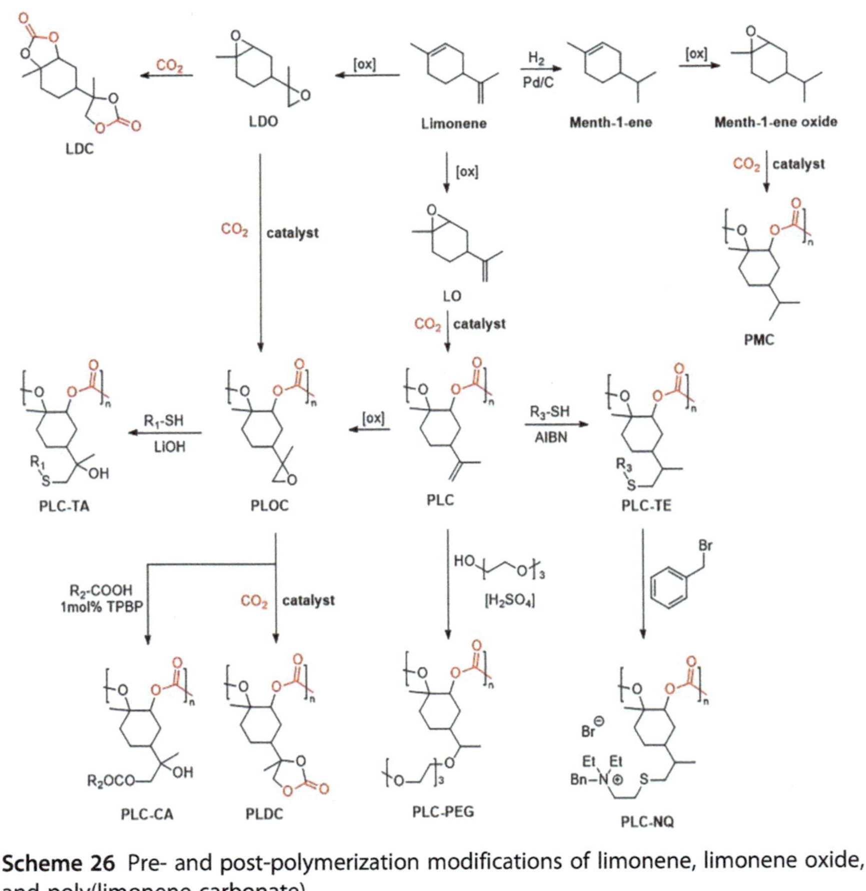

Scheme 26 Pre- and post-polymerization modifications of limonene, limonene oxide, and poly(limonene carbonate).

Other terpene-derived epoxides can be used along with CO_2 as substrates for ROCOP processes to generate fully bio-renewable polymers. For instance, α-pinene oxide and menthen-2-ene oxide have been used as bio-derived substrate to produce polycarbonate materials although these processes are not as developed as the ROCOP processes of LO and CO_2. To the best of our knowledge, there are no academic reports on the successful ROCOP of α-pinene oxide with CO_2. The only report on the ROCOP reaction of α-pinene oxide and CO_2 is collected in a Chinese patent that briefly describes the synthesis of poly(α-pinene carbonate) catalyzed by the combination of a chromium catalyst and PPNCl (Scheme 27) *(181)*.

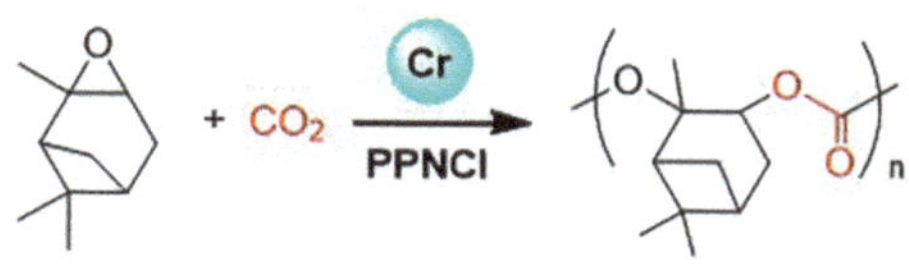

Scheme 27 ROCOP of α-pinene and CO_2.

Despite all the interesting properties described for PLC that have broadened the scope of applications for this polymer derived from renewable resources and CO_2 *(179,182)*, the main drawbacks are its relatively low thermal decomposition temperature and the possibility of undesired crosslinking and oxidation reactions during processing due to the presence of double bonds. Recently, a new and more thermally stable fully bio-renewable polycarbonate (PMen2C) has been reported by ROCOP of menth-2-ene oxide and CO_2 using β-diiminate acetate zinc catalyst **cat14** under mild reaction conditions (25 °C and 25 bar) (Scheme 28) *(161)*. This polycarbonate has shown to be a potential sustainable engineering polymeric material due to its excellent thermal properties which are unusual for aliphatic polycarbonates and of utmost importance for melt processing. Poly(menth-2-ene carbonate) showed a high T_g value of up to 144 °C, an exceptional thermal stability ($T_{5\%}$) of up to 308 °C, which is significantly higher than PCHC, ($T_{5\%}$=283 °C, M_n=52,900), PLOC ($T_{5\%}$=229 °C, M_n=114,000) and its reduced form obtained from menth-1-ene oxide *(180)*.

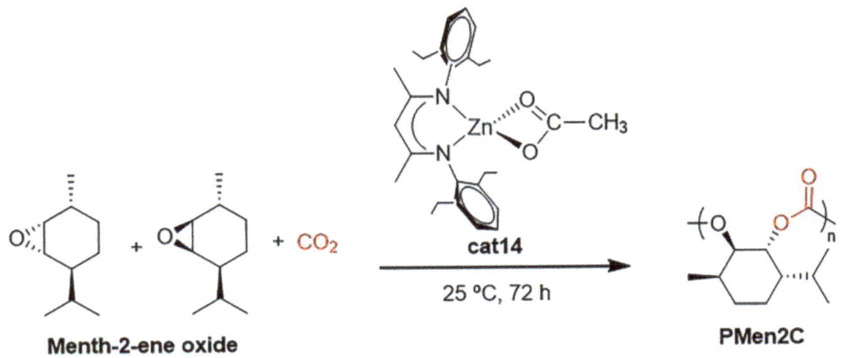

Scheme 28 ROCOP of menth-2-ene oxide and CO_2.

4. Outlook and conclusions

The resources necessary to manufacture chemical products, minerals and materials derived from petroleum are becoming scarce on the planet and is causing an increase in the economic and environmental costs of most raw materials. As an important contribution to a sustainable future, the chemicals industry and its products must adapt to a circular economy through the implementation of processes to eliminate waste, optimize and save resources as well as recycle and reuse products in order to preserve the environment *(183)*. Moreover, most products should be made from renewable feedstocks in a circular economy in order to reduce pressure on natural resources *(184)*. These efforts will contribute to achieving a sustainable development by optimizing the planet's resources to ensure their lasting protection. However, the biggest challenge will be changing the way products are designed, which will have to be led by scientists who are devoting considerable efforts to figure out how chemicals and their underpinning synthetic chemistry fit into a circular economy.

The literature has shown that virtually all current classes of chemical compounds and a great variety of new products can be synthesized from renewable resources using catalytic processes. The burning question is, to what extent these routes will be implemented in industry on a large scale to make a significant impact. When it comes to existing products, these processes must compete in cost with fossil-based products. To assess the sustainability of a chemical process, its influence on the economic, social and ecological spheres must be studied *(185)*. In the last decade, excellent works have been published on the development of more ecological and sustainable compounds and polymers.

In this chapter, we have covered some excellent examples for CO_2 fixation and bio-derived feedstocks conversion into commercially interesting fully renewable products such as cyclic carbonates and polycarbonate-type polymers. The analysis has not been focused on the use of CO_2 as a measure for the global mitigation of CO_2 but on the identification of opportunities of exploiting and using renewable raw materials that avoid the use of fossil fuel-based resources in the transition toward a more sustainable future production. Therefore, the challenge of developing a CO_2-based chemistry is not only motivated by the amount of CO_2 emissions that can be avoided, but also by the importance of developing new processes and synthetic routes to improve the carbon balance and other factors that define the impact on the environment during the production of chemical compounds in accordance with the Principles of Green Chemistry *(186)*.

Some remarkable advances have been made on the design of completely renewable cyclic carbonates; a research field which has advanced enormously in recent years and the development of suitable catalysts has been the key. These catalysts are evolving toward the Principles of Green Chemistry, and they should preferably be inexpensive, readily available, scalable, and sustainable in terms of their components. The incorporation of carbon dioxide into bio-based raw materials such as terpenes, platform molecules derived from sugar, glycerol, fatty acids or vegetable oils offers an alternative solution to convert low-value materials into high-value functional molecules or materials which can find application as additives, solvents or polymer precursors.

Although the progress is clear, there are still several aspects that still need to be improved. For instance, most of catalytic processes developed to date rely on the use of purified substrates, while most commercially available bio-renewable precursors on a large scale contain impurities that may affect the stability, activity and reusability of the catalytic system involved. Therefore, catalysts must be further developed in order to not only be capable of combining high (chemo)selectivity and catalytic activity, but also be compatible with chemical precursors obtained from waste streams. In addition, it is necessary to consider the cost of the design of the catalytic system to implement the catalytic process at industrial scale. Therefore, future research should be directed toward the expansion of the bio-derived cyclic carbonates scope, which are obtained from CO_2 and fully bio-renewable precursors, which can find novel industrial applications. To achieve this challenge, the development of more efficient and selective catalytic systems that can cope with impurities is needed.

The development of heterogeneous, or at least recyclable, catalysts with strong reactivity profiles will be required in order to meet the expected increasing demand for fully bio-renewable organic molecules and polymeric materials. There is still no extensive knowledge on the development of heterogeneous catalysts for the chemical fixation of CO_2 into epoxidized bio-precursors, and no viable catalyst has been developed so far for its application in large-scale production. This will be a challenge to address in the coming years to satisfy the expected fully bio-based cyclic carbonate market.

On the other hand, a key factor to accelerate and facilitate the transition from fossil fuel-derived materials to more sustainable ones trying to meet the circular economy requirements is to develop new production routes to chemically fix CO_2 into polymers. In this chapter, the most relevant synthetic routes to prepare fully bio-based polycarbonates using CO_2 as a renewable feedstock have been described. Two chemical transformations have been discussed: (i) the polymerization of CO_2-based building blocks such as cyclic carbonates, (ii) direct copolymerization of CO_2 with appropriate co-monomers such as epoxides.

The preparation of CO_2-based polymers with high CO_2 content can be achieved by the direct copolymerization of CO_2 with various bio-comonomers. This ring-opening copolymerization process is the most widespread method for preparing fully bio-renewable CO_2-based polymers. The success of this approach largely depends on the catalyst design. Currently, there are few catalytic systems, generally based on metals, that showed to be highly efficient and selective for the utilization of this type of bio-precursors, but novel breakthroughs are expected in the near future. Some aliphatic polycarbonate polyols derived from ethylene and propylene oxide have been commercialized by some chemical companies. These polycarbonate polyols materials have found application for polyurethane production.

The copolymerization of CO_2 with bio-derived epoxides is a potentially relevant process for the large-scale production of fully biodegradable polymers; however, there are important challenges to solve: (i) very few bio-renewable monomers have been activated so far, and (ii) only two families of metal catalysts have shown good catalytic activity for these bio-derived precursors. Much research is being devoted to investigating new pathways to overcome these limitations and push this attractive carbon dioxide utilization technology toward realistic utility. Finally, this overview on the current state-of-the-art of carbon dioxide utilization for the synthesis of polycarbonates has shown that CO_2-based technologies can also provide access to totally bio-renewable polymers with new functionalities and structures, which will contribute prospectively to the exploration and development of interesting new applications.

References

1. World Energy Statistics and Balances - Data product - IEA. https://www.iea.org/data-and-statistics/data-product/world-energy-statistics-and-balances (accessed December 22, 2021). 2021.
2. Greenhouse Gas Emissions from Energy - Data product - IEA. https://www.iea.org/data-and-statistics/data-product/greenhouse-gas-emissions-from-energy (accessed December 22, 2021). 2021.
3. Summary for Policymakers — Global Warming of 1.5 °C. https://www.ipcc.ch/sr15/chapter/spm/ (accessed December 22, 2021). 2021.
4. Ghiat, I.; Al-Ansari, T. *J. CO_2 Util.* **2021**, *45*, 101432. https://doi.org/10.1016/J.JCOU.2020.101432.
5. Cuéllar-Franca, R. M.; Azapagic, A. *J. CO_2 Util.* **2015**, *9*, 82–102. https://doi.org/10.1016/J.JCOU.2014.12.001.
6. Aresta, M.; Dibenedetto, A.; Angelini, A. *Chem. Rev.* **2013**, *114*, 1709–1742. https://doi.org/10.1021/CR4002758.
7. Aresta, M.; Dibenedetto, A.; Quaranta, E. *Reaction Mechanisms in Carbon Dioxide Conversion*; Springer: Berlin, Heidelberg, 2016; pp. 1–34. https://doi.org/10.1007/978-3-662-46831-9_1.
8. COP26 Outcomes - UN Climate Change Conference (COP26) at the SEC – Glasgow 2021. https://ukcop26.org/the-conference/cop26-outcomes/ (accessed December 23, 2021). 2021.
9. CO2 emissions – Global Energy Review 2021 – Analysis - IEA. https://www.iea.org/reports/global-energy-review-2021/co2-emissions (accessed December 23, 2021). 2021.
10. A European Green Deal | European Commission. https://ec.europa.eu/info/strategy/priorities-2019-2024/european-green-deal_en (accessed December 23, 2021). 2021.
11. Dunn, P. J., Hii, K. K., Krische, M. J., Williams, M. T., Eds. *Sustainable Catalysis : Challenges and Practices for the Pharmaceutical and Fine Chemical Industries*; John Wiley & Sons, Inc: Hoboken, New Jersey, US, 2013.
12. Artz, J.; Müller, T. E.; Thenert, K.; Kleinekorte, J.; Meys, R.; Sternberg, A.; Bardow, A.; Leitner, W. *Chem. Rev.* **2017**, *118*, 434–504. https://doi.org/10.1021/ACS.CHEMREV.7B00435.
13. Zhang, X.; Fevre, M.; Jones, G. O.; Waymouth, R. M. *Chem. Rev.* **2017**, *118*, 839–885. https://doi.org/10.1021/ACS.CHEMREV.7B00329.
14. North, M., Ed. *Sustainable Catalysis: With Non-endangered Metals, Part 1, Royal Society of Chemistry, Cambridge*; 2015. https://doi.org/https://doi.org/10.1039/9781782622116.
15. Tripathi, N.; Hills, C. D.; Singh, R. S.; Atkinson, C. J. *NPJ Clim. Atmos. Sci.* **2019**, *2*, 35. https://doi.org/10.1038/s41612-019-0093-5.
16. Takkellapati, S.; Li, T.; Gonzalez, M. A. *Clean Technol. Environ. Foreign Policy* **2018**, *20*, 1615–1630. https://doi.org/10.1007/S10098-018-1568-5/FIGURES/1.
17. Sheldon, R. A. *Green Chem.* **2014**, *16*, 950–963. https://doi.org/10.1039/C3GC41935E.

18. Protti, S., Palmieri, A., Eds. *Sustainable Organic Synthesis*; Royal Society of Chemistry: Cambridge, 2021. https://doi.org/10.1039/9781839164842.
19. Aresta, M., Ed. *Carbon Dioxide as Chemical Feedstock*; John Wiley & Sons, Inc: Hoboken, New Jersey, 2010. https://doi.org/10.1002/9783527629916.
20. Styring, P.; Quadrelli, E. A.; Armstrong, K. *Carbon Dioxide Utilisation: Closing the Carbon Cycle*, 1st Edition; Elsevier: Amsterdam, 2014. https://doi.org/10.1016/C2012-0-02814-1.
21. Arakawa, H.; Aresta, M.; Armor, J. N.; Barteau, M. A.; Beckman, E. J.; Bell, A. T.; Bercaw, J. E.; Creutz, C.; Dinjus, E.; Dixon, D. A.; Domen, K.; DuBois, D. L.; Eckert, J.; Fujita, E.; Gibson, D. H.; Goddard, W. A.; Goodman, D. W.; Keller, J.; Kubas, G. J.; Kung, H. H.; Lyons, J. E.; Manzer, L. E.; Marks, T. J.; Morokuma, K.; Nicholas, K. M.; Periana, R.; Que, L.; Rostrup-Nielson, J.; Sachtler, W. M. H.; Schmidt, L. D.; Sen, A.; Somorjai, G. A.; Stair, P. C.; Stults, B. R.; Tumas, W. *Chem. Rev.* **2001**, *101*, 953–996. https://doi.org/10.1021/CR000018S.
22. Schwarz, H. *Coord. Chem. Rev.* **2017**, *334*, 112–123. https://doi.org/10.1016/J.CCR.2016.03.009.
23. Álvarez, A.; Borges, M.; Corral-Pérez, J. J.; Olcina, J. G.; Hu, L.; Cornu, D.; Huang, R.; Stoian, D.; Urakawa, A. *ChemPhysChem* **2017**, *18*, 3135–3141. https://doi.org/10.1002/CPHC.201700782.
24. Paparo, A.; Okuda, J. *Coord. Chem. Rev.* **2017**, *334*, 136–149. https://doi.org/10.1016/j.ccr.2016.06.005.
25. Bozell, J. J.; Petersen, G. R. *Green Chem.* **2010**, *12*, 539–554. https://doi.org/10.1039/B922014C.
26. Centi, G.; Perathoner, S. *Green Carbon Dioxide: Advances in CO_2 Utilization*; John Wiley & Sons, Inc: Hoboken, New Jersey, US, 2014. https://doi.org/10.1002/9781118831922.
27. Sabri, M. A.; Al Jitan, S.; Bahamon, D.; Vega, L. F.; Palmisano, G. *Sci. Total Environ.* **2021**, *790*, 148081. https://doi.org/10.1016/J.SCITOTENV.2021.148081.
28. Burkart, M. D.; Hazari, N.; Tway, C. L.; Zeitler, E. L. *ACS Catal.* **2019**, *9*, 7937–7956. https://doi.org/10.1021/ACSCATAL.9B02113.
29. Yuan, Z.; Eden, M. R.; Gani, R. *Ind. Eng. Chem. Res.* **2016**, *55*, 3383–3419. https://doi.org/10.1021/ACS.IECR.5B03277.
30. Sancho-Sanz, I.; Korili, S. A.; Gil, A. *Catal. Rev.* **2021**,. https://doi.org/10.1080/01614940.2021.1968197.
31. Saeidi, S.; Najari, S.; Hessel, V.; Wilson, K.; Keil, F. J.; Concepción, P.; Suib, S. L.; Rodrigues, A. E. *Prog. Energy Combust. Sci.* **2021**, *85*, 100905. https://doi.org/10.1016/J.PECS.2021.100905.
32. Do, T. N.; You, C.; Kim, J. *Energ. Environ. Sci.* **2022**, *15*, 169–184. https://doi.org/10.1039/D1EE01444G.
33. Dieterich, V.; Buttler, A.; Hanel, A.; Spliethoff, H.; Fendt, S. *Energ. Environ. Sci.* **2020**, *13*, 3207–3252. https://doi.org/10.1039/D0EE01187H.
34. Mbatha, S.; Everson, R. C.; Musyoka, N. M.; Langmi, H. W.; Lanzini, A.; Brilman, W. *Energy Fuel* **2021**, *5*, 3490–3569. https://doi.org/10.1039/D1SE00635E.
35. Pérez-Fortes, M.; Schöneberger, J. C.; Boulamanti, A.; Harrison, G.; Tzimas, E. *Int. J. Hydrogen Energy* **2016**, *41*, 16444–16462. https://doi.org/10.1016/J.IJHYDENE.2016.05.199.
36. Li, M.; Sun, Z.; Hu, Y. H. *J. Mater. Chem. A* **2021**, *9*, 12495–12520. https://doi.org/10.1039/D1TA00440A.
37. Modak, A.; Bhanja, P.; Dutta, S.; Chowdhury, B.; Bhaumik, A. *Green Chem.* **2020**, *22*, 4002–4033. https://doi.org/10.1039/D0GC01092H.

38. Sable, D. A.; Vadagaonkar, K. S.; Kapdi, A. R.; Bhanage, B. M. *Org. Biomol. Chem.* **2021**, *19*, 5725–5757. https://doi.org/10.1039/D1OB00755F.
39. Zevenhoven, R.; Eloneva, S.; Teir, S. *Catal. Today* **2006**, *115*, 73–79. https://doi.org/10.1016/J.CATTOD.2006.02.020.
40. Tabanelli, T.; Bonincontro, D.; Albonetti, S.; Cavani, F. *Stud. Surf. Sci. Catal.* **2019**, *178*, 125–144. https://doi.org/10.1016/B978-0-444-64127-4.00007-0.
41. Omae, I. *Coord. Chem. Rev.* **2012**, *256*, 1384–1405. https://doi.org/10.1016/J.CCR.2012.03.017.
42. Fukuoka, S.; Tojo, M.; Hachiya, H.; Aminaka, M.; Hasegawa, K. *Polym. J.* **2007**, *39*, 91–114. https://doi.org/10.1295/polymj.pj2006140.
43. Pescarmona, P. P. *Curr. Opin. Green Sustain. Chem.* **2021**, *29*, 100457. https://doi.org/10.1016/J.COGSC.2021.100457.
44. Grignard, B.; Gennen, S.; Jérôme, C.; Kleij, A. W.; Detrembleur, C. *Chem. Soc. Rev.* **2019**, *48*, 4466–4514. https://doi.org/10.1039/C9CS00047J.
45. Song, B.; Qin, A.; Tang, B. Z. *Cell Rep. Phys. Sci.* **2022**, 100719. https://doi.org/10.1016/J.XCRP.2021.100719.
46. Ye, S.; Wang, S.; Lin, L.; Xiao, M.; Meng, Y. *Adv. Ind. Eng. Polym. Res.* **2019**, *2*, 143–160. https://doi.org/10.1016/J.AIEPR.2019.09.004.
47. Poland, S. J.; Darensbourg, D. J. *Green Chem.* **2017**, *19*, 4990–5011. https://doi.org/10.1039/C7GC02560B.
48. Huang, J.; Worch, J. C.; Dove, A. P.; Coulembier, O. *ChemSusChem* **2020**, *13*, 469–487. https://doi.org/10.1002/CSSC.201902719.
49. Mülhaupt, R. *Macromol. Chem. Phys.* **2013**, *214*, 159–174. https://doi.org/10.1002/MACP.201200439.
50. Clark, J. H.; Budarin, V.; Deswarte, F. E. I.; Hardy, J. J. E.; Kerton, F. M.; Hunt, A. J.; Luque, R.; Macquarrie, D. J.; Milkowski, K.; Rodriguez, A.; Samuel, O.; Tavener, S. J.; White, R. J.; Wilson, A. J. *Green Chem.* **2006**, *8*, 853–860. https://doi.org/10.1039/B604483M.
51. Philp, J. C.; Ritchie, R. J.; Allan, J. E. M. *Trends Biotechnol.* **2013**, *31*, 219–222. https://doi.org/10.1016/J.TIBTECH.2012.12.007.
52. Lange, L.; Connor, K. O.; Arason, S.; Bundgård-Jørgensen, U.; Canalis, A.; Carrez, D.; Gallagher, J.; Gøtke, N.; Huyghe, C.; Jarry, B.; Llorente, P.; Marinova, M.; Martins, L. O.; Mengal, P.; Paiano, P.; Panoutsou, C.; Rodrigues, L.; Stengel, D. B.; van der Meer, Y.; Vieira, H. *Front. Bioeng. Biotechnol.* **2021**, *8*, 1456. https://doi.org/10.3389/FBIOE.2020.619066/BIBTEX.
53. Farmer, T. J.; Mascal, M. *Introduction to Chemicals from Biomass*; John Wiley & Sons: Chichester, UK, 2014. Chapter 4 https://doi.org/10.1002/9781118714478.CH4.
54. Ventura, M.; Marinas, A.; Domine, M. E. *Top. Catal.* **2020**, *63*, 846–865. https://doi.org/10.1007/S11244-020-01309-9.
55. Serrano-Ruiz, J. C.; Luque, R.; Sepúlveda-Escribano, A. *Chem. Soc. Rev.* **2011**, *40*, 5266–5281. https://doi.org/10.1039/C1CS15131B.
56. Saravanamurugan, S., Li, H., Riisager, A., Pandey, A., Eds. *Biomass, Biofuels, Biochemicals in Recent Advances in Development of Platform Chemicals*; Elsevier: Amsterdam, 2019.
57. Stadler, B. M.; Wulf, C.; Werner, T.; Tin, S.; De Vries, J. G. *ACS Catal.* **2019**, *9*, 8012–8067. https://doi.org/10.1021/ACSCATAL.9B01665.
58. Werpy, T.; Petersen, G. *Top Value Added Chemicals from Biomass: Volume I - - Results of Screening for Potential Candidates from Sugars and Synthesis Gas, United States, 2004.* https://doi.org/10.2172/15008859.
59. Aomchad, V.; Cristòfol, À.; Della Monica, F.; Limburg, B.; D'Elia, V.; Kleij, A. W. *Green Chem.* **2021**, *23*, 1077–1113. https://doi.org/10.1039/D0GC03824E.

60. Lamparelli, D. H.; Capacchione, C. *Catalysts* **2021**, *11*, 961. https://doi.org/10.3390/CATAL11080961.
61. Shaikh, A.-A. G.; Sivaram, S. *Chem. Rev.* **1996**, *96*, 951–976. https://doi.org/10.1021/cr950067i.
62. Schäffner, B.; Schäffner, F.; Verevkin, S. P.; Börner, A. *Chem. Rev.* **2010**, *110*, 4554–4581. https://doi.org/10.1021/cr900393d.
63. Philipp, M.; Bhandary, R.; Groche, F. J.; Schönhoff, M.; Rieger, B. *Electrochim. Acta* **2015**, *173*, 687–697. https://doi.org/10.1016/j.electacta.2015.05.108.
64. Ouhib, F.; Meabe, L.; Mahmoud, A.; Grignard, B.; Thomassin, J. M.; Boschini, F.; Zhu, H.; Forsyth, M.; Mecerreyes, D.; Detrembleur, C. *ACS Appl. Polym. Mater.* **2020**, *2*, 922–931. https://doi.org/10.1021/ACSAPM.9B01130.
65. Sonnati, M. O.; Amigoni, S.; Taffin De Givenchy, E. P.; Darmanin, T.; Choulet, O.; Guittard, F. *Green Chem.* **2013**, *15*, 283–306. https://doi.org/10.1039/c2gc36525a.
66. Han, Z.; Rong, L.; Wu, J.; Zhang, L.; Wang, Z.; Ding, K. *Angew. Chem. Int. Ed.* **2012**, *51*, 13041–13045. https://doi.org/10.1002/anie.201207781.
67. Liu, H.; Huang, Z.; Han, Z.; Ding, K.; Liu, H.; Xia, C.; Chen, J. *Green Chem.* **2015**, *17*, 4281–4290. https://doi.org/10.1039/c5gc00810g.
68. Guo, W.; Gónzalez-Fabra, J.; Bandeira, N. A. G.; Bo, C.; Kleij, A. W. *Angew. Chem. Int. Ed.* **2015**, *54*, 11686–11690. https://doi.org/10.1002/anie.201504956.
69. Blain, M.; Yau, H.; Jean-Gérard, L.; Auvergne, R.; Benazet, D.; Schreiner, P. R.; Caillol, S.; Andrioletti, B. *ChemSusChem* **2016**, *9*, 2269–2272. https://doi.org/10.1002/cssc.201600778.
70. Khan, A.; Zheng, R.; Kan, Y.; Ye, J.; Xing, J.; Zhang, Y. J. *Angew. Chem. Int. Ed.* **2014**, *53*, 6439–6442. https://doi.org/10.1002/anie.201403754.
71. Boot-Handford, M. E.; Abanades, J. C.; Anthony, E. J.; Blunt, M. J.; Brandani, S.; Mac Dowell, N.; Fernández, J. R.; Ferrari, M. C.; Gross, R.; Hallett, J. P.; Haszeldine, R. S.; Heptonstall, P.; Lyngfelt, A.; Makuch, Z.; Mangano, E.; Porter, R. T. J.; Pourkashanian, M.; Rochelle, G. T.; Shah, N.; Yao, J. G.; Fennell, P. S. *Energ. Environ. Sci.* **2014**, 7, 130–189. https://doi.org/10.1039/c3ee42350f.
72. Nemirowsky, J. *J. Für Prakt. Chemie.* **1883**, *28*, 439–440. https://doi.org/10.1002/prac.18830280136.
73. Selva, M.; Caretto, A.; Noè, M.; Perosa, A. *Org. Biomol. Chem.* **2014**, *12*, 4143–4155. https://doi.org/10.1039/c4ob00655k.
74. Indran, V. P.; Haji Saud, A. S.; Maniam, G. P.; Yusoff, M. M.; Taufiq-Yap, Y. H.; Mohd, M. H. *RSC Adv.* **2016**, *6*, 34877–34884. https://doi.org/10.1039/c5ra26286k.
75. Doro, F.; Winnertz, P.; Leitner, W.; Prokofieva, A.; Müller, T. E. *Green Chem.* **2011**, *13*, 292–295. https://doi.org/10.1039/c0gc00817f.
76. Reithofer, M. R.; Sum, Y. N.; Zhang, Y. *Green Chem.* **2013**, *15*, 2086–2090. https://doi.org/10.1039/c3gc40790j.
77. Kumar, S.; Singhal, N.; Singh, R. K.; Gupta, P.; Singh, R.; Jain, S. L. *Dalton Trans.* **2015**, *44*, 11860–11866. https://doi.org/10.1039/c5dt01012h.
78. Hu, J.; Ma, J.; Zhu, Q.; Qian, Q.; Han, H.; Mei, Q.; Han, B. *Green Chem.* **2016**, *18*, 382–385. https://doi.org/10.1039/c5gc01870f.
79. North, M. *ARKIVOC* **2012**, 610–628. https://doi.org/10.3998/ark.5550190.0013.115.
80. Shaikh, R. R.; Pornpraprom, S.; D'Elia, V. *ACS Catal.* **2018**, *8*, 419–450. https://doi.org/10.1021/acscatal.7b03580.
81. Kamphuis, A. J.; Picchioni, F.; Pescarmona, P. P. *Green Chem.* **2019**, *21*, 406–448. https://doi.org/10.1039/c8gc03086c.
82. Asahi Kasei Corporation. *Asahi Kasei Licenses Technology Package to Manufacture High-Purity Ethylene Carbonate and Dimethyl Carbonate Using CO_2 as Main Feedstock*; 2021 (News | Asahi Kasei. https://www.asahi-kasei.com/news/2021/e210701.html (accessed December 12, 2021).

83. Breitmaier, E., Ed. *Terpenes: Flavors, Fragrances, Pharmaca, Pheromones*; Elsevier: Amsterdam, 2006. https://doi.org/10.1002/9783527609949.
84. Brown, G. D. *Phytochemistry* **1994**, *35*, 425–433. https://doi.org/10.1016/S0031-9422(00)94775-7.
85. Virot, M.; Tomao, V.; Ginies, C.; Visinoni, F.; Chemat, F. *J. Chromatogr. A* **2008**, *1196–1197*, 147–152. https://doi.org/10.1016/J.CHROMA.2008.04.035.
86. Fiorani, G.; Stuck, M.; Martín, C.; Belmonte, M. M.; Martin, E.; Escudero-Adán, E. C.; Kleij, A. W. *ChemSusChem* **2016**, *9*, 1304–1311. https://doi.org/10.1002/cssc.201600238.
87. Martínez, J.; Fernández-Baeza, J.; Sánchez-Barba, L. F.; Castro-Osma, J. A.; Lara-Sánchez, A.; Otero, A. *ChemSusChem* **2017**, *10*, 2886–2890. https://doi.org/10.1002/cssc.201700898.
88. Longwitz, L.; Steinbauer, J.; Spannenberg, A.; Werner, T. *ACS Catal.* **2018**, *8*, 665–672. https://doi.org/10.1021/acscatal.7b03367.
89. de la Cruz-Martínez, F.; Buchaca, M. M. D. S.; Martínez, J.; Fernández-Baeza, J.; Sánchez-Barba, L. F.; Rodríguez-Diéguez, A.; Castro-Osma, J. A.; Lara-Sánchez, A. *ACS Sustain. Chem. Eng.* **2019**, 7, 20126–20138. https://doi.org/10.1021/acssuschemeng.9b06016.
90. Navarro, M.; Sánchez-Barba, L. F.; Garcés, A.; Fernández-Baeza, J.; Fernández, I.; Lara-Sánchez, A.; Rodríguez, A. M. *Catal. Sci. Technol.* **2020**, *10*, 3265–3278. https://doi.org/10.1039/D0CY00593B.
91. Morikawa, H.; Minamoto, M.; Gorou, Y.; Yamaguchi, J. I.; Morinaga, H.; Motokucho, S. *Bull. Chem. Soc. Jpn.* **2017**, *91*, 92–94. https://doi.org/10.1246/BCSJ.20170300.
92. Bähr, M.; Bitto, A.; Mülhaupt, R. *Green Chem.* **2012**, *14*, 1447–1454. https://doi.org/10.1039/c2gc35099h.
93. Schimpf, V.; Ritter, B. S.; Weis, P.; Parison, K.; Mülhaupt, R. *Macromolecules* **2017**, *50*, 944–955. https://doi.org/10.1021/acs.macromol.6b02460.
94. Xiong, X.; Yu, I. K. M.; Tsang, D. C. W.; Bolan, N. S.; Sik Ok, Y.; Igalavithana, A. D.; Kirkham, M. B.; Kim, K. H.; Vikrant, K. *Chem. Eng. J.* **2019**, *375*, 121983. https://doi.org/10.1016/J.CEJ.2019.121983.
95. Zhou, Y.; Zhao, W.; Lai, Y.; Zhang, B.; Zhang, D. *Front. Plant Sci.* **2020**, *11*, 1315. https://doi.org/10.3389/FPLS.2020.01315/BIBTEX.
96. Poussard, L.; Mariage, J.; Grignard, B.; Detrembleur, C.; Jéroîme, C.; Calberg, C.; Heinrichs, B.; De Winter, J.; Gerbaux, P.; Raquez, J. M.; Bonnaud, L.; Dubois, P. *Macromolecules* **2016**, *49*, 2162–2171. https://doi.org/10.1021/ACS.MACROMOL.5B02467.
97. Hu, S.; Chen, X.; Torkelson, J. M. *ACS Sustain. Chem. Eng.* **2019**, 7, 10025–10034. https://doi.org/10.1021/acssuschemeng.9b01239.
98. Liu, X.; Yang, X.; Wang, S.; Wang, S.; Wang, Z.; Liu, S.; Xu, X.; Liu, H.; Song, Z. *ACS Sustain. Chem. Eng.* **2021**, *9*, 4175–4184. https://doi.org/10.1021/ACSSUSCHEMENG.1C00033.
99. Miloslavskiy, D.; Gotlib, E.; Figovsky, O.; Pashin, D. *Int. Lett. Chem. Phys. Astron.* **2014**, *27*, 20–29. https://doi.org/10.18052/WWW.SCIPRESS.COM/ILCPA.27.20.
100. Doll, K. M.; Erhan, S. Z. *Green Chem.* **2005**, 7, 849–854. https://doi.org/10.1039/B511014A.
101. Yu, A. Z.; Setien, R. A.; Sahouani, J. M.; Docken, J.; Webster, D. C. *J. Coatings Technol. Res.* **2019**, *16*, 41–57. https://doi.org/10.1007/S11998-018-0135-7/FIGURES/13.
102. Guzmán, A. F.; Echeverri, D. A.; Rios, L. A. *J. Chem. Technol. Biotechnol.* **2017**, *92*, 1104–1110. https://doi.org/10.1002/JCTB.5104.
103. Mazo, P.; Rios, L. *J. Am. Oil Chem. Soc.* **2013**, *90*, 725–730. https://doi.org/10.1007/S11746-013-2214-3/TABLES/2.

104. Mazo, P. C.; Rios, L. A. *Chem. Eng. J.* **2012**, *210*, 333–338. https://doi.org/10.1016/J.CEJ.2012.08.099.
105. Mahendran, A. R.; Aust, N.; Wuzella, G.; Müller, U.; Kandelbauer, A. *J. Polym. Environ.* **2012**, *20*, 926–931. https://doi.org/10.1007/S10924-012-0491-9/FIGURES/6.
106. Büttner, H.; Steinbauer, J.; Wulf, C.; Dindaroglu, M.; Schmalz, H. G.; Werner, T. *ChemSusChem* **2017**, *10*, 1076–1079. https://doi.org/10.1002/CSSC.201601163.
107. Natongchai, W.; Pornpraprom, S.; D'Elia, V. *J. Org. Chem.* **2020**, *9*, 801–810. https://doi.org/10.1002/AJOC.202000154.
108. Grignard, B.; Thomassin, J. M.; Gennen, S.; Poussard, L.; Bonnaud, L.; Raquez, J. M.; Dubois, P.; Tran, M. P.; Park, C. B.; Jerome, C.; Detrembleur, C. *Green Chem.* **2016**, *18*, 2206–2215. https://doi.org/10.1039/C5GC02723C.
109. Büttner, H.; Grimmer, C.; Steinbauer, J.; Werner, T. *ACS Sustain. Chem. Eng.* **2016**, *4*, 4805–4814. https://doi.org/10.1021/ACSSUSCHEMENG.6B01092/SUPPL_FILE/SC6B01092_LIVESLIDES.MP4.
110. Martínez, J.; de la Cruz-Martínez, F.; de Sarasa Buchaca, M. M.; Caballero, M. P.; Ojeda-Amador, R. M.; Salvador, M. D.; Fregapane, G.; Tejeda, J.; Castro-Osma, J. A.; Lara-Sánchez, A. *J. Environ. Chem. Eng.* **2021**, *9*, 105464. https://doi.org/10.1016/J.JECE.2021.105464.
111. Li, Z.; Zhao, Y.; Yan, S.; Wang, X.; Kang, M.; Wang, J.; Xiang, H. *Catal. Lett.* **2008**, *123*, 246–251. https://doi.org/10.1007/S10562-008-9414-8/TABLES/3.
112. Jalilian, M.; Yeganeh, H.; Haghighi, M. N. *Polym. Adv. Technol.* **2010**, *21*, 118–127. https://doi.org/10.1002/PAT.1406.
113. Hu, Y.; Steinbauer, J.; Stefanow, V.; Spannenberg, A.; Werner, T. *ACS Sustain. Chem. Eng.* **2019**, 7, 13257–13269. https://doi.org/10.1021/ACSSUSCHEMENG.9B02502.
114. Farhadian, A.; Gol Afshani, M. B.; Babaei Miyardan, A.; Nabid, M. R.; Safari, N. *ChemistrySelect* **2017**, *2*, 1431–1435. https://doi.org/10.1002/SLCT.201601891.
115. Gregory, G. L.; Jenisch, L. M.; Charles, B.; Kociok-Köhn, G.; Buchard, A. *Macromolecules* **2016**, *49*, 7165–7169. https://doi.org/10.1021/ACS.MACROMOL.6B01492/SUPPL_FILE/MA6B01492_SI_002.CIF.
116. Gregory, G. L.; Kociok-Köhn, G.; Buchard, A. *Polym. Chem.* **2017**, *8*, 2093–2104. https://doi.org/10.1039/C7PY00236J.
117. Gustafson, T. P.; Lonnecker, A. T.; Heo, G. S.; Zhang, S.; Dove, A. P.; Wooley, K. L. *Biomacromolecules* **2013**, *14*, 3346–3353. https://doi.org/10.1021/BM4010832.
118. Chen, X.; Gross, R. A. *Macromolecules* **1999**, *32*, 308–314. https://doi.org/10.1021/MA980895O.
119. Mikami, K.; Lonnecker, A. T.; Gustafson, T. P.; Zinnel, N. F.; Pai, P. J.; Russell, D. H.; Wooley, K. L. *J. Am. Chem. Soc.* **2013**, *135*, 6826–6829. https://doi.org/10.1021/JA402319M/SUPPL_FILE/JA402319M_SI_004.CIF.
120. Gregory, G. L.; Ulmann, M.; Buchard, A. *RSC Adv.* **2015**, *5*, 39404–39408. https://doi.org/10.1039/C5RA07290E.
121. McGuire, T. M.; López-Vidal, E. M.; Gregory, G. L.; Buchard, A. *J. CO_2 Util.* **2018**, *27*, 283–288. https://doi.org/10.1016/J.JCOU.2018.08.009.
122. Gregory, G. L.; Hierons, E. M.; Kociok-Köhn, G.; Sharma, R. I.; Buchard, A. *Polym. Chem.* **2017**, *8*, 1714–1721. https://doi.org/10.1039/C7PY00118E.
123. Pati, D.; Feng, X.; Hadjichristidis, N.; Gnanou, Y. *Macromolecules* **2017**, *50*, 1362–1370. https://doi.org/10.1021/ACS.MACROMOL.6B02527.
124. Pati, D.; Feng, X.; Hadjichristidis, N.; Gnanou, Y. *J. CO_2 Util.* **2018**, *24*, 564–571. https://doi.org/10.1016/J.JCOU.2018.02.008.
125. Plastics in a circular economy | European Commission. https://ec.europa.eu/info/research-and-innovation/research-area/environment/circular-economy/plastics-circular-economy_en (accessed January 2, 2022).

126. Haider, T. P.; Völker, C.; Kramm, J.; Landfester, K.; Wurm, F. R. *Angew. Chem. Int. Ed.* **2019**, *58*, 50–62. https://doi.org/10.1002/ANIE.201805766.
127. Ye, S.; Xiang, X.; Wang, S.; Han, D.; Xiao, M.; Meng, Y. *ACS Sustain. Chem. Eng.* **2020**, *8*, 1923–1932. https://doi.org/10.1021/ACSSUSCHEMENG.9B06294.
128. Song, B.; Bai, T.; Xu, X.; Chen, X.; Liu, D.; Guo, J.; Qin, A.; Ling, J.; Tang, B. Z. *Macromolecules* **2019**, *52*, 5546–5554. https://doi.org/10.1021/ACS.MACROMOL.9B00898.
129. Sulley, G. S.; Gregory, G. L.; Chen, T. T. D.; Peña Carrodeguas, L.; Trott, G.; Santmarti, A.; Lee, K. Y.; Terrill, N. J.; Williams, C. K. *J. Am. Chem. Soc.* **2020**, *142*, 4367–4378. https://doi.org/10.1021/JACS.9B13106.
130. Gennen, S.; Grignard, B.; Tassaing, T.; Jérôme, C.; Detrembleur, C. *Angew. Chem. Int. Ed.* **2017**, *56*, 10394–10398. https://doi.org/10.1002/anie.201704467.
131. Zhu, Y.; Romain, C.; Williams, C. K. *Nature* **2016**, *540*, 354–362. https://doi.org/10.1038/nature21001.
132. Bachmann, M.; Kätelhön, A.; Winter, B.; Meys, R.; Müller, L. J.; Bardow, A. *Faraday Discuss.* **2021**, *230*, 227–246. https://doi.org/10.1039/D0FD00134A.
133. Yu, Y.; Fang, L. M.; Liu, Y.; Lu, X. B. *ACS Catal.* **2021**, *11*, 8349–8357. https://doi.org/10.1021/ACSCATAL.1C01376.
134. Hersh, S. N.; Choi, K. Y. *J. Appl. Polym. Sci.* **1990**, *41*, 1033–1046. https://doi.org/10.1002/APP.1990.070410514.
135. Kim, Y.; Choi, K. Y.; Chamberlin, T. A. *Ind. Eng. Chem. Res.* **2002**, *31*, 2118–2127. https://doi.org/10.1021/IE00009A008.
136. Woo, B. G. *Ind. Eng. Chem. Res.* **2001**, *40*, 1312–1319. https://doi.org/10.1021/IE000908Y.
137. Fukuoka, S.; Kawamura, M.; Komiya, K.; Tojo, M.; Hachiya, H.; Hasegawa, K.; Aminaka, M.; Okamoto, H.; Fukawa, I.; Konno, S. *Green Chem.* **2003**, *5*, 497–507. https://doi.org/10.1039/B304963A.
138. Singh, A. *Encycl. Polym. Nanomater.* **2015**, 1793–1796. https://doi.org/10.1007/978-3-642-29648-2_419.
139. T. Watanabe and S. Fukuoka, n.d. Jap. Patent 57–176932 A1 (Dec. 30, 1981); 64-3181B2 (Jan. 19, 1989), assigned to Asahi Kasei Chemicals Corp.
140. Langanke, J.; Wolf, A.; Hofmann, J.; Böhm, K.; Subhani, M. A.; Müller, T. E.; Leitner, W.; Gürtler, C. *Green Chem.* **2014**, *16*, 1865–1870. https://doi.org/10.1039/C3GC41788C.
141. Xu, J.; Feng, E.; Song, J. *J. Appl. Polym. Sci.* **2014**, *131*. https://doi.org/10.1002/APP.39822.
142. CO_2 as a raw material, Covestro AG. https://www.covestro.com/en/sustainability/flagship-solutions/co2-as-a-raw-material (accessed January 5, 2022).
143. Poly(Alkylene Carbonates) 2022 Polymerization Reaction - Empower Materials. https://www.empowermaterials.com/polyalkylene-carbonates-basic-polymerization-reaction-2 (accessed January 5, 2022).
144. Luinstra, G. A.; Borchardt, E. In *Synthetic Biodegradable Polymers*; Rieger, B., Künkel, A., Coates, G., Reichardt, R., Dinjus, E., Zevaco, T., Eds.; Advances in Polymer Science, 245; Springer: Berlin, Heidelberg, 2011; pp. 29–48. https://doi.org/10.1007/12_2011_126.
145. Feng, J.; Zhuo, R. X.; Zhang, X. Z. *Prog. Polym. Sci.* **2012**, *37*, 211–236. https://doi.org/10.1016/J.PROGPOLYMSCI.2011.07.008.
146. Tian, H.; Tang, Z.; Zhuang, X.; Chen, X.; Jing, X. *Prog. Polym. Sci.* **2012**, *37*, 237–280. https://doi.org/10.1016/J.PROGPOLYMSCI.2011.06.004.
147. Artham, T.; Doble, M. *Macromol. Biosci.* **2008**, *8*, 14–24. https://doi.org/10.1002/MABI.200700106.
148. Zhu, K. J.; Hendren, R. W.; Jensen, K.; Pitt, C. G. *Macromolecules* **2002**, *24*, 1736–1740. https://doi.org/10.1021/MA00008A008.

149. Yu, W.; Maynard, E.; Chiaradia, V.; Arno, M. C.; Dove, A. P. *Chem. Rev.* **2021**, *121*, 10865–10907. https://doi.org/10.1021/acs.chemrev.0c00883.
150. Lee, J. C.; Litt, M. H. *Macromolecules* **2000**, *33*, 1618–1627. https://doi.org/10.1021/MA9914321.
151. Clements, J. H. *Ind. Eng. Chem. Res.* **2003**, *42*, 663–674. https://doi.org/10.1021/IE020678I.
152. Guerin, W.; Diallo, A. K.; Kirilov, E.; Helou, M.; Slawinski, M.; Brusson, J. M.; Carpentier, J. F.; Guillaume, S. M. *Macromolecules* **2014**, *47*, 4230–4235. https://doi.org/10.1021/MA5009397.
153. Agarwal, S.; Naumann, N.; Xie, X. *Macromolecules* **2002**, *35*, 7713–7717. https://doi.org/10.1021/MA020584K.
154. Guerin, W.; Helou, M.; Slawinski, M.; Brusson, J. M.; Carpentier, J. F.; Guillaume, S. M. *Polym. Chem.* **2015**, *6*, 1972–1985. https://doi.org/10.1039/C4PY01660B.
155. Haba, O.; Tomizuka, H.; Endo, T. *Macromolecules* **2005**, *38*, 3562–3563. https://doi.org/10.1021/MA0476745.
156. Bhat, G. A.; Luo, M.; Darensbourg, D. J. *Green Chem.* **2020**, *22*, 7707–7724. https://doi.org/10.1039/D0GC03219K.
157. Tempelaar, S.; Mespouille, L.; Coulembier, O.; Dubois, P.; Dove, A. P. *Chem. Soc. Rev.* **2013**, *42*, 1312–1336. https://doi.org/10.1039/C2CS35268K.
158. Mespouille, L.; Coulembier, O.; Kawalec, M.; Dove, A. P.; Dubois, P. *Prog. Polym. Sci.* **2014**, *39*, 1144–1164. https://doi.org/10.1016/J.PROGPOLYMSCI.2014.02.003.
159. Maquilón, C.; Della Monica, F.; Limburg, B.; Kleij, A. W. *Adv. Synth. Catal.* **2021**, *363*, 4033–4040. https://doi.org/10.1002/ADSC.202100654.
160. Tran, D. K.; Rashad, A. Z.; Darensbourg, D. J.; Wooley, K. L. *Polym. Chem.* **2021**, *12*, 5271–5278. https://doi.org/10.1039/D1PY00784J.
161. Wambach, A.; Agarwal, S.; Greiner, A. *ACS Sustain. Chem. Eng.* **2020**, *8*, 14690–14693. https://doi.org/10.1021/ACSSUSCHEMENG.0C04335.
162. Inoue, S.; Koinuma, H.; Tsuruta, T. *J. Polym. Sci. Part B Polym. Lett.* **1969**, 7, 287–292. https://doi.org/10.1002/POL.1969.110070408.
163. Novomer, https://www.novomer.com/ (accessed January 5, 2022).
164. Andrea, K. A.; Kerton, F. M. *ACS Catal.* **2019**, *9*, 1799–1809. https://doi.org/10.1021/ACSCATAL.8B04282.
165. Yang, G. W.; Zhang, Y. Y.; Xie, R.; Wu, G. P. *J. Am. Chem. Soc.* **2020**, *142*, 12245–12255. https://doi.org/10.1021/JACS.0C03651.
166. Yang, G. W.; Xu, C. K.; Xie, R.; Zhang, Y. Y.; Zhu, X. F.; Wu, G. P. *J. Am. Chem. Soc.* **2021**, *143*, 3455–3465. https://doi.org/10.1021/JACS.0C12425.
167. Lu, X. B.; Darensbourg, D. J. *Chem. Soc. Rev.* **2012**, *41*, 1462–1484. https://doi.org/10.1039/C1CS15142H.
168. Firdaus, M.; Montero De Espinosa, L.; Meier, M. A. R. *Macromolecules* **2011**, *44*, 7253–7262. https://doi.org/10.1021/MA201544E.
169. Zhao, J.; Schlaad, H. *Adv. Polym. Sci.* **2011**, *253*, 151–190. https://doi.org/10.1007/12_2012_166.
170. Ciriminna, R.; Lomeli Rodriguez, M.; Demma Carà, P.; Lopez Sanchez, J. A.; Pagliaro, M. *Chem. Commun.* **2014**, *50*, 15288–15296. https://doi.org/10.1039/C4CC06147K.
171. Hauenstein, O.; Reiter, M.; Agarwal, S.; Rieger, B.; Greiner, A. *Green Chem.* **2016**, *18*, 760–770. https://doi.org/10.1039/C5GC01694K.
172. Byrne, C. M.; Allen, S. D.; Lobkovsky, E. B.; Coates, G. W. *J. Am. Chem. Soc.* **2004**, *126*, 11404–11405. https://doi.org/10.1021/JA0472580.
173. Cheng, M.; Moore, D. R.; Reczek, J. J.; Chamberlain, B. M.; Lobkovsky, E. B.; Coates, G. W. *J. Am. Chem. Soc.* **2001**, *123*, 8738–8749. https://doi.org/10.1021/JA003850N.

174. Auriemma, F.; De Rosa, C.; Di Caprio, M. R.; Di Girolamo, R.; Ellis, W. C.; Coates, G. W. *Angew. Chem. Int. Ed.* **2015**, *54*, 1215–1218. https://doi.org/10.1002/ANIE.201410211.
175. Li, C.; Sablong, R. J.; Koning, C. E. *Angew. Chem. Int. Ed.* **2016**, *55*, 11572–11576. https://doi.org/10.1002/anie.201604674.
176. Reiter, M.; Vagin, S.; Kronast, A.; Jandl, C.; Rieger, B. *Chem. Sci.* **2017**, *8*, 1876–1882. https://doi.org/10.1039/C6SC04477H.
177. Peña Carrodeguas, L.; González-Fabra, J.; Castro-Gómez, F.; Bo, C.; Kleij, A. W. *Chem. A Eur. J.* **2015**, *21*, 6115–6122. https://doi.org/10.1002/CHEM.201406334.
178. Li, C.; Sablong, R. J.; Koning, C. E. *Eur. Polym. J.* **2015**, *67*, 449–458. https://doi.org/10.1016/J.EURPOLYMJ.2015.01.003.
179. Hauenstein, O.; Mushfequr Rahman, M.; Elsayed, M.; Krause-Rehberg, R.; Agarwal, S.; Abetz, V.; Greiner, A. *Adv. Mater. Technol.* **2017**, *2*, 1700026. https://doi.org/10.1002/ADMT.201700026.
180. Hauenstein, O.; Agarwal, S.; Greiner, A. *Nat. Commun.* **2016**, 7, 11862. https://doi.org/10.1038/ncomms11862.
181. Kunming University. *Method for Preparing Polycarbonate by Copolymerizing Carbon Dioxide and Alpha-Pinene Derivatives*; Chinese patent CN103333329B, Kunming University of Science and Technology, 2013.
182. Stößer, T.; Li, C.; Unruangsri, J.; Saini, P. K.; Sablong, R. J.; Meier, M. A. R.; Williams, C. K.; Koning, C. *Polym. Chem.* **2017**, *8*, 6099–6105. https://doi.org/10.1039/C7PY01223C.
183. Circular economy action plan. https://ec.europa.eu/environment/strategy/circular-economy-action-plan_es (accessed January 10, 2022).
184. Clark, J. H. *Green Chem.* **2019**, *21*, 1168–1170. https://doi.org/10.1039/C9GC90021G.
185. Sheldon, R. A. *Green Chem.* **2017**, *19*, 18–43. https://doi.org/10.1039/C6GC02157C.
186. Anastas, P.; Eghbali, N. *Chem. Soc. Rev.* **2010**, *39*, 301–312. https://doi.org/10.1039/B918763B.

About the authors

Felipe de la Cruz-Martínez was born in Badajoz (Spain) in 1990. He studied chemistry at University of Extremadura and then he moved at the University of Castilla-La Mancha where he completed his PhD studies on the carbon dioxide utilization for the synthesis of industrial valuable products under the supervision of Prof. Agustín Lara-Sánchez and Dr. José A. Castro-Osma. He also did a PhD stay at the University of York under the supervision of Prof. Michael North focused on the synthesis of CO_2-based hydroxyurethanes. Since July 2021, he is working as a postdoctoral fellow in the Dr. Jesús Campos group developing bimetallic pairs for the activation of small molecules.

José Antonio Castro-Osma is a lecturer at the School of Pharmacy of the Universidad de Castilla-La Mancha. After completing his PhD studies at the Universidad de Castilla-La Mancha in 2013 under the supervision of Prof. Antonio Otero and Dr. Agustín Lara-Sánchez, he completed a postdoctoral position with Prof. Michael North at the Green Chemistry Centre of Excellence at the University of York. His current research is focused on the development of new catalytic processes for the transformation of CO_2 and biomass-derived chemical feedstocks into high-value added chemical products and polymers.

Agustín Lara-Sánchez was born in La Solana (Ciudad Real), Spain, in 1972. He obtained his graduate degree in 1995 from the University of Castilla-La Mancha and his Doctorate in 1999 with Prof. Antonio Otero and Dr. Juan Fernández. From 2000 to 2001, he worked as a Marie Curie postdoctoral fellow at the University of East Anglia with Prof. Manfred Bochmann researching in the design of catalysts for the polymerization of olefins and the synthesis of biodegradable polymers. From 2002 to 2019 he was a lecturer at the University of Castilla-La Mancha. In September 2019, he was appointed Full Professor in Inorganic Chemistry at the University of Castilla-La Mancha. He has authored more than 100 publications in reviewed journals. His current research interests have been directed toward the study of catalytic processes for the synthesis of products of industrial interest such as cyclic carbonates and biodegradable polymers from CO_2 and renewable resources.

MIX
Papier | Fördert gute Waldnutzung
FSC® C083411

Kolibri 360 GmbH
Ferdinand-Jühlke-Straße 7
99095 Erfurt, Deutschland
produktsicherheit@kolibri360.de